Wisconsin's Natural Communities

Wisconsin's Natural Communities

How to Recognize Them, Where to Find Them

Randy Hoffman

THE UNIVERSITY OF WISCONSIN PRESS

The University of Wisconsin Press
1930 Monroe Street, 3rd Floor
Madison, Wisconsin 53711-2059
uwpress.wisc.edu

3 Henrietta Street
London WC2E 8LU, England
eurospanbookstore.com

Text design by Mira Nenonen

Printed in the United States of America

Library of Congress Cataloging-in-Publication Data

Hoffman, Randolph M.
Wisconsin's natural communities: how to recognize them, where to find them / Randolph M. Hoffman
p. cm.
Includes bibliographical references (p.).
ISBN 0-299-17080-2 (cloth: alk. paper)—
ISBN 0-299-17084-5 (paper: alk. paper)
1. Biotic communities—Wisconsin. 2. Natural areas—Wisconsin—Guidebooks. 3. Wisconsin—Guidebooks. I. Title.
QH105.W6 H64 2000
577.8′2′09775—dc21
00-011124

ISBN 978-0-299-17083-7 (e-book)

for Laura and Eric

Contents

Tables

Acknowledgments

A lifelong curiosity about the natural world together with my parents, Norman and Beverly Hoffman, who inspired in me an awe of nature and instilled an urge to give something back to society, provided the motivation to write this book. Without these formative influences, I might never have undertaken this project.

In my early twenties, I had an obsession with birds that gradually led to an awareness of other forms of life. I found myself needing to know all I could about everything in nature. Plant forays, grubbing through rotten logs, and muddling shallow waters became my pleasure activities, and slowly, like putting together the pieces of a puzzle, I gained an understanding of natural communities. This book is my attempt to share that accumulated knowledge. Nature is so complex, however, that one person cannot possibly know everything, and, most assuredly, omissions of characteristic species occur in every chapter.

I owe a tremendous debt of gratitude to the many artists who have given me permission to use their work in this book. I especially want to thank my good friend Thomas Meyer for his photographic talents and for his permission to use a number of his prints. Throughout my career, I have loved the work of Jim McEvoy. Every time I look at one of his drawings, I am impressed by the way Jim captures the essence of the relationship between a species and its habitat. I am thrilled that he has allowed me to use some of his work in this book. A special thank you must be given to Eric Epstein for his permission to use his photographs and especially for hundreds of hours of lively ecological discussions. Randall Smith, Kathy Kirk, Bill Smith, Mike Mossman, Julie Fox Martin, Georgine Price and Les Ferge also provided exceptional photographs and graphics that made the book complete. I am grateful to the Ohio Department of Natural Resources, the Missouri Department of Conservation, the USDA Forest Service, and, especially, the hundreds of dedicated employees at the Wisconsin Department of Natural Resources for graphics, photographs, and knowledge offered to me for this project.

Many people contributed to the written portion of the book. John Krause must be acknowledged for coaxing me to discuss my book with the University of Wisconsin Press. Patricia Duyfhuisen and Yvette LaPierre edited, with great skill, my scientific technical language into something understandable to the general reader. I also want to thank Bill Brooks, John Bates, Ken Lange, Les Ferge and two anonymous reviewers for giving constructive comments on how to improve the manuscript.

Finally, and most important, I want to thank my son, Eric, for creating the maps in the book—his computer skills continue to astonish me; my spouse, JoAnn, for transcribing my poor handwriting into legible text and for her unwavering support in this venture; and my daughter, Laura, for her ability to keep her dad focused on the important things when the workload seemed overbearing. I also want to thank nature, for challenging all my senses and being so much more entertaining and gratifying than the digital world could ever envision.

Wisconsin's Natural Communities

Introduction

Wisconsin has tremendous natural beauty, and many people enjoy its bounty. Visitors travel thousands of miles and residents stay home to vacation in Wisconsin. Fortunately, Wisconsin still has abundant resources available for its residents and friends to enjoy. More than 15,000 lakes and countless miles of rivers and shores attract water lovers. Still others seek out the whispering, soothing effect of a northern pine forest. Others prefer the glacial landscapes of the kettle moraine or the pastoral hills of western Wisconsin. But a common thread among all these travelers is the desire to connect with and know more about the natural world.

This book presents a framework for understanding the natural world of Wisconsin and is written for amateur naturalists, hunters, bird watchers, hikers, campers, anglers, and others who appreciate the natural world and want a deeper understanding of what makes a healthy natural community. It is also intended for landowners or land managers who want to manage their land in the best way possible for the health of the natural community. The book is designed to help you understand the workings of these natural communities and to provide a foundation for recognizing the interconnections between different species in a community and between species and their habitats.

Three decades of field study, a life's passion for understanding the natural world, and a feeling of responsibility to pass my knowledge to others has led me to write this book. It is meant for others with a similar passion to learn about the natural world and an innate need to provide a better world for future generations.

The book is divided into two parts: Part 1 describes the natural communities found in Wisconsin, and part 2 identifies specific sites where you can experience good examples of those communities.

Each chapter in part 1 describes a distinct natural community in the state, identifying it by its most recognizable features, for example, "southern oak forest" (some communities, such as calcareous fen or mesic prairie, are identified by scientific nomenclature simply because no other name is recognized). The primary "indicators" of the community have been listed to help you identify it. Additional information is given on the natural history of the community, or how topography, climate, and disturbance can affect species composition. Any boxed material describes interesting ecological interactions that occur in the community.

The community description is followed by a list of activities that you can undertake to better understand the community. The activities, which range from simple to quite complex, could serve as the field portion of classroom studies or as guidelines for the amateur naturalist intent on studying the natural world. Following some of the activities are references to sources in the bibliography that you can consult for more information. The characteristic plant, fungus, and animal species found in the community are listed at the end of the chapter.

Part 2 describes 50 sites in Wisconsin where you can study healthy examples of each natural community. Each site description lists the important species and natural features of the site and includes a map of the area. Information on rare species is kept general because unfortunately some people still plunder the natural world without regard for species rarity.

Common species names are used throughout in an attempt to make the book accessible to as many readers as possible. The common names I have used for several groups of species, such as vascular plants, moths, and so forth, are those found (with their scientific counterparts) in the "group specific" reference books that have been marked with an asterisk (*) in the bibliography (for example, all the common moth names come from *A Field Guide to the Moths of Eastern North America* by Charles Covell). Species belonging to those groups, such as beetles and mussels, whose common names do not come from a single source are listed (with both common and scientific names) in the appendix. Excellent introductory works on Wisconsin's natural history are listed in the bibliography, and a glossary of less-familiar words and phrases follows the appendix.

To put all this information in context, it's important to understand that Wisconsin has a phenomenal diversity of life, though this book only addresses the larger more common species, and that the proportion of diversity on any piece of land is in constant change due to processes from the past (natural foundations) and present (natural processes and human activity).

Diversity of Life

Diversity refers to the aggregation of all the species in the state, their populations, and interactions between both relatives and other species. It is generally agreed that diversity is essential to the health of an ecosystem because species interactions create a relative stability. Plants capture sunlight and produce energy. Grazers eat the plants. Other animals eat the grazers. Decomposers return nutrients to the soil. All these interactions and more helped shape the face of the land that the settlers found, which provided a basis for our state's society.

Wisconsin lies at the crossroads of northern conifer forest, eastern hardwood forest, and western prairie. Each region has distinct plant and animal communities that converge in Wisconsin, providing great species diversity. Although we don't know precisely how many species of plants and animals there are in Wisconsin, we do know that there are between 50,000 and a 100,000 species (not including microbes). This estimate comes from a 1997 Society for Conservation Biology seminar at which various biologists quantified the species diversity of Wisconsin. The estimated numbers of plant and animal species are given in table 1.

This book lists over 2,000 of those species, some of which are familiar to all of us, while others are known by only a few professionals. Although the study of Wisconsin's ecosystem is an enormous undertaking, and we may never fully understand how the different elements work together, the knowledge of species diversity and the search for understanding enriches us.

Many species are known from only a few specimens, and new species discovered in Wisconsin are described nearly every year. Some species may be critical to the function of a natural community. Others most certainly contain chemical compounds that could be important to human health. Because we know so little about the majority of the species living in our state, we should pause to think carefully about land-use decisions and how those decisions may permanently alter the natural balance.

Natural Foundations

Within Wisconsin's borders, just under 36,000,000 acres support natural communities and species that have changed dramatically over time. A few billion years ago, the landscape could have been molten rock, and one billion years ago it could have been ocean floor. More recently, about 15,000 years ago, the landscape was 85 percent ice and 15 percent tundra. A few thou-

Table 1. Species estimates from a 1997 Society for Conservation Biology seminar

Groups	*Number of known species*
Mammals	66
Birds	252 breeding species
Birds	>400 recorded in the state
Reptiles and amphibians	53
Fish	153 (13 extirpated)
Freshwater mussels	81
Snails	246
Spiders	600
Other invertebrates	16,000 to 25,000
Insects	20,000 to 40,000
Lichens	>700
Fungi	12,000 to 30,000
Algae	1300
Mosses/Liverworts	530
Plants	2,642 (1,991 native and 651 exotic)
Total	53,000 to 100,000 species

sand years can manifest dramatic changes. The first Americans hunted the wooly mammoth to extinction. Later descendents of these first Americans farmed and gathered food, which changed the land. European immigrants cleared the forests, plowed the prairies, and drained the marshes.

Conservation biologists often refer to the condition of the land at the time of the first major influx of European immigrants, which began around 1830, as pre-European settlement. Up until around 1850 when the land survey started, few areas of the state had been settled by European immigrants. These areas did not have platted lands, and the federal government needed to have some type of system to prevent multiple claims on the same piece of property. During the mid-1800s, the government surveyed nearly all of Wisconsin. These land surveyors laid out a grid across the state that formed the basis for townships and all future land transactions. Fortunately, the land surveyors documented the witness trees, location of lakes and streams, and general condition of the soil for agriculture. These notes give us a snapshot view of the state's natural condition during that period. An interpretation of these land survey records helped in the development of the range maps used in part 1 of this book. Table 2 shows the approximate acreage of the natural communities in the mid-1800s and in the late 1900s.

While we can't predict the future of species diversity in our natural world, it is likely that the common species found in the more widespread natural communities, such as agricultural and urban communities and northern hardwood–hemlock forest, will continue to survive and thrive. However, we may lose even the common species in some remnant communities, such as mesic prairie and oak opening. By understanding the habitat needs of both common and rare species, we can take action to prevent loss of diversity.

Natural Processes and Human Activity

Natural processes are a dynamic of the natural world. Many processes occur over long time frames, such as the advance and retreat of a glacier. Others are subtle, such as fungi breaking down plant material. Still others are dramatic, such as fire sweeping through and radically

Table 2. Comparison of acres for communities in this book

Natural Community	*Mid-1800s*	*Late 1900s*[a]
Southern oak forest	970,000	1,365,000
Southern red oak–mixed forest	415,000	585,000
Sugar maple–basswood forest	3,432,000	50,000
Floodplain forest	420,000	225,000
Dry pine forest	340,000	505,000
White pine–hardwood forest	1,930,000	610,000
Northern hardwood–hemlock forest	11,750,000	7,700,000
Forested swamps	560,000	500,000
Northern forested bogs and fens	1,680,000	1,200,000
Great Lakes shoreland forest	670,000	420,000
Dry prairie	105,000	6,500
Dry-mesic prairie	630,000	1,000
Mesic prairie	840,000	300
Wet-mesic prairie	420,000	2,500
Wet prairie	105,000	200
Calcareous fen	500	300
Southern sedge meadow	1,000,000	200,000[b]
Northern sedge meadow	115,000	100,000
Open bog	110,000	100,000
Oak opening	5,500,000	500
Oak barrens	1,800,000	325,000
Pine barrens	2,340,000	110,000
Sand barrens	—	2,000
Shrub communities and lake dunes	?	?
Lakes	800,000	970,000
Agricultural and abandoned land	?	19,027,000
Cities, residences	?	1,500,000
Roads	—	420,000

[a] The late 1900s acreage reflects the total coverage of the dominant species. The condition of those acres depends on past land use history, and most of those acres listed have a simplified association of species. The common species listed under the community should be present, but the uncommon or rare species listed under the individual site sections may not be present. Areas containing a full range of species composition and interactions are found at only a fraction of the forested and wetland communities. Finally, the agricultural–urban areas have the most nonnative species, which were absent prior to pre-European settlement.

[b] Most of the southern sedge meadow land is now dominated by invasive exotic species.

changing a landscape. It's these large events that cause large shifts in natural community composition.

- Remnants of the Ice Age are with us today. Landforms and soils in most of the state resulted from the effects the ice had on the land, such as the rugged kettle moraine topography or the rich loam soil on the flat plains around Antigo. Even the retreat of the glacier left lasting effects. As the ice retreated northward, the vegetation changed from tundra to spruce–fir forest, then to pine forest, and then, in southern Wisconsin, to oak forest. In certain places where the conditions are right, relicts of the former natural communities persist in small pockets.
- Severe storms, tornadoes, and downbursts can change the composition of species. A jumble of downed trees left in the wake of one of these storms prevents deer from eating

herbs and seedlings, giving an advantage to those protected plants. Also, a once-shaded forest is now open to full sunlight. Species adapted to low light levels cannot tolerate the bright sun and die. Exposed soil around the tipped tree root may provide conditions favorable for aspen establishment.

- Widespread drought can diminish populations of dominant species and change the species composition of a natural community.
- Floods can scour banks, uproot hundreds of trees, and deposit silt in new locations. Deep channels are cleared of the material deposited during low-water times, thereby reestablishing a lost part of the ebb and flow of the river. Deposits of new seeds can enhance the restored habitat.
- Insect or disease outbreaks can affect great areas. Jack pine budworms can dramatically change the landscape by eating nearly all the jack pine needles, which kills the trees. As a result, more sunlight reaches the ground, which encourages sun-loving groundlayer species and discourages shade-tolerant species. Also, the dead trees are susceptible to catastrophic fire, which can bring about radical changes in species composition.
- Prairies and savannas may be subject to grazing by large animals, such as deer and elk. Grazing produces a mosaic of vegetation heights, with some tall ungrazed grasses and some grasses grazed close to the ground. The height of vegetation must be important to some species because many grassland birds will choose either short grass areas, mid-grass areas, or tall grass as preferred habitat.
- Fire reduces the amount of woody material, provides nutrients (ash) for new growth, and directly kills some animals.

Activity by humans can also result in dramatic changes to the land. Road or parking-lot construction, for example, reduces species diversity to very low levels because few species can live on asphalt and cement. Less dramatic changes, such as conversion of forest land to a backyard, enhance species such as American robins and earthworms that are able to tolerate the conversion. Frequent and prolonged activities like annual plowing or mowing limit the number of species able to use the site. Trees and shrubs cannot take hold, which prohibits the land from growing the full complement of species it is capable of growing.

How to Use the Book

Below each chapter title in part 1 is a list of two to five indicator species that will help you identify the natural community. These are the dominant (most frequently observed) plants in the community. Most of the trees are easy to recognize, but some of the prairie and wetland species may require learning new identification skills. You can identify unknown species by referring to the plant identification books in the bibliography. Next refer to the species lists at the end of the chapter. As you study a site, you should find a majority of the species listed. In addition, check the list of Wisconsin's common species at the end of this introduction. These are species that can be found in almost all natural communities. Using these lists in combination should provide a fairly complete resource of the dominant parts of the natural community, although not all the species will be at every site. Of course, you will find many, many other species, both common and rare, but that's the excitement of discovery. The community description provides a basis for understanding the interactions between these species. Range maps, which are interpreted from the original mid-1800s land survey records by comparing the recorded species with natural community composition today, help you locate each community.

Part 2 identifies 50 sites in the state where you can experience exemplary natural communities. The sites are separated into five regions—west, northwest, northeast, southeast, and south central—and each site description includes a map to the area. Following the title of each site is a list of natural communities found at the site. A description of the site and its features is followed by a list of specialties: rare species and natural features that are particularly important at the site. This list is valuable to those wishing to observe the rarest parts of our natural world. Less rare but nevertheless interesting species found at the site are listed at the end of the chapter, and occasionally large concentrations of common species are identified, as a guide for those wishing to view natural phenomena.

Let's assume you own an oak forest in southern Wisconsin, and you want to learn more about managing it. First you would determine the natural community by identifying the common plants. You would then refer to the characteristic species listed at the end of the appropriate natural community chapter in part 1 to better understand the complexity of life in your woods. You could then take note of the community's range and list of exemplary sites. Plan a visit to one or more of the sites, using the site maps and descriptions provided in part 2. This field trip would allow you to compare your land to a high quality community of the same type. The specialties lists found with the site descriptions in part 2 highlight rare species that you might want to search for on your own property. Finally, you could scan over the activities section to see if any of the activities interest you. Let me encourage you to always document your land's ecological heritage and to pass that information along to the next generation.

Common Species

The following species are found in many natural communities. Even though disturbances caused by human activities limit the number of species on a particular site, some species, such as killdeer and dandelions, are able to live and thrive in these modified habitats because they have little competition from other species. These highly adaptable species are found in agricultural and urban settings, as well as in natural communities across the state. The daggers (†) in all species lists throughout this book indicate invasive exotic species.

Common Plants

bella honeysuckle†
black raspberry
box elder
bull thistle†
butter and eggs
common dandelion†
common milkweed
common plantain
common ragweed
crabgrass
field chickweed
field mustard
Kentucky bluegrass†
lamb's quarters
Missouri gooseberry
mouse-ear chickweed
mullein
pineapple weed
quackgrass†
Queen Anne's lace
reed canary grass†
shepard's purse
smooth brome†
staghorn sumac
tall stinging nettle†
tartarian honeysuckle†
timothy
Virginia creeper
white mustard

white sweet clover†
yellow sweet clover†

Mushrooms

arched earthstar
artist's conk
brick tops
Caesar's fiberhead
carbon balls
chicken mushroom
common mycena
crested coral
crested polypore
eyelash cup
fawn mushroom
giant clitocybe
gypsy
honey mushroom
lilac fiberhead
meadow mushroom
orange mock oyster
oyster mushroom
pear-shaped puffball
powder-cap amanita
puffball agaric
recurved cup
red and black russula
red-cracked bolete
ribbed-stalked cup
ringed tubaria
scaly pholiota
scaly-tooth
scarlet cup
scarlet waxy cap
smoky polypore
smooth thimble-cap
spring polypore
tinder polypore
turkey-tail
white coral
white fiberhead
yellow-cracked bolete
yellow fairy cups

Butterflies and Moths

cabbage white†
codling moth
common sulphur
eastern tailed blue
eastern tent caterpillar
fall cankerworm
Isabella tiger moth
monarch
orange sulphur†
red admiral
sod webworm
yellow bear moth

Wide-Ranging Insects

(The following list of insects includes cosmopolitan species that don't have a preference for any particular habitat and predatory species that must pursue prey in many different habitats. These insects are part of nearly every natural community.)

alternate shield-back
American ambush bug
American bumble bee
American carrion beetle
anchor bug
anchor shield-back
antelope stag beetle
aphodine dung beetle
ash-gray blister beetle
bald-faced hornet
bee fly
big-headed fly
black corsair
black horse fly
bluebone beetle
blue bottle fly
box elder bug
broad-headed bug
burrowing wasp
candy-stripe leafhopper
Carolina locust
chinch bug†
chocolate serica
chrysanthemum lace bug
clematis blister beetle
cluster fly
common firefly
contort drynid
convergent lady beetle
cornfield ant
damsel bug
deer fly
digger wasp
drone fly
dung scavenger
eastern sand wasp
elm sawfly
European ground beetle
European leafhopper
eye gnat
face fly
fancy ant hister beetle
fancy dung beetle
field cricket
flat net-wing
flesh fly
flower fly
four-spotted collops
garden carrion beetle
giant cicada killer
giant ichneumon
golden chalcid
goldenrod plant bug
goldenrod soldier beetle
gold-necked burying beetle
green lacewing
green stink bug
hairy checkered beetle
hairy rove beetle
hieroglyph beetle
honey bee
horned tree fungus beetle
horse shoe crab beetle
house cricket
house fly

humpbacked fly
larder beetle
large milkweed bug
latrine fly
leafcutting bee
leaf-footed bug
little house fly
long-legged fly
louse fly
maculated dung beetle
maculated lady beetle
mantidfly
March fly
masked hunter
minute pirate bug
mydas fly
nine-spotted lady beetle
northern katydid
northern yellow jacket
painted rove beetle
paper wasp
parasitic rove beetle
parenthesis lady beetle
pavement ant
pedunculate ground beetle
Pennsylvania ambush bug
Pennsylvania hister beetle
picnic beetle
pigeon horntail
red flat bark beetle
red-legged locust
red-lined carrion beetle
red-necked click beetle
robber fly
scooped scarab
secondary screw-worm
shiny fungus beetle
silky ant
six-spotted leafhopper
small milkweed bug
spined assassin bug
spotted June beetle
squash bug
tarnished plant bug
thirteen-spotted lady beetle
three-banded lady beetle
tumble bug
twice-stabbed stink bug
two-banded fungus beetle
two-lined soldier beetle
two-spotted lady beetle
two-spotted sap beetle
two-spotted stink bug
two-striped planthopper
undulated lady beetle
vinegar fly
wasp-like clerid
water scavenger beetle
wheel bug
white-footed ichneumon
yellow bumble bee
yellow jacket

Reptiles and Amphibians

American toad
eastern garter snake
leopard frog

Birds

American goldfinch
American robin
blue jay
brown-headed cowbird
common crow
common grackle
downy woodpecker
European starling†
great-horned owl
house sparrow†
house wren
mourning dove
red-tailed hawk
red-winged blackbird
ring-necked pheasant
rock dove†
song sparrow
white-breasted nuthatch

Mammals

cottontail
feral cats†
feral dogs†
gray squirrel
house mouse†
Norway rat†
opossum
raccoon
striped skunk
white-tailed deer

Communities & Characteristic Species

Forests

Wisconsin forests at *the Millennium: An Assessment* (Finan 2000) documents nearly 16,000,000 acres of forest land in the state, a significant increase over the 13,000,000 acres of the early 1930s when forest resource exploitation reached its zenith in what is referred to as the "cut-over" period. On the surface we seem to be doing quite well, even in comparison to the 22,000,000 acres of forest land recorded by the land surveyors in the mid-1800s.

Forest assessment statistics place values on various aspects of the forest resource, such as the shade and aesthetic values of urban forests and the economics of plantations. Many forest values, however, are human-centered, such as revenue from timber sales, hunting sites, and recreation areas, and have little to do with ecology and the understanding of natural communities.

Wisconsin's forests have been managed for thousands of years, beginning with Native Americans. Most of the tribes living in heavily forested areas were hunter-gatherers, but they did have small gardens. They set fire to clear plots or improve blueberry grounds. In southern and western Wisconsin, more-agrarian tribes burned prairies and savannas annually. These fires most assuredly burned into the forest, affecting its composition. In addition, some areas were spared from fire, such as favored sugar-mapling areas, which also had an effect on species composition.

A combination of natural and these human-generated disturbances resulted in the forests that the European settlers found. Since then, a combination of resource exploitation, market-driven lumbering, forest management for sustained yield, and a small amount of ecological management has resulted in the forests that we have today. Many tracts of forest are now isolated by agriculture and development. These woods have plants and animals more indicative of Europe and Asia than North America.

Today's forests, while getting older compared to the 1930s, are much younger than those of the previous several thousand years. Seldom do we allow our forests to age beyond 100 years before harvest. Centuries ago, 80 to 90 percent of the sugar maple forests had trees more than 150 years old, and 20 to 40 percent of the white pine forests had trees more than 200 years old. Old-growth forests have many characteristics not found in younger forests, which may have consequences for the survival of some species. For example, in older forests wind storms would blow trees down every year. Today's forests with younger trees are less susceptible to wind. Moreover, in centuries past the technology was unavailable to salvage wind-blown trees, but tree salvage after storms is standard now. The state-endangered American pine marten, a member of the weasel family, does best in forests with numerous dead falls and snag trees.

An understanding of the full range of natural forest values, such as maintaining biological diversity, soil generation, watershed, and ground water recharge, will, I hope, give future generations more management options. Forest ecology does not function on political or property lines. We need to understand that forests can be very open or densely tangled, have patches of young trees or large areas of old decaying trees, and, in most cases and from an ecological standpoint, be affected positively by fire. Also, forests need to be viewed from a landscape perspective. Species composition and forest structure shift across the landscape over the course of time, and we should not manage a piece of land in an identical manner in perpetuity.

Southern Oak Forest–Oak Woodland

Indicators: Black Oak, White Oak, Bur Oak, Black Cherry, Red Oak

Ecology

The oak forests of Wisconsin hold a special place for many people. Oaks have qualities many Midwesterners embrace, such as durability, hardiness, and longevity. Others see value in oak forests as a source of lumber or as habitat for wildlife.

The southern oak forest–oak woodland community is widespread, but the number and type of oaks and other plant species in a given forest depend on a variety of factors, including the type of soil and the moisture and nutrients it contains, and the slope of the land and its orientation to mid-afternoon sun. Fire is another variable that affects the plant species found in the forest. For 3,000 years before European settlement, Native Americans used fire extensively to clear and control undergrowth, which encouraged savanna communities with scattered oaks. After European settlement, fire gradually fell out of favor as a management tool, and as a result, most of Wisconsin's savannas and even some prairies converted to oak forests. In the mid-1800s Wisconsin's oak forests covered approximately 970,000 acres. Currently there are about 1,365,000 acres of oak forest in the state. Oak forest acreage has reached its zenith and is now gradually declining.

Southern oak forests contain a mix of oak species with abundant shrubs, including oak

Fire

Wildfire has been influencing the Wisconsin landscape for several thousand years. For more than 3,000 years, Native Americans burned the land regularly. They set fires primarily in late summer or fall, and occasionally in spring, and the fires were allowed to burn until the weather put them out. Native Americans set fires for a variety of reasons, including encircling game, destroying an enemy village or protecting village sites from fire set by enemies, and clearing land for agricultural purposes. In very wet years, some areas did not burn, while in dry years nearly everything burned. Plants and animals that adapted to fire thrive in its wake. Those that didn't, such as sugar maple, spring beauty, or spotted salamander, either perished or migrated to safe areas.

Frequent fires tend to prevent tree establishment. Fire every 1 to 5 years usually allows prairies or meadows, and fire every 5 to 20 years allows savanna communities. Less frequent fire allows development of oak and pine forest because the thick bark on oaks and pines effectively insulates the growing parts during infrequent fires. These communities have adapted to survive fire. For example, most of a prairie plant consists of a dense root system that lies uderground, safe from fire. Insects that survive fire can experience rapid population growth; even with just a few surviving individuals, recovery is possible.

Southern oak forest (illustration by Randall Smith; used by permission)

saplings, in the understory. Oak woodlands, on the other hand, have an oak canopy but very few shrubs and saplings in the understory.

Range

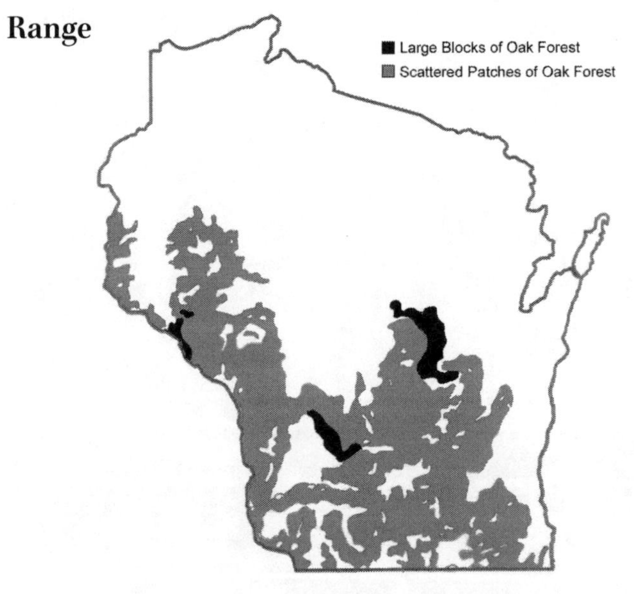

Southern Oak Forest–Oak Woodland

The map indicates areas where extensive southern oak forest and oak woodland communities occur. In southwestern Wis-

Red Oaks

Members of the red oak group (red, black, and Hill's oaks) require two years for acorns to develop. Each spring the oak trees bloom and are pollinated by the wind. The developing acorns mature and drop in the second year. The seed crop varies year to year, with an abundant crop usually occurring once every three to five years. The variable acorn production benefits oaks because a reliable acorn crop year after year would allow acorn-feeding insects, such as acorn weevils and beetles, to sustain optimum populations.

Acorns are too heavy to permit wide distribution on their own. Squirrels and chipmunks bury acorns in a scattered fashion near existing trees, but how do acorns turn up far away? Acorns are spread primarily by two bird species: red-headed woodpeckers and blue jays. A red-headed woodpecker, a specialist feeder on acorns, caches them for use during the winter. Woodpeckers always drop a few acorns as they transport them, and a few fall from the caches, too. High over-winter populations of red-headed woodpeck-ers occur during years of high acorn production.

The most important transporter of acorns, however, is the blue jay. In the fall blue jays carry acorns several miles to bury them for winter use. Studies show that blue jays can transport as much as 50 percent of the acorns produced in a year an average distance of about one mile (Green 1995). The acorns that are not found and eaten in the winter will then germinate the next spring.

consin, oak forests grow primarily on hillsides over limestone. Because the soil depth, sunlight, and moisture vary up and down a hill, the composition of an oak forest on a hillside can vary over short distances. Red oak dominates on the deep soils near the base of the hill, while black oak is abundant on the thin soils near the crest of a hill. White oaks are often found on flat ridge tops.

You can see examples of southern oak forests at Devil's Lake State Park, Governor Dodge State Park, Nelson Dewey State Park, Perrot State Park, and Wyalusing State Park.

While all oak communities have features in common, landforms and soils change dramatically across the state, and local conditions favor different types of oak forest and woodland. Sandy oak forests, for example, have tree and groundlayer species such as black oak, white oak, Pennsylvania sedge, and hog peanut, all of which do best in well-drained sandy soils. This oak forest community can sustain itself for centuries. Sandy oak forests occur in Polk, Dunn, Chippewa, Eau Claire, Trempealeau, Jackson, Monroe, Juneau, Sauk, Adams, Marquette, Green Lake, Waushara, and Waupaca Counties, as well as several sites in the Black River State Forest.

Glacial moraine oak forests originated on the gravelly till of the glaciated portion of southern Wisconsin. They are similar to sandy oak forests but also contain plants that thrive in more nutrient-rich soils, such as wild geranium and common enchanter's nightshade. Glacial moraine oak forests occur in Fond du Lac, Dodge, Waukesha, Walworth, Racine, and Kenosha Counties. You can find good examples in the North Kettle Moraine, South Kettle Moraine, and Upper Mukwonago River.

Prairie oak groves were once common in the southern Wisconsin portion of the prairie peninsula. Settlers recognized these groves as an important part of the southern Wisconsin landscape, as exemplified by community names such as Union Grove, Cottage Grove, and Patch Grove. Most prairie oak groves are on the leeward side of natural firebreaks, such as rivers or lakes, which stop most fires, letting only a small percent through to burn the groves. The natural topography of the land allows more humus to build up, and groves form on the deep, rich soil. Prairie oak groves develop only in areas dominated by prairie and scattered bur oak savanna. A few groves remain in Grant, Lafayette, Green, Rock, Iowa, Dane, and Columbia Counties. You can visit an example in Green County Preserves.

Swamp white oak forests and woodlands grow in floodplains where sand and gravel deposits provide drainage after spring floods. The annual spring flooding and summer drying undoubtedly help determine species composition. Floodplain oak forests and woodlands were once common but now are rare on sandy terraces near the lower Wisconsin and lower Chippewa Rivers because the lock and dam system on the Mississippi River inundated all these floodplain oak woodlands. Fortunately, two excellent examples remain in the Lake Pepin and Avoca Prairie areas.

Oak woodlands are fire-shaped communities that unfortunately may be functionally extinct as a natural community. Intact functioning oak woodlands could some day be seen again if enlightened people begin managing large tracts of oak woods with fire. In these woodlands, oaks make up 60 to 90 percent of the canopy. The oak trunks are longer than those that grow in the open savanna, and they have open branching higher up, which gives a bent or twisted appearance to the canopy. Wildflowers, grasses, and occasional shrubs create a dense groundlayer. Relatives of open prairie species, such as pale-leaved sunflower, side-flowering aster, Short's aster, poke milkweed, and silky wild rye, are abundant. Oak woodlands also have some species normally associated with moist prairies, such as Culver's root and shooting star, which can tolerate some shade.

You can find artifacts of this nearly lost community in narrow strips along roadsides in Grant, Crawford, Green, Sauk, Buffalo, and Pepin Counties. A few acres of oak woodland remain at Devil's Lake State Park, Lower Wisconsin River, and Green County Preserves.

Activities

- Observe the oak canopy and estimate the extent of insect damage (chewed, discolored, or bumpy areas) on the leaves (average is 5 to 30 percent). Assuming the damage is within a normal range, anticipate the effects of no or little insect damage or massive defoliation. The effects can be seen in tree growth, additional predators, and the amount of sunlight reaching the forest floor.
- Observe succession. What are the trees in the understory? Are the small seedlings the same species as the saplings?
- Visit an oak forest in winter and draw a tree form. Sketch the branching pattern. Many large low branches indicate that there was more space between trees at one time. A mostly straight trunk bur oak with gnarly and bending branches 30 to 50 feet up may indicate a quality oak woodland.
- Develop a list of species found in different patches in the forest. Choose sunlit areas, dense shade areas, and edges of the woods. A comparison with the oak savanna species may help in making management decisions. Oak savanna may require clearing of the underbrush and applying fire to restore the savanna community.
- Take a flashlight into an oak woods in late summer. Carefully watch for walkingsticks. If you happen upon one of these nighttime feeders, give it a little scare and watch it assume a stick position. Describe the purpose for such an action.

Characteristic Species

Plants

American hazelnut
blue marsh violet
bracken fern
bracted tick-trefoil
common blackberry
common enchanter's nightshade
false Solomon's seal
gray dogwood
hairy sweet cicely
hog-peanut
lopseed
pale-leaved sunflower
rattlesnake fern
shining bedstraw
spreading dogbane
starry false Solomon's seal

- tall agrimony
- tall blue lettuce
- wild geranium
- wild sarsaparilla
- wild strawberry

Mosses and Lichens

- blister lichen
- common tree moss
- dog lichen
- feathered neckera
- southern leucodon
- urn moss
- wrinkled shield lichen

Mushrooms

- beafsteak polypore
- bear's head tooth
- black velvet bolete
- blusher
- chanterelle
- citron amanita
- crowded parchment
- early spring entoloma
- false turkey-tail
- fragrant armillaria
- Frost's bolete
- funnel clitocybe
- hen of the woods
- jack o'lantern
- old man of the woods
- smooth chanterelle
- splash cups
- warted oak polypore

Butterflies and Moths

- acorn moth
- banded hairstreak
- banded tussock-moth
- black-patched graylet
- common wood nymph
- consort
- Edward's hairstreak
- fawn sallow
- filbertworm moth
- fragile white carpet
- giant swallowtail
- greater wax moth
- hobomok skipper
- LeConte's hapola
- little wood satyr
- mourning underwing
- oak skeletonizer
- orange-patched smoky moth
- orange-tipped oakworm
- pink-striped oakworm
- red-humped caterpillar
- red twin-spot
- residua underwing
- scarlet-winged lichen-moth
- spiny oakworm
- striped hairstreak
- tearful underwing
- widow underwing

Beetles

- antelope beetle
- ant-like flower beetle
- banded net-wing
- black firefly
- black oak weevil
- caterpillar hunter
- eyed click beetle
- faded click beetle
- hazelnut weevil
- hermit flower beetle
- horned passolus
- lurid flat-headed borer
- oak branch borer
- oak timberworm
- oak twig pruner
- raspberry cane borer
- red-necked cane borer
- red-shouldered longhorn
- scriptured leaf beetle
- seed-eating ground beetle
- square weevil
- stag beetle
- tile-horned prionus
- two-lined flower longhorn
- white oak borer

Insects

- American grasshopper
- broad-winged katydid
- brown stink bug
- common meadow katydid
- common walkingstick
- fork-tailed bush katydid
- green-striped grasshopper
- jumping oak gall
- oak apple gall
- oak lace bug
- oak treehopper
- ornate pygmy grasshopper
- partridge scolops
- Scudder's grasshopper
- short-legged shield bearer
- snowy tree cricket
- sprinkled grasshopper
- Walsh's grasshopper

Spiders

- ant mimic
- branch-tip spider
- daring jumping spider
- filmy dome spider
- parson spider
- purse-web spider

Reptiles and Amphibians

- brown snake
- eastern gray treefrog
- tiger salamander

Birds

- Baltimore oriole
- eastern towhee
- gray catbird
- indigo bunting
- northern cardinal
- red-eyed vireo
- red-headed woodpecker
- rose-breasted grosbeak
- yellow-billed cuckoo

Mammals

- coyote
- eastern chipmunk
- fox squirrel
- red fox
- white-footed mouse
- woodchuck

Southern Red Oak–Mixed Forest

Indicators: Red Oak, White Oak, Basswood, Sugar Maple

Ecology

The southern red oak–mixed forest develops where deeper, more nutrient-rich soils favor a mix of oaks and moist-soil tree species. Red and white oaks dominate, along with red maple and shagbark hickory on drier sites and sugar maple and basswood on moister sites. The southern red oak–mixed forest expanded greatly when fire suppression became common after European settlement, which allowed oak seedlings and saplings to reach maturity. Other species, such as American elm and white ash, also began to thrive due to fire suppression. As the forest developed, shade-tolerant species like sugar maple grew and competed favorably with other tree seedlings and saplings.

Sugar maple dominance is accelerating in red oak–mixed forests due in part to the tree's tolerance of low light levels, but land management influences these forests more. Throughout a good portion of the twentieth century, grazing and the annual burning of pastures helped oaks maintain dominance on many sites because fire kills young maples. However, fire suppression, removal of cattle from woods, and overharvesting economically valuable red oak, hickory, and black cherry accelerated the growth of species able to withstand heavy shading. High prices for red oak lumber continue to tempt landowners to harvest all the red oak in their forests. When the oak is removed, species such as basswood, white ash, American elm, and red maple become dominant. These trees thrive as understory species under oaks and rapidly grow toward the sun when an oak is removed.

Ephemeral Ponds

Each spring, ephemeral, or temporary, ponds form in the southern red oak–mixed forest. Many plants, such as several species of beggarticks and smartweeds, grow around these ponds and nowhere else in the forest and provide structures on which amphibians and invertebrates can attach their eggs.

Wood frogs are the first frog species to breed in the spring, and they prefer these ephemeral ponds. The males call to attract females (their call somewhat resembles a quack), and they mate as the female lays her eggs on submerged plants and branches. Wood frogs prefer the smaller ponds that dry up in summer because they usually don't contain salamander larvae, which love to eat wood-frog eggs. Salamander larvae live in many of the larger ponds that last until late summer.

Much smaller ephemeral ponds form when water runs down a tree trunk and accumulates in cavities within a living tree. These "ponds" can produce their own ecosystems. Tree-hole mosquitoes, for example, live out their larval stage nowhere else. One tree-hole species is a vector for La Crosse encephalitis.

Southern red oak–mixed forest (illustration by Jim McEvoy; courtesy of Wisconsin Department of Natural Resources)

Cerulean Warbler

The cerulean warbler is most closely associated with southern red oak forest or red oak–mixed forest. This state-threatened species, which takes its name from the color of the male's plumage, is a canopy dweller, preferring red oaks. They are not found in dry black oak savannas or any of the northern forests with conifers, but occasionally small numbers are found in floodplain and sugar maple–basswood forests.

These birds are decreasing for many reasons, including destruction of their winter habitat in rainforests of Columbia, Equador, and Peru and loss of large tracts of old oak–mixed forests for breeding. Cerulean warblers seldom nest in trees less than 60 years old and prefer trees more than 100 years old. They also require large tracts of forest of more than 100 acres to nest successfully and are rarely found on sites less than 40 acres (Wisconsin Department of Natural Resources Surveys, unpublished).

Cerulean warbler (photograph courtesy of Ohio Department of Natural Resources)

Southern Red Oak–Mixed Forest

Range

The best remaining red oak–mixed forest sites are those on nutrient-rich, moist soils in areas that burned regularly in the past, which permitted excellent oak growth and kept competing species in check. Without regular and continued fire disturbance or some type of simulated fire effects, such as grazing, the remaining red oak–mixed forests might not be here 50 years from now. The best conditions for the development of this community are found in western Wisconsin, southern Wisconsin prairie groves, and in the Kettle Moraine State Forests, especially the North Unit.

Excellent examples of southern red oak–mixed forests can be found at Baraboo Hills, Cadiz Springs, Devil's Lake State Park, Governor Dodge State Park, Interstate State Park, Nelson Dewey State Park, North Kettle Moraine, Perrot State Park, South Kettle Moraine, Upper Mukwonago River, Wildcat Mountain State Park, and Wyalusing State Park.

Activities

- Slope and aspect greatly influence species composition, which is easily observed in the field. List the species found on the south-facing slope of a woods in the driftless area, and then list the species found on a north-facing slope. Compare the lists and explain why they are different. Hint:

a slope facing directly into the afternoon sun will be hotter and drier.

- Map the tree species of a study site of your choice. Geographic Positioning Systems (GPS) make for very accurate mapping, or simply sketch the tree patterns from the ground. The maps can show pattern, clumps, or scattering of trees. Describe how other species may use these patterns. For example, the hickory tussock-moth would be more likely observed in a grove of hickories than oak- or maple-dominated areas (Forman and Godron 1986).
- Document the plant species found in woodlots isolated from other woods. Can the plants reestablish if the woodlot is blown down or completely logged and the direct sunlight kills the delicate herbs? Where will the seeds come from to reestablish these herbs (Forman and Godron 1986: 108–9)?
- Observe squirrels and chipmunks in the fall as they gather and store acorns. What effects might this activity, called scatter hoarding, have on the oaks and other tree species?
- Visit isolated woodlots at night to observe southern flying squirrels. Are they found in all sizes of woodlot or is there a minimum size? Does distance from other woodlots have an effect on their presence?

Characteristic Species

Plants

arrow-leaved aster
bellwort
bloodroot
blue cohosh
blue marsh violet
bottlebrush grass
climbing bittersweet
clustered black snakeroot
common carrion flower
common enchanter's nightshade
downy arrow-wood
early meadow-rue
elm-leaved goldenrod
false Solomon's seal
gray dogwood
hairy sweet cicely
hog peanut
interrupted fern
jack-in-the-pulpit
lady fern
lion's foot
long-awned wood-grass
lopseed
maidenhair fern
May apple
pale-leaved sunflower
Pennsylvania sedge
pointed tick-trefoil
rattlesnake fern
shining bedstraw
spikenard
summer grape
upright carrion flower
wild geranium
wild sarsaparilla
wood anemone

Mosses and Lichens

dog lichen
many-fruited dog lichen
plume lichen
smooth rock tripe
spreading leather

Mushrooms

bear lentinus
black trumpet
bracelet cort
cinnabar cort
cinnabar-red polypore
club-shaped stinkhorn

common split gill
crimped gill
jellied false coral
moose antlers
Ravenel's stinkhorn
silvery-violet cort
veined cup
walnut mycena
yellow patches

Butterflies and Moths

Abbot's sphinx
achemon sphinx
alternate woodling
American angle-shades
American idia
bad-wing
black-blotched schizura
black zale
blinded sphinx
broad-lined sallow
brown-lined sallow
Canadian melanolophia
carpenter worm
chain-dotted geometer
clymene hapola
copper underwing
crinkled flannel moth
drab brown wave
eastern tiger swallowtail
eight-spotted forester
fawn sallow
five-spotted hawk moth
galium sphinx
grapevine epimenis
gray-edged bomolocha
gray scoopwing
green arches
hickory tussock-moth
horned spanworm
horrid zale
imperial moth
lesser grape vine looper
lined ruby tiger moth
lobelia dagger
luna moth
lynx flower moth
medium dagger
nessus sphinx
obscure underwing
pink patched looper
promethea moth
question mark
small engrailed
straight-lined looper
unicorn caterpillar
white-dotted prominent
white-lined sphinx
white-striped black
widow underwing
wild cherry sphinx
yellow gray underwing

Beetles

ash-gray lady beetle
bark hister beetle
black firefly
cylindrical hardwood borer
dark brown tiger beetle
darkling beetle
eyed click beetle
hairy May beetle
horned passolus
May beetle
mole-like flower beetle
narrow-necked click beetle
notch-mouthed ground beetle
oak branch borer
oak timberworm
painted-hickory borer
pinching bug beetle
red oak borer
scriptured leaf beetle
seed-eating ground beetle
six-spotted tiger beetle
slender ground beetle
square weevil
stag beetle
tan bark borer
twig pruner
two-lined chestnut borer
two-lined flower longhorn
woodland ground beetle

Insects

Carolina ground cricket
currant fruit fly
four-lined plant bug
oak apple gall
oak lace bug
oak treehopper
oystershell scale
potter wasp
sprinkled grasshopper
walnut husk fly
Walsh's grasshopper

Spiders

branch-tip spider
brown widow
daring jumping spider
filmy dome spider
forest wolf spider
hammock spider
house spider
marbled spider
northern widow
parson spider
purse-web spider
spined micrathena

Landsnails

armed pupa snail
fraternal pill snail
great zonite shell
hairy pill snail

modest pupa snail
ripe forest snail
tree zonite shell

Reptiles and Amphibians
blue-spotted salamander
brown snake
eastern gray treefrog
spring peeper
tiger salamander

Birds
American redstart
blue-gray gnatcatcher
eastern wood-pewee
great crested flycatcher
least flycatcher
northern cardinal
red-eyed vireo
red-headed woodpecker
rose-breasted grosbeak
wood thrush
yellow-billed cuckoo
yellow-throated vireo

Mammals
eastern chipmunk
fox squirrel
red fox
short-tailed shrew
white-footed mouse
woodchuck

Sugar Maple–Basswood Forest

Indicators: Sugar Maple, Basswood, White Ash, Bitternut Hickory, Beech (Eastern Wisconsin Only)

Ecology

Before European settlement, the east central Wisconsin landscape contained large areas of old-growth sugar maple–beech–basswood or sugar maple–basswood forest. The sugar maple–basswood–beech forest was most extensive north and east of Lake Winnebago and south through central Manitowoc and Sheboygan Counties and in eastern Washington and western Ozaukee Counties. Beech reaches the western extent of its range in eastern Wisconsin and drops out as a part of the sugar maple–basswood forest in the rest of the state.

Both historic areas and remnants remaining today of sugar maple–basswood forest are not contiguous. These patches developed in protected areas in an otherwise fire-influenced landscape. Large forested tracts existed before extensive European settlement in Green County, an area centered in the central Kickapoo River Valley, and a northwest area in western Pepin and Dunn Counties and in eastern Pierce and St. Croix Counties. Though the sugar maple–

Sugar maple–basswood forest (photograph by Mike Mossman; courtesy of Wisconsin Department of Natural Resources)

Old-growth Forest

Old-growth does not refer to big trees or an end point in development. It's a term describing the ecosystem that develops as individual trees in the forest die from old age. Trees attaining great age play host to numerous shelf fungi. Many species of these fungi live on decaying material either in parts of the standing tree or everywhere within a fallen trunk. The trees may be large, but not all old-growth trees are large trees. An ecosystem that has been developing longer than the life span of the oldest trees is definitely an old ecosystem, but small patches may be disturbed somewhere in the forest. For instance, any disturbance that comes along—a tornado, forest fire, or death of a single tree—causes different events to occur at the point of disturbance. Most likely younger trees of the same species as the surrounding forest will grow in the disturbed area. The ecology and species composition of the gap will probably be different because it will contain species that do better in more sunlight. As the individuals in the gap become older, however, the natural processes and species composition will become closer and closer to the surrounding forest. Wind-blown light seeds of aspen and birch or oak acorns transported by blue jays can survive in the abundant sunlight, but they will eventually be replaced by sugar maple seedlings that do better under the shade of the newcomers. Taken as a large area, the forest remains quite stable, with a small amount of early succession species, a lot more of mid-succession species, a majority of late-succession species, and a small amount of ancient, virtually undisturbed forest.

Bird eating a land snail (illustration by author)

Forest Snails

Bird populations, especially those that winter in the tropics, have shown downward trends for many years. Numerous explanations have been offered, including loss of breeding habitat, loss of wintering habitat, loss of migration stop-over habitat, nest predation, nest parasitism by cowbirds, and habitat simplification.

Several studies in Europe have revealed a previously uninvestigated area of concern. In many habitats, calcium is deficient, hard to find, or concentrated in small patches. Birds need calcium for building strong egg shells—a familiar theme, for we nearly lost our bald eagles and peregrine falcons to eggshell thinning.

The research indicates that female forest birds, even seed eaters, actively search out and consume landsnails immediately prior to nesting. The thickness of many eggshells depends on how successful these birds were at finding snails (Graveland et al. 1994). Habitats with few snails have much less productivity than snail-rich habitats. Moist habitats and older forests have more snails than drier and younger forests. Some species of snails and birds may need the moisture conditions of an old forest over the long term to be sustainable.

basswood forests are extensive, they are still islands in a matrix of prairie and savanna. Because of their isolation, these islands of sugar maple–basswood forests each show regional variation in species composition.

The east central forests formed on the leeward sides of hills where the topography slowed or stopped fire. Near Lake Michigan, beech became a co-dominant in this forest. The presence of beech has profound effects on the forest because of a host of species that associate with it, such as beechdrops, a parasitic plant living only on beech roots. These associate species require beech as their sole food source (Niering and Olmstead 1979). These interrelationships enforce the notion of natural complexity because every species has several more species with which it interacts.

Range

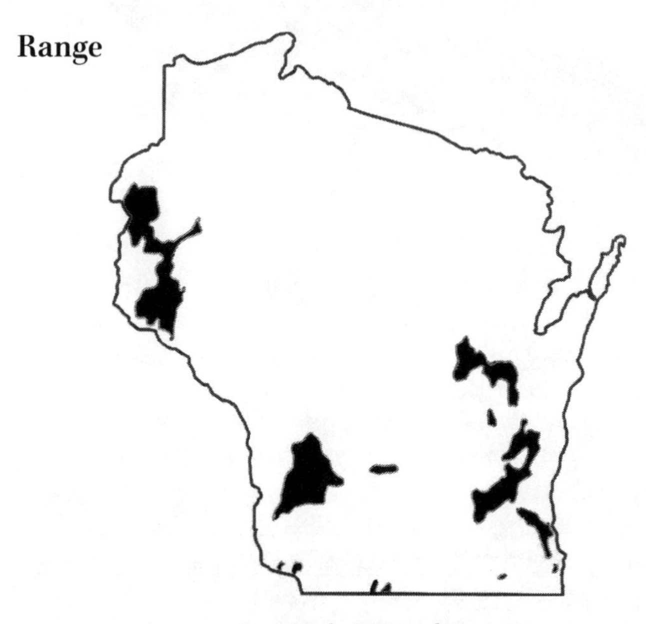

Sugar Maple–Basswood Forest

Two small areas of sugar maple–basswood forest occur in Green and Grant Counties, true islands in a sea of prairie and savanna. These forests are notable because they contain many species that reach their northern range limit here. Species diversity, however, is not quite as rich in these two areas as in other sugar maple–basswood forests because they are very isolated, and many species simply cannot cross the intervening land.

A large area of mixed sugar maple–basswood–oak forest with a significant white pine component occurs in the heart of the driftless area. The Kickapoo River may have been the most influential factor in this forest's development because the deep valleys and the river acted as a huge firebreak. East of the Kickapoo River, in Crawford, Vernon, Richland, and portions of Sauk Counties, is a large area where sugar maple–basswood forests are more abundant and have more plant species than the other forests discussed so far. Many species in the driftless area are relicts of post-glacial periods, which gives the area a smattering of northern species, such as eastern hemlock, yellow birch, and starflower.

A big block of sugar maple–basswood forest occurs in

eastern St. Croix, Polk, and Pierce Counties and in western Dunn and Pepin Counties. There is less diversity of tree species in this forest than in others, but the forest has the most diverse groundlayer species of any southern sugar maple–basswood forest in Wisconsin. Similar to the Big Woods of Minnesota, this forest owes its diversity to geography and subsurface geology. It lies at the crossroads of prairie–savanna, northern forest, and southern forest, so many species have range limits near here. Because this forest grows on nutrient-rich, shallow soil over limestone, many plants thrive that would not survive on more nutrient-poor soils.

These forests were self-perpetuating and originally covered nearly 3,450,000 acres in Wisconsin. Through fragmentation and conversion to rich farmland, only about 50,000 acres of this forest type remain. The shaded areas on the map show the present sugar maple–basswood forests, although conversion to farmland reduced the forests in eastern Wisconsin and in Green County since the data were gathered. Many remaining sites have the appropriate tree species but do not have the full complement of groundlayer species. Because these forests are small, they are susceptible to invasion by a few exotic species, resulting in a very poor herb layer, much different from the original forest.

The state's flora has lost several species that require this habitat. Current examples of old-growth maple–basswood forest can be found at Green County Preserves, Interstate State Park, Maribel Caves, North Kettle Moraine, and Wyalusing State Park.

Activities

- List the plant species found within 3 to 5 feet of a developed trail, and those more than 25 feet away from the trail. Do the same exercise on a deer trail. Compare the lists and identify the exotic species. Are exotic species the reason for the differences in the list?
- Observe interior forest herbs in summer and fall. Open ripened pods and examine the seeds. List ways the seeds could make it to an adjacent wooded area (Kricher 1988).
- Measure the depth of the leaf litter. Dig a trench perpendicular to the surface to get a leaf litter profile. View the process of decomposition and its effect on soil formation and maintenance. You should be able to observe different layers of decomposition with color and texture changes (Kricher 1988).

- Visit a medium- to large-sized beech tree in September. Look near the tree for the flowering stalks of beechdrops. Once you find them, map the extent of the beechdrops. Describe their relationship to the beeches (Niering and Olmstead 1979).

Characteristic Species

Plants

annual bedstraw
bloodroot
blue cohosh
blue marsh violet
downy Solomon's seal
early meadow-rue
hairy sweet cicely
large-flowered trillium
maidenhair fern
May apple
Pennsylvania sedge
rattlesnake fern
sharp-lobed hepatica
spring beauty
Virginia waterleaf
wild leek
wood-nettle
yellow wood violet

Mosses and Lichens

blister lichen
broom moss
common beaked moss
common cedar moss
common fern moss
feathered neckera
flat-stemmed entodon
giant byrum
knight's plume
knothole moss
mealy goblet lichen
mountain fern moss
northern leucodon
round stemmed entodon
silvery byrum
slender cedar moss
sod lichen
spreading leather
wavy broom moss
whip fork moss
white moss
wiry fern moss
wrinkled shield lichen

Mushrooms

bearded tooth
blue mycena
dead man's fingers
elegant stinkhorn
flat crep
golden pholiota
hairy rubber cup
half-free morel
jelly leaf
maple argocybe
pinwheel marasmius
small chanterelle
white morel

Butterflies and Moths

ashen pinion
basswood leafroller
black-blotched schizura
black-etched prominent
blinded sphinx
clymene hapola
curve-lined acontia
eyed baileya
galium sphinx
harvester
imperial moth
Io moth
linden bark borer
linden looper
maple zale
ochre dagger
olive angle shades
polyphemus moth
question mark
saddled prominent
white-stripped black
yellow-banded underwing
yellowhorn

Beetles

banded fungus beetle
banded sap beetle
basswood leaf miner
blackish tiger beetle
darkling beetle
eastern snail-eater
flat ground beetle
fuscous May beetle
linden borer
living beech borer
May beetle
odor-of-leather beetle
paederine rove beetle
pole borer
red sap beetle
short-winged blister beetle
six-spotted sap beetle
small ironclad beetle
snail-eating ground beetle
snout beetle
sugar maple borer

- twig pruner
- velvety bark beetle
- white-banded fungus weevil

Insects

- basswood lace bug
- Carolina ground cricket
- red carpenter ant
- slender-legged camel cricket
- snowflea
- snow scorpion

Spiders

- brown recluse
- daring jumping spider
- elegant crab spider
- filmy dome spider
- forest wolf spider
- house spider
- northern widow
- spined micrathena

Landsnails

- allied labyrinth snail
- apple seed snail
- armed pupa snail
- five-toothed pupa snail
- fraternal pill snail
- great zonite shell
- Holzinger's pupa snail
- minute disk snail
- modest pupa snail
- moss pupa snail
- open disk snail
- ripe forest snail
- striped forest snail
- three-toothed forest snail
- tiny harp snail
- toothless pupa snail
- tree zonite shell
- white lipped forest snail

Reptiles and Amphibians

- blue-spotted salamander
- eastern garter snake
- northern ring necked snake
- spring peeper
- wood frog

Birds

- eastern wood-pewee
- great crested flycatcher
- hairy woodpecker
- ovenbird
- pileated woodpecker
- red-bellied woodpecker
- red-eyed vireo
- scarlet tanager
- veery
- wood thrush

Mammals

- eastern chipmunk
- masked shrew
- white-footed mouse

Floodplain Forest and Ash Swamps

Indicators: Silver Maple, Green Ash, Swamp White Oak, Black Willow, Cottonwood, River Birch

Ecology

Floodplain forests occur mostly along medium to large rivers and develop on a variety of soils, including gravel, fine sands, and deep silts. A few sites have deposits of clay that prevent water from draining into the soil, resulting in ponds perched above the level of the river. The eroded soils of the watershed are transported by the flooding river, suspended in the flood waters, and precipitated out as sand, silt, or gravel. Repeated flooding creates canopy openings, as the flood waters damage individual trees by undermining the root structures of certain species, such as silver maple and river birch. Floodplain forests usually have high water only in spring or after periods of heavy rainfall.

Species composition in any one spot depends greatly on the regularity, duration, and intensity of the floods, which erode some areas of the floodplain and deposit soils in others. Soil deposits determine combinations of tree species. On soils where silt has accumulated, silver maple, green ash, red maple, and American elm (before most of the elms were lost to Dutch elm disease) are the most abundant trees. Black willow and cottonwoods can grow anywhere but frequently occur near the watercourses or on exposed sandbars. Along the Chippewa, St. Croix, and Wisconsin Rivers, river birch forms a significant portion of the canopy. In well-drained areas, swamp white oak, yellowbud hickory, butternut, or hackberry can be very plentiful. Additional canopy species are basswood, red elm, and black ash.

Range

The map shows the areas of floodplain forest in the state. Ash swamps, however, occur in patches too small to map at this scale. Ash swamps are found mostly along Lake Michigan

Floodplain Forest

Elm Ecology

Elm trees, with their flexible branches swooping to form umbrella-like crowns, were a major component in floodplain forests. Three species of elm (slippery, American, and rock) were all affected by Dutch elm disease, which arrived in the 1950s and continues today, and most of the elms have disappeared from the forest. The loss of a dominant species such as elm is bound to have had an effect on the forest structure and composition because each species has so many other species that associate with it. In some instances, one species is so dependent on another that it cannot survive without it.

When looking at floodplain forests today, a naturalist can see standing dead elms that are somewhat protected from decay and many fallen trunks that are rapidly being transformed by decomposers, such as elm borers, which now have plenty of habitat. Mycologists often focus on the same trees because some species of morel mushrooms thrive on the decaying elm roots.

But what about the species that use the living trees? How has the elm tree population crash affected other species, such as the elm casebearer, the elm spanworm, the elm sphinx, or the elm leaf beetle? We don't know because these species have not been studied. The elm borer is probably doing fine right now because it feeds on those standing dead elms. But what about the elm calligrapha that bores under the living bark of larger trees? And how have the Baltimore orioles coped? Their favorite nesting trees were most definitely elms. Like the oriole, some species can utilize other trees for nesting, but a few species are dependent on the elm trees for their survival. When one species is lost, most of the time several other species are also lost.

Dead elm (illustration by Jim McEvoy; courtesy of Wisconsin Department of Natural Resources)

Morel mushroom (photograph by Thomas A. Meyer; used by permission)

border counties in the low areas of landscapes carved out by advancing and retreating glaciers. They depend on constant water and highly organic soils, usually on an impervious clay layer. Floodplain species, such as green ash and American elm, dominate the canopy in these low, wet areas, although silver maple occasionally becomes dominant. Groundlayer plants are often sparse because in most years surface water lasts well into summer.

Floodplain forests are most diverse near the Illinois border. Traveling north, species drop out at a slow but perceptible rate. The largest and most intact remaining floodplain forests are found along the lower Wisconsin, lower Chippewa, St. Croix, Milwaukee, Yellow, Black, lower Wolf, and Jump Rivers and at scattered locations on the Mississippi River.

These forests are still intact to varying degrees, mainly because humans, even with all their ingenuity, haven't found a better "use" for the land. With the exceptions of reservoirs, poorly planned drainage schemes, and ruts from logging operations, floodplain forests have been minimally altered, although some hardwood swamps have been drained. In the mid-1800s floodplain forests and hardwood swamps covered about 420,000 acres; today they cover approximately 255,000 acres.

Excellent examples of floodplain communities can be seen at Avon Bottoms, Interstate State Park, Lake Pepin, Lower Wisconsin River Valley, Necedah National Wildlife Refuge, North Kettle Moraine (especially hardwood swamps), Upper Mukwonago River, White River Marsh, Wildcat Mountain State Park, and Wyalusing State Park.

Activities

- Visit a floodplain forest daily in spring and list the species and numbers of migratory birds. Make a list when the floodplain has experienced a spring flood, leaving numerous pools of water, and one when the floodplain is dry. Compare the differences in migrant bird use.
- Visit sandy islands on the Wisconsin or Chippewa Rivers in late summer. List the different species found as the island gets drier. Slightly higher portions of the islands will dry earlier and have a longer period for plant growth and thus different species composition. Notice the patterns of plant species; they often form concentric rings around the island according to elevation.
- Find a dead tree with the bark still attached and mark it so you can find it again. Identify and map the location of a few

mushroom species nearby. Measure the distance and direction the mushrooms are from the tree's base. Return for several years and document the movement of the mushroom species in distance and direction from the tree. Describe the ecological interaction (Kricher 1988: 279–84).

- Find a tree hole with water in it. Take a sample of the water and place the contents under magnification. Observe the diversity of life.

Characteristic Species

Plants

blue marsh violet
clearweed
common elderberry
common water-horehound
false nettle
fringed loosestrife
germander
green dragon

Green dragon (courtesy of Wisconsin Department of Natural Resources)

honewort
hop sedge
jack-in-the-pulpit
late goldenrod
mad-dog skullcap
orange jewelweed
riverbank grape
sensitive fern
Virginia wild rye
white grass
wood-nettle

Mosses and Lichens

blister lichen
broom moss
common beaked moss
common cedar moss
common fern moss
common tree moss
feather moss
flat-stemmed entodon
mealy goblet lichen
nodding moss
northern leucodon
pointed mnium
silvery byrum
sod lichen
southern leucodon
spreading leather
velvet tree apron moss
wavy broom moss
whip fork moss
wrinkled shield lichen

Mushrooms

ash-tree bolete
birch polypore
bone polypore
clinker polypore
destructive pholiota
elm oyster
little nest polypore
scaly-inky cap
velvet foot
yellow-unicorn entoloma
yellow morel

Butterflies and Moths

Abbot's sphinx
achemon sphinx
American ermine-moth
ash borer
ashen pinion
bella moth
big poplar sphinx
birch casebearer
birch dagger
black-barred brown
black-etched prominent
blinded sphinx
blind-eyed sphinx
Canadian melanophila
Carolina sphinx
carpenterworm
cecropia moth
clymene hapola
cottonwood dagger
curve-lined owlet
darling underwing
dock rustic
Doris tiger moth

double-toothed prominent
eastern tiger swallowtail
eight-spotted forester
elm casebearer
elm spanworm
elm sphinx
five-spotted hawk moth
giant swallowtail
grapevine epimenis
great ash sphinx
Grote's sallow
hackberry butterfly
hag moth
Io moth
laurel sphinx
lesser grapevine looper
little wood satyr
moonseed moth
mourning cloak
nessus sphinx
ochre dagger
one-eyed sphinx
pepper and salt geometer
polyphemus moth
promethea moth
red-humped carpenter moth
scribbler
small engrailed
small-eyed sphinx
spear-marked black
spotted apatelodes
spotted tiger-moth
stout spanworm
sweetheart underwing
thin-lined owlet
Thoreau's flower moth
unclear dagger
unicorn caterpillar
viceroy
waved sphinx
white-lined sphinx
white underwing

Beetles
bark hister beetle
big-jawed rove beetle
black elm bark weevil
black firefly
bombardier beetle
broad rove beetle
bronze tiger beetle
brown prionid
cottonwood borer
cottonwood leaf beetle
cross-toothed rove beetle
eastern snail-eater
elm borer
elm calligrapha
elm leaf beetle
faded click beetle
flat ground beetle
golden net-wing
goldsmith beetle
grape flea beetle
grape rootworm
hackberry engraver
long-necked longhorn
marginated flower longhorn
native elm bark beetle
obscure rove beetle
pinching bug beetle
pole borer
red elm bark weevil
red-necked darkling beetle
red-shouldered fungus beetle
red-spotted fungus beetle
rustic borer
sculptured sap beetle
striated ground beetle
timber borer
two-spotted bark beetle
willow leaf beetle
woodland ground beetle

Insects
black carpenter ant
elm lace bug
elm leaf miner
grape leafhopper
hackberry lace bug
large-headed grasshopper
oystershell scale
red carpenter ant
red plant bug
slender-legged camel cricket
snow scorpion
tree-hole mosquito
willow plant bug
window-pane fly

Spiders
brown widow
daring jumping spider
elegant crab spider
forest wolf spider
hammock spider
house spider
marbled spider
parson spider
three-banded crab spider
two-spotted spider mite

Landsnails
amber retinella
apple seed snail
armed pupa snail
brown pupa snail
Foster's forest snail
great zonite shell
handsome vallonia
Lea's pill snail
minute disk snail
modest pupa snail
parallel disk snail
ripe forest snail
shining zonite shell

- Singley's disk snail
- small zonite shell
- southern pill snail
- striped forest snail
- three-toothed forest snail
- tiny harp snail
- toothless pupa snail
- tree bark pupa snail
- white-lipped forest snail
- white swamp snail
- white zonite shell

Reptiles and Amphibians

- blue-spotted salamander
- central newt
- eastern gray treefrog
- eastern hognose snake
- fox snake
- green frog
- northern water snake
- spring peeper
- wood frog

Birds

- American redstart
- Baltimore oriole
- barred owl
- blue-gray gnatcatcher
- blue-winged warbler
- eastern wood-pewee
- great crested flycatcher
- hairy woodpecker
- least flycatcher
- ovenbird
- pileated woodpecker
- red-bellied woodpecker
- rose-breasted grosbeak
- veery
- warbling vireo
- wood duck
- wood thrush
- yellow-throated vireo

Mammals

- beaver
- coyote
- eastern chipmunk
- masked shrew
- red fox
- short-tailed shrew

Dry Pine Forest

Indicators: Red Pine, Jack Pine, Hill's Oak, Large-Toothed Aspen, White Pine

Ecology

Although soils and climate determine where dry pine forests can potentially grow, fire determines where the forests actually develop. To develop as forests, red and white pines need protection from intense fires long enough to develop thick bark (40 to 50 years). The east and north sides of lakes and larger streams provide protection from frequent conflagrations. Other areas, such as steep north-facing slopes or islands and peninsulas in bogs and wetlands, offer enough protection for development of dry pine forest.

Jack pine needs fire to replace itself naturally because its cones usually open after being heated by fire. Also, a fire-scorched landscape is ideal for germinating the light wind-borne seeds of red and white pine. Because the seedbed is no longer prepared by fire, naturally regenerated pine forests have become very rare.

Exploitation of red and white pine was intensive in the early days of logging. After harvest, immense slash accumulated, and massive fires scoured the area. The removal of most white

Burnt Wood Community

Fire prevention is a crucial activity in Wisconsin. Countless lives, structures, and income-producing trees have been spared due to our fire prevention and control efforts. Control and prevention of fire has tremendous societal benefits, but in some areas we may need to consider the adverse effects of fire suppression on fire-influenced ecosystems.

In nature, no one act is simple and complexity rules. For example, fire benefits many species in a pine forest, in addition to the pines which rely on fire for regeneration. After a burn, species like crows, ravens, jays, foxes, and weasels scavenge the area for food. Olive-sided flycatchers often perch on the dead snags and sally for flying insects. Black-backed woodpeckers inhabit burned conifer areas, sometimes in dense numbers, for several years after a fire to feed on insects found under the bark and in the wood. They forage for grubs and larvae, which feed on the dead trees, by scaling off large pieces of bark and lapping up the exposed insects. Even humans will search recently burned sites for mushrooms, such as pink burn cup.

Some insect species breed and lay eggs only on the bark of recently burned conifers. About 40 species of insects, mostly beetles, fly to recently fire-killed conifers to lay their eggs. One group of these insects in the metallic wood-boring beetle family have antennae that sense infrared radiation emitted by the fires (Hart 1998). They then fly to the fire to beat their competition for this resource. It isn't known how far away they can sense this radiation, but individuals have shown up at a burn site more than 60 miles from the nearest conifers.

Dry pine forest (photograph by Eric Epstein; used by permission)

and red pine prior to the fires left few trees to serve as seed sources. Aspen and white birch seeds were more available, and the forest changed in composition. This turnover in composition continues today because the forests are intensely managed for aspen pulpwood. Although red pines are still a major part of the dry pine forest, they are mostly grown in dense plantation rows which block most of the sunlight and leave the forest floor with few groundlayer species.

Sawyers

Sometimes on a calm quiet morning with few birds singing, scraping sounds can be heard in the pine woods. The gnawing sound carries only a few meters and is much like the sound of a wood awl cutting into a block of pine. This is the sound of pine sawyers doing their business. No, not loggers with chain saws or lumberjacks with crosscuts—these are the grubs of several species of long-horned beetles eating the insides of the pine trees. The most common are the Carolina sawyer and the white-spotted sawyer. The sound is made by the mouth parts of the soft larvae scraping a piece of pine wood off for consumption.

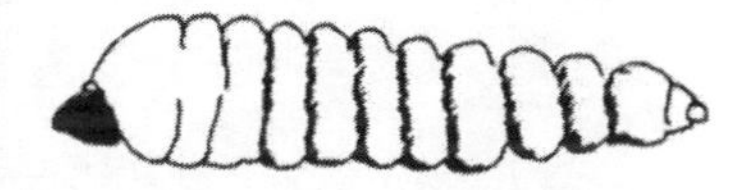

Pine sawyer larva (illustration by Jim McEvoy; courtesy of Wisconsin Department of Natural Resources)

Dry Pine Forest

Range

Most of Wisconsin's pine forests developed on four areas of sandy glacial deposits. If viewed from a satellite, the pine forests appear as islands of different vegetation imbedded in a sea of northern hardwood forest. These islands are found in portions of Marinette, Vilas, Oneida, and northern Lincoln Counties, and in a strip from Polk County diagonally through most of Burnett, southern Douglas, and northern Washburn Counties, to nearly the tip of the Bayfield Peninsula. Pine forests are also found in the northern portion of glacial Lake Wisconsin, northern Adams and Juneau Counties, southeastern Jackson County, and southwestern Wood County.

Examples of dry pine forest are found at Black River State Forest, Moquah Barrens, Necedah National Wildlife Refuge, and Vilas County. There is a virgin northern pine forest at Lucius Woods County Park in the Douglas County Sharptail Barrens section. An excellent example of a dry pine forest can be seen on Stockton Island in the Apostle Islands. A southern relict pine stand with all three pine species occurs at Governor Dodge State Park.

Activities

- One way to age a red pine is to count the whorls of limbs. A red pine adds one row of branches every year. A young tree's age can be easily estimated, but older trees lose their bottom branches and the estimation process becomes more difficult.

- When the media report insect outbreaks or you notice higher than normal numbers of insects, spend some time at a pine forest watching the birds for their response to the abundant food source.
- Enter an old stand of red pine and dig through the litter of needles. Upon reaching bare soil, examine the different layers of needle decomposition and describe the decomposition process (Kricher 1988).
- Visit an area that was burned by a large wildfire, whether a few days or a few decades ago (records of old wildfires are kept by fire control officials). Observe the plant response to the fire, such as flowering and emergence of numerous mushroom species. Observe the effect fire has on the pine forest community, such as new pine seedlings or birds using the blackened snags for perches. Compare the burned to the unburned areas and note the differences (Curtis 1959; Forman and Godron 1986).

Characteristic Species

Plants

American hazelnut
barren strawberry
beaked hazelnut
big-leaved aster
bracken fern
bush honeysuckle
Canada mayflower
dwarf raspberry
early low blueberry
false Solomon's seal
pipsissewa
spreading dogbane
starflower
wild sarsaparilla
wintergreen
wood anemone

Mosses and Lichens

antler lichen
British soldiers
common feather moss
cornucopia cladonia
curd lichen
empty-cup lichen
flabby lichen
ladder lichen
many-fruited dog lichen
matted byrum
mealy goblet lichen
pitted cetraria
puffed shield lichen
reindeer lichen
water measuring cord moss
yellow pine lichen
yellow wall lichen

Mushrooms

conifer false morel
crustlike cup
deadly cort
dirty milky
dusky waxy cap
elegant polypore
false chanterelle
family collybia
fetid armillaria
irregular earth tongue
orange jelly
peppery bolete
rooting cauliflower mushroom
rosy polypore
slippery jack
yellow pholiota
yellow rabbit ears
yellow tuning fork

Butterflies and Moths

bicolored moth
big poplar sphinx
black zigzag
brown collared dart
Canadian sphinx

Canadian tiger swallowtail
carpenter moth
catocaline dart
Compton's tortoise shell
Comstock's sallow
connubial underwing
darling underwing
decorated owlet
dreamy dusky wing
Esther moth
fillet dart
five-lined gray
hobomok skipper
jack pine budworm
LeConte's hapola
manto tussock
neighbor
northern pine sphinx
orange sallow
pine elfin
pink-striped oakworm
red-fronted emerald
red-humped caterpillar
six-spotted gray
snowberry clearwing
sordid underwing
virgin moth
white underwing

Beetles

aspen leaf beetle
black-horned pine borer
black oak weevil
bronze poplar borer
Carolina sawyer
eastern raspberry fruitworm
fire beetle
goldsmith beetle
hazelnut weevil
northern pine weevil
pine engraver
pine heartwood borer
pine weevil
poplar borer
red-necked cane borer
red pine flat-headed borer
red turpentine beetle
rough flower beetle
six-spined engraver
white-spotted sawyer

Insects

black carpenter ant
broad-winged katydid
brown stink bug
coral-winged grasshopper
huckleberry grasshopper
mining bee
oak treehopper
ornate pygmy grasshopper
oystershell scale
pine spittlebug
poplar plant bug
red carpenter ant
sanded pygmy grasshopper
Saratoga spittlebug
spined soldier bug
sweat bee

Spiders

bowl and doily spider
daring jumping spider
eastern wood tick
elegant crab spider
forest wolf spider
hammock spider
inconspicuous crab spider
metaphid jumping spider
purse-web spider

Reptiles and Amphibians

blue-spotted salamander
fox snake

Birds

blackburnian warbler
chipping sparrow
dark-eyed junco
eastern wood-pewee
hermit thrush
ovenbird
pine warbler (large pines only)
red-breasted nuthatch
yellow-rumped warbler

Mammals

coyote
least chipmunk
red fox
red squirrel
white-footed mouse
woodchuck

White Pine–Hardwood Forest

Indicators: White Pine, Red Maple, Red Oak, Paper Birch, Red Pine

Ecology

White pine–hardwood forests were common throughout northern Wisconsin at the time of European settlement, covering nearly 2,000,000 acres. The primary trees of this natural community were white pine, red pine, red oak, and red maple, in various combinations. The most diverse combinations of tree species formed on sandy soils with more organic nutrients and good water drainage. On sandy soils, red pines were more numerous. On rich loamy soils, hardwoods played a more important role, although occasionally a pine–hardwood forest can develop on richer soils after a major fire. Long-lived pine–hardwood communities developed in areas with a history of infrequent fires and soils falling in the midrange between droughty sands and rich loam. Aspen and paper birch were also part of the pre-European settlement forest, competing with the pines and oaks after catastrophic fires. Pines and hardwoods interspersed with these short-lived species, forming a complex forest of different patches and ages.

Around the mid-1880s, the pine barons began exploiting the forests, nearly eliminating the white pines. Follow-up logging of the hardwoods, along with intense fires that sterilized the soils, converted much of this community to an early succession forest dominated by aspen and paper birch. Although fire normally benefits pine reproduction, not many pines were left to provide seed after the forests were logged.

Today's forests barely resemble the diverse pine–hardwood forest of 150 years ago, which developed over centuries with young trees, mid-aged trees, and 400-year-old monarchs forming the canopy. Today, demand for aspen and pine pulp, white-tailed deer, and ruffed grouse encourages land managers and growers to manage the forests accordingly. This focused management emphasizes young trees, which are cut for products, so old large trees do not develop. One or two tree species usually dominate in these forest stands. The forest community is less diverse in these managed forests because species that do best in older forests are missing.

Large tracts of old-growth pine–hardwood forests, with their more diverse structure, composition, and function, are very rare. Restoring large areas to older, more diverse pine–hardwood forest is unlikely to happen because of the immense pressure on the forest for products.

Nest with a View

White pines can grow to be the tallest trees in Wisconsin and can live for several hundred years. These monarchs of the forest also often become lightning rods in thunderstorms and are struck again and again. This repeated assault leaves many of the tallest trees with thin, spindly crowns.

White pines provide a living space for two well-known species, the bald eagle and the osprey. The height of the trees and the open crowns caused by lightening strikes or by bark girdling by porcupines offer an ideal platform for huge stick nests. Sometimes this preference backfires when the nest is struck by lightning, an example of the risks species face at the hand of nature.

White pine–hardwood forest (illustration by Jim McEvoy; courtesy of Wisconsin Department of Natural Resources)

Range

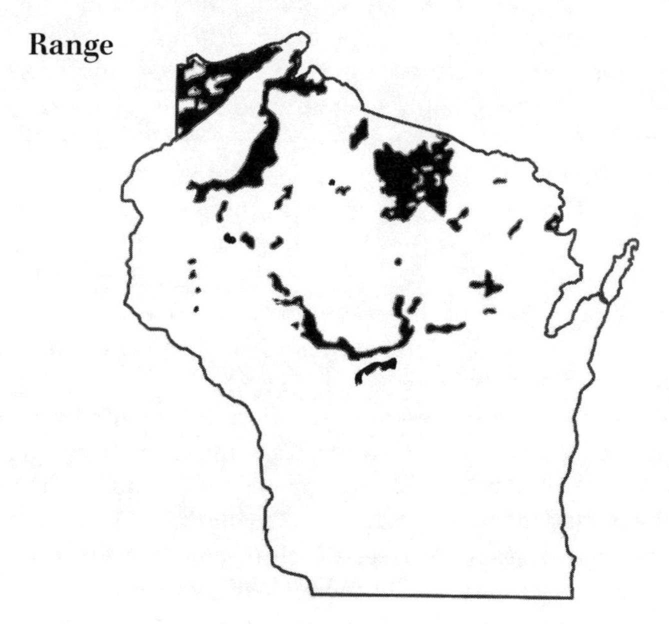

White Pine–Hardwood Forest

Stands of white pine still persist in southwestern Wisconsin on areas too small to be mapped, occasionally in combination with red pine and jack pine. They are relicts of an age when conifer forests covered the area. These relicts persist on sandstone cliffs with shallow soils because the position of the rock allows ample light to reach the ground for seed germination. A small number of northern herbs, such as starflower, Canada mayflower, and partridgeberry, grow in the groundlayer, which varies with each site. A gradual reduction of northern herbs occurs as you go south.

Examples of white pine–hardwood forests can be found in the Apostle Islands, Black River State Forest, Blue Hills, Copper Falls State Park, Lakewood Area, Moose Junction, Vilas County, Whitefish Dunes State Park, and Wisconsin Point.

Relict stands can be found in the Baraboo Hills and Governor Dodge State Park.

Activities

- Find a forest of white pine and hardwoods. Lay out a grid 25 meters by 25 meters (or yards), count the trees larger than two inches in diameter in the plot and list them in a table according to the most numerous. Repeat the process elsewhere in the forest. If one species comprises most the individuals, then it is very dominant (Pratt 1995: 103–4).
- Look for cavities in dead and living trees. Observe the cavity entrance for activity. As many as 85 bird species use tree cavities for nesting. Count the number of tree holes in the same plots as above. Describe the value to the forest ecosystem of tree cavities (Green 1995).
- Find a basswood, birch, or pine tree with rows of holes in the bark and observe. The holes were most likely drilled by yellow-bellied sapsuckers. Each tree is visited on a regular basis by the sapsuckers. At other times, various species use this sugary food source or prey upon those so attracted. Visitors include hummingbirds, bees, ants, flies, spiders, flying squirrels, and many more.
- Observe foraging flocks of birds. You should notice that several species are participants in these mixed foraging flocks, sometimes called guilds. Observe their foraging behavior. Do they concentrate their activities on a few species of plants?

Characteristic Species

Plants

beaked hazelnut
big-leaved aster
bluebead
bracken fern
bush honeysuckle
Canada mayflower
downy Solomon's seal
early low blueberry
false Solomon's seal
ground-pine
large-leaved shinleaf
maple-leaved arrow-wood
partridgeberry
rough-leaved ricegrass
starflower
swamp fly honeysuckle
sweet-scented bedstraw
twisted stalk
wild sarsaparilla
wood anemone

Mosses and Lichens

blister lichen
common cedar moss
crisped lichen
curd lichen
dog lichen
feather moss
flabby lichen
lung lichen
many-fruited dog lichen
swollen lichen
water measuring cord moss

Mushrooms

bay bolete
chicken-fat suillus

conifer-cone baeospora
dark bolete
fibril trich
gray-veil amanita
painted suillus
pinecone tooth
tree ear
velvety fairy fan
yellow-red gill polypore

Butterflies and Moths

alternate woodling
American angle shades
banded purple
brown-lined sallow
Compton's tortoise shell
figure-seven moth
galium sphinx
gray comma
gray scoopwing
green arches
green comma
half-wing
harvester
imperial moth
Indian skipper
mournful thyris
pearly eye
pink-edged sulphur
question mark
spotted thyris
stalk borer

Beetles

black-horned pine borer
notch-mouthed ground beetle
pinching bug beetle
pine heartwood borer
pine stump borer
red-horned grain beetle
red oak borer
red sap beetle
ribbed pine borer
ten-lined June beetle
white pine weevil

Insects

big-leaf aster plant bug
black carpenter ant
four-lined plant bug
mining bee
pine spittlebug
poplar plant bug
potter wasp
red carpenter ant

Spiders

daring jumping spider
elegant crab spider
filmy dome spider
forest wolf spider
inconspicuous crab spider
marbled spider
metaphid jumping spider
purse-web spider
three-banded crab spider

Landsnails

fraternal pill snail
modest pupa snail
open disk snail
tiny harp snail
tree zonite shell

Reptiles and Amphibians

blue-spotted salamander
fox snake

Birds

Baltimore oriole
black and white warbler
black-billed cuckoo
broad-winged hawk
chestnut-sided warbler
eastern wood-pewee
hairy woodpecker
ovenbird
red-eyed vireo
rose-breasted grosbeak
ruffed grouse
scarlet tanager

Mammals

coyote
eastern chipmunk
least chipmunk
red squirrel
white-footed mouse
woodchuck

Northern Hardwood–Hemlock Forest

Indicators: Hemlock, Sugar Maple, Yellow Birch, Basswood, Beech (Eastern Wisconsin Only)

The northern hardwood–hemlock forest was the largest feature of Wisconsin's landscape before European settlement, covering one-third of the state's area or nearly 14,500,000 acres. At the turn of the twenty-first century, this forest type is still significant. Hardwood or hardwood–hemlock forests still cover nearly 7,700,000 acres in the northern part of the state, which ranks second only to agricultural lands in percentage of landscape covered.

Statewide, this forest has different patterns of tree distribution. The forests of the northeast are the most diverse, having all Wisconsin's northern hardwoods in the canopy. Of the five major tree species (sugar maple, basswood, yellow birch, hemlock, and beech), three drop out as one travels west. Beech drops out of the canopy at the Wolf River, hemlock drops out in Sawyer and Bayfield Counties, while yellow birch persists to the St. Croix River. Farther west only sugar maple and basswood persist. In the western counties, the canopy becomes very similar to the trees listed in the Sugar Maple–Basswood Forest chapter, though the groundlayer is much different. More northerly species, such as Carolina spring beauty, small Bishop's cap, and small enchanter's nightshade, replace the spring beauty, Bishop's cap and enchanter's nightshade found in the southern sugar maple–basswood forest.

Changes occur across the landscape, but surprisingly, much greater change can happen across short distances. For example, a walk across the few hundred feet comprising a drumlin can reveal different species from one side to the other and different species at the base and the top. Northern hardwood forest composition can have any one of several tree combinations, depending on soils, topography, moisture (including drainage patterns), and distance from seed-bearing trees.

Range

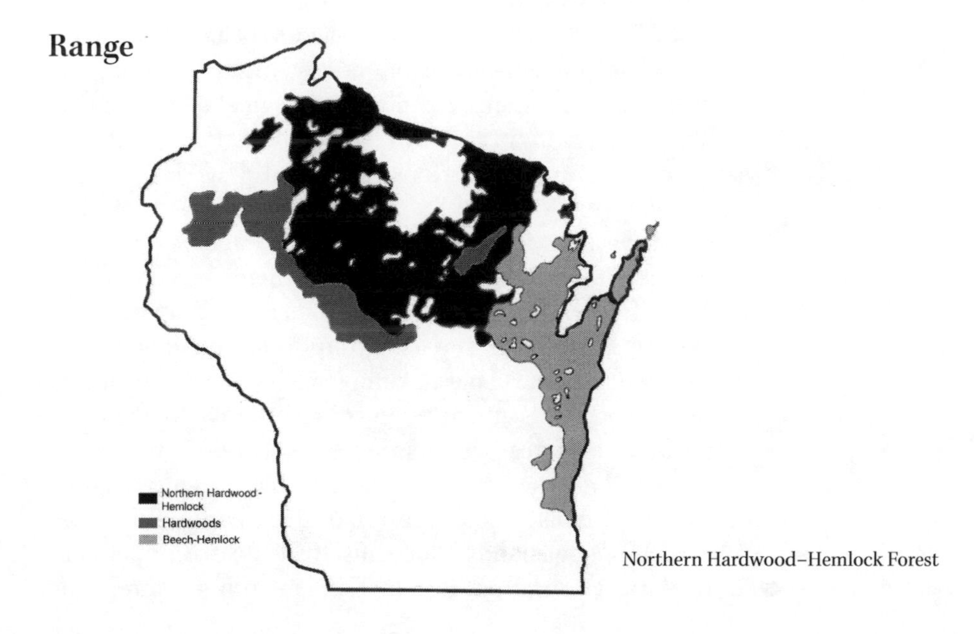

Northern Hardwood–Hemlock Forest

Old-growth northern hardwood–hemock ecosystem (courtesy of USDA Forest Service)

The map shows areas of the state with a distinct hardwood dominance. Sugar maple, white ash, and yellow birch, with occasional hemlock and red oak, are often the most abundant tree species. A large area in the center of Marathon County has a forest of this type, and to the west, forests in Polk, Barron, and Washburn Counties have sugar maple, yellow birch, and basswood as dominant species.

The northern hardwood–hemlock forests were much different before European settlement. These forests covered large tracts with towering white pines that seemed almost primeval to the European settlers. In reality, these 1850s forests were not all ancient trees; many areas did not have white pines, and as much as 10 percent of the forest was young trees.

Today's hemlock forest is a mere shadow of its former abundance. Hemlock does best on poorly drained areas near swamps or on lakeshores. Hemlock was in great demand in the early years of logging because the leather industry used the bark as a source of tannin. The extensive removal of the seed-bearing trees, changes in local conditions, and shifts to a warmer, drier climate have all had negative impacts on hemlock.

As with other northern forest types, there are a few relicts in the driftless area. These small stands are mostly hemlock and yellow birch, with a sparse groundlayer made up of distinctively northern species. Examples of relicts can be found in the Baraboo Hills and at Wildcat Mountain State Park.

Because this forest type is so large, most areas in the northwestern and northeastern regions have a representative mix of this forest type. Good examples of these communities are found at Apostle Islands, Blue Hills, Brule River, Copper Falls State Park, Harrington Beach, Kohler–Andrae State Park,

Lakewood Area, Marathon County, Maribel Caves, Moose Junction, Newport State Park, Northern Bayfield County, Point Beach State Forest, Rock–Washington Islands, Three Lakes, Vilas County, and Whitefish Dunes State Park.

Activities

- Visit one of the sites listed above and sketch the structure of the forest. Include woody debris lying on the forest floor, standing dead trees, and mounds of soil that formed when a tree tipped over. Observe the plants growing on the exposed soils of the mound. Yellow birch is a common sapling on such mounds, as is white pine.
- Find an old yellow birch tree with shelf fungi growing near the base or a few feet up the trunk. Mark the tree and return for several years. Measure the advance of shelf fungi up the trunk and the growth of each individual fungus (Kricher 1995).
- Observe several hemlock logs lying on the ground. List the tree species growing on the logs. Compare the species on different logs and try to explain the differences. A hemlock log can take more than 100 years to decompose, and species using downed logs change as decomposition proceeds. Once a hemlock falls, it still provides habitat for well over 100 years (Maser and Trappe 1984).
- Go to a hemlock stand after a severe thunderstorm. Search the area for downed trees and observe the lichens growing on the upper branches and trunk. Describe their shape, abundance, and associated species. Many species of lichen only grow on these old trees, and they provide habitat for invertebrates of many species, which in turn provide a superb source of food for birds (Pettersson et al. 1995).
- Visit a hemlock relict in the southern part of the state and list the plant species found there. Compare the species list with that of a northern hemlock forest. Develop a sapling list for the plants growing under the hemlocks and one for adjacent hardwoods, then compare them (Curtis 1959).

Characteristic Species

Plants

big-leaved aster
bluebead
blue marsh violet
Canada mayflower
downy Solomon's seal
hairy sweet cicely
lady fern
large-flowered trillium
partridgeberry
shining clubmoss
small Bishop's cap

starflower
stiff clubmoss
swamp fly honeysuckle
sweet-scented bedstraw
tall white violet
twisted stalk
white baneberry
wild sarsaparilla
wood anemone
yellow wood violet

Mosses and Lichens

blister lichen
blue-dot lichen
bristly lichen
broom moss
common beaked moss
common fern moss
common tree apron moss
dog lichen
feathered neckera
flat-stemmed entodon
frayed lichen
giant byrum
knight's plume
lung lichen
many-fruited dog lichen
mealy goblet lichen
mountain fern moss
powder gun moss
puffed shield lichen
round-stemmed entodon
script lichen
silvery byrum
slender cedar moss
spreading leather
wavy broom moss
whip fork moss
white moss
wrinkled shield lichen
yellow-green pored lichen

Mushrooms

angel's wings
birch polypore
bitter bolete
bleeding conifer parchment
blue-pored polypore
comb tooth
death cap
flat-topped coral
hairy parchment
hemlock varnish shelf
king bolete
lichen agaric
lilac-brown bolete
mossy maple polypore
northern tooth
orange sponge polypore
pink-fringed milky
powdery sulfur bolete
red mouth bolete
strangulated amanita
strap-shaped coral
turpentine waxy cap
winter polypore
wood clitocybe
yellow-footed chanterelle
yellow oyster mop
yellow-tipped coral

Butterflies and Moths

ashen pinion
birch casebearer
blinded sphinx
Clemen's sphinx
common tan wave
curve-lined acontia
drab-brown wave
eyed baileya
galium sphinx
gray comma
green arches
green comma
hemlock looper
horned spanworm
Io moth
lemon plagodis
linden looper
luna moth
maple zale
pale-winged gray
polyphemus moth
saddled prominent
small-eyed sphinx
spear-marked black
wanton pinion
yellow-banded underwing
yellowhorn

Beetles

banded sap beetle
basswood leaf miner
cross-toothed rove beetle
golden net-wing
hairy May beetle
hemlock borer
linden borer
living beech borer
May beetle
obscure rove beetle
paederine rove beetle
pinching bug beetle
red-necked darkling beetle
short-winged blister beetle
six-spotted sap beetle
six-spotted tiger beetle
small ironclad beetle
stumpy longhorn
sugar maple borer
ten-lined June beetle

Insects

basswood lace bug
big-leaf aster plant bug
island grasshopper
pine spittlebug
poplar plant bug
spotted camel cricket

Spiders

- daring jumping spider
- filmy dome spider
- forest wolf spider
- northern widow
- three-banded crab spider
- zebra spider

Landsnails

- allied labyrinth snail
- armed pupa snail
- fraternal pill snail
- hairy pill snail
- handsome vallonia
- Holzinger's pupa snail
- minute disk snail
- modest pupa snail
- open disk snail
- ripe forest snail
- tiny harp snail
- tree zonite shell

Reptiles and Amphibians

- blue-spotted salamander
- northern red-bellied snake
- red-backed salamander
- spotted salamander
- spring peeper
- wood frog

Birds

- American redstart
- Baltimore oriole
- barred owl
- black-billed cuckoo
- broad-winged hawk
- chestnut-sided warbler
- eastern wood-pewee
- hairy woodpecker
- least flycatcher
- northern flicker
- ovenbird
- red-eyed vireo
- rose-breasted grosbeak
- ruffed grouse
- scarlet tanager
- veery
- wood thrush
- yellow-bellied sapsucker

Mammals

- cinereous shrew
- coyote
- deer mouse
- least chipmunk
- red fox
- red squirrel
- short-tailed shrew
- snowshoe hare

Forested Swamps

Indicators: White Cedar, Black Ash, White Pine

Ecology

Forested swamps develop where trees grow on wet soil saturated by neutral or alkaline water. There are three types in Wisconsin: white cedar swamps, black ash swamps, and white pine flatwoods. Different local conditions influence species composition, and changes can be dramatic over short distances.

WHITE CEDAR SWAMPS

White cedar swamps are usually the most familiar and recognizable of the forested swamps. Water flow and chemistry, soils, slope, and past history all play an important part in determining the present forest composition. Seeps and ground-water flow keep areas bathed in nutrient-rich water, which allows the cedars to become established. If the water flow continues, cedar swamps can be very long-lived—some trees can be hundreds of years old.

Mosses, flowering plants, and low-lying woody plants are abundant. Goldthread, creeping snowberry, twinflower, bunchberry, and dwarf raspberry are found in nearly every white cedar swamp.

The white cedar swamps have undergone some logging, but many areas were not harvested by early loggers, and an ancient ecological integrity persists. White cedar swamps form a significant part of our old-growth landscape, and the limited logging and old-growth remnants may be why these swamps still harbor many rare plants and animals.

Orchid Habitats

An ancient white cedar swamp reveals microhabitats more easily than any other natural community. The cedar trees are never straight and tall. They are always falling over at various angles, and some are lying flat to the ground, yet still very much alive. Mosses carpet the ground, but moss species change over the course of inches. Watery depressions are everywhere. Streams and rivulets course in the open and then disappear, sometimes flowing right under the roots of a large cedar. All of these ingredients create a multitude of small habitats.

Each of these small areas seems to have its own complement of orchid species. The rare calypso orchid is usually found in the higher drier mosses. The much more common club-spur orchid and heart-leaved twayblade are found below the raised cedar trunks and above the small pools. Adder's tongue orchids are almost exclusively found at the edges of the water in the depressions. Northern green orchids are most commonly found along rivulets. Dwarf rattlesnake plantain is most commonly found on the mosses covering flattened trunks or tree bases. The most diverse orchid populations are in cedar swamps with a mineral-rich water source and open areas for the species needing more sunlight.

Forested swamp (photograph by Thomas A. Meyer; used by permission)

Low areas found in northeast Wisconsin have the most diverse composition of any swamp. White cedar can also be found on lowlands bordering streams or in upland areas containing active seeps. In western Wisconsin, white cedar becomes increasingly less common to the point of being rare along the St. Croix River.

HARDWOOD SWAMPS

Hardwood-dominated swamps are the other extensive type of forested swamp. They develop primarily in lake beds and ponds that dry by midsummer and in floodplains. The moving water of the floodplains or the drying action in lake beds and ponds prevents water-absorbing plants, especially sphagnum mosses, from becoming established. The absence of these water-retaining plants dries the soil in late summer, which inhibits conifer survival. Dominant species are black ash and red maple. Prior to the invasion of Dutch elm disease, American elm was also dominant in the canopy. Less important canopy trees include yellow birch, hemlock, and white cedar. These ash-dominated areas have a much different groundlayer from white cedar swamps. Sphagnum mosses are rare in these situations, unlike white cedar swamps. The ground is covered with species such as yellow jewelweed and several mint species that can tolerate seasonal flooding.

Hardwood swamps are more commonly found in the northwestern part of the state. Large areas in Douglas County have ash-dominated forests. The Empire Swamp Road (see the Moose Junction chapter) is an excellent place to observe a large area of black ash swamp.

WHITE PINE FLATWOODS

The white pine forest of eastern Jackson County, southwestern Wood County, and northeastern Monroe County is a conifer-dominated relict. These forests, sometimes called flatwoods, developed on wet sands perched on an impervious clay layer. The dominant species are white pine and red maple. The characteristic groundlayer species are sphagnum mosses, winterberry, cinnamon fern, and species far removed from their normal range, such as Massachusetts fern and long sedge.

Range

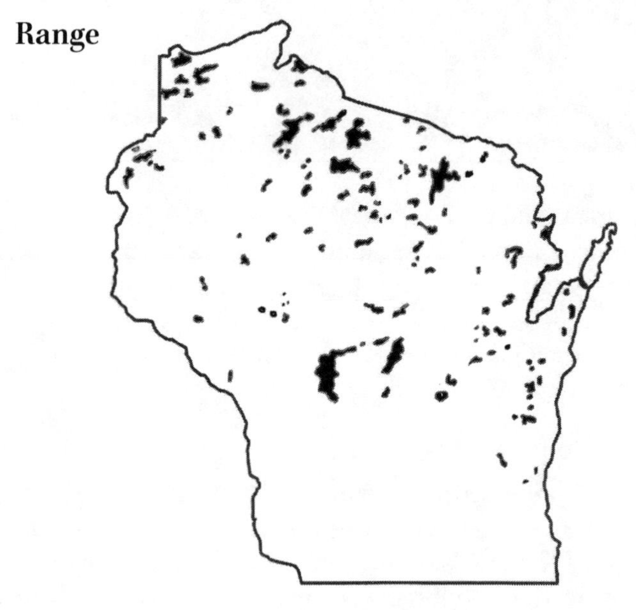

Large Forested Wetlands

The map shows all the large forested wetlands in the state, including swamps, bogs, and fens (the communities often are so intertwined that separation at this scale isn't possible). As much as 500,000 acres of white cedar swamp, black ash swamp, and white pine flatwoods remain scattered throughout northern, central, and eastern Wisconsin. Good stands can be found at the Apostle Islands, Black River State Forest (relicts), Brule River, Harrington Beach State Park, Lakewood

Area, Maribel Caves, Newport State Park, Northern Bayfield County, North Kettle Moraine (relicts), Ridges Sanctuary, Three Lakes, and Vilas County.

Activities

- Carefully observe woodpecker-drilled cavities on white cedar tree trunks. These trees have many wood-boring insects that birds eat, but what do the wood-boring insects eat? Discolored wood usually indicates the presence of ambrosia beetles. Tunnels drilled by the beetles are infected with ambrosia fungi spores, which in turn is eaten by the beetle larvae. Some beetle species eat only one species of ambrosia fungi (Hickin 1963).
- Gather a whitened branch from a black ash tree. Observe the white rot fungi and the apparent slight differences in texture and color of these fungi. As many as 12 species can live on a branch of a few feet in length. Many will partition the branch into segments, with each species only occupying a small area (Carroll and Wicklow 1992).
- Enter a white cedar swamp in midsummer. Carefully observe flying insects and look for a phantom crane fly that appears like a huge, slow-flying mosquito with black and white striped, outstretched legs.
- When observing the white cedar swamp's flowering plants, notice the tops of the plants, especially orchids. Have they been bitten off? White-tailed deer consider orchids, especially lady's slippers, to be their "ice cream." A simple enclosure can be built to protect orchids. The structure requires 12 three-foot-long pieces of lathe or ½-inch-width boards, screws or nails, and string. Attach the lathe together into a square, then attach four pieces perpendicular to the square, as if you were making a cube. Then attach four more pieces to square-off the cube, but secure it one foot away from the ends so it can be pushed into the soil. Tie the string one foot apart around the sides and over the top and place over the orchids. Deer will not put their heads through the openings, and the orchids can successfully reproduce.

Characteristic Species

Plants

bunchberry
Canada mayflower
cinnamon fern
creeping snowberry
crested wood-fern
dwarf raspberry
fancy wood-fern
goldthread
oak fern
sensitive fern
small Bishop's cap

small enchanter's nightshade
starflower
swamp red currant
sweet-scented bedstraw
three-seeded sedge
twinflower
winterberry

Mosses and Lichens

common tree moss
dog lichen
feather moss
frayed lichen
limp tufted lichen
lung lichen
many-fruited dog lichen
mealy goblet lichen
mountain fern moss
nodding moss
old man's beard
puffed shield lichen
red collar moss (deer yards)
round stemmed entodon
silvery byrum
squamalate lichen
studded leather lichen
tufted-hair lichen
twig lichen
velvet tree apron moss
wavy broom moss
whip fork moss
white moss

Mushrooms

ash-tree bolete
blood-red cort
blue-toothed entoloma
chanterelle waxy cap
emetic russula
golden waxy cap
little brown collybia
orange earth tongue
orange moss agaric
pungent cort
red-gilled cort
swamp beacon
trumpet chanterelle
violet cort
white spongy polypore
wrinkled thimble-cap

Butterflies and Moths

aphrodite fritillary
Atlantis fritillary
birch casebearer
black-banded carpet
black zigzag
Carolina sphinx
Clemen's sphinx
Comstock's sallow
cranberry spanworm
currant spanworm
double-banded carpet
five-spotted hawk moth
garden tiger moth
gray comma
great ash sphinx
Grote's sallow
hop merchant
Io moth
jocose sallow
laurel sphinx
lemon plagodis
many-lined carpet
osmunda borer
pale-winged gray
promethea moth
question mark
sharp-angled carpet
small-eyed sphinx
subflexus straw
white-striped black

Beetles

banded longhorn
black-horned pine borer
eastern spruce bark beetle
fire beetle
hemlock borer
ribbed pine borer
strawberry root weevil
stumpy longhorn
white-spotted sawyer

Insects

spotted camel cricket

Spiders

bowl and doily spider
daring jumping spider
hemlock cobweb weaver
three-banded crab spider

Landsnails

allied labyrinth snail
amber retinella
apple seed snail
Lea's pill snail
striped forest snail
three-toothed forest snail
tiny harp snail
white swamp snail

Reptiles and Amphibians

eastern gray treefrog
red-backed salamander
red-bellied snake
spotted salamander
spring peeper
wood frog

Birds

black and white warbler
blue-headed vireo
chipping sparrow
hermit thrush
magnolia warbler
Nashville warbler
northern parula
northern waterthrush
olive-sided flycatcher

ovenbird
pileated woodpecker
pine siskin
purple finch
raven
red-breasted nuthatch
veery
white-throated sparrow
yellow-rumped warbler

Mammals

coyote
eastern chipmunk
least chipmunk
masked shrew
red fox
red squirrel
short-tailed shrew
white-footed mouse

Northern Forested Bogs and Fens

Indicators: Black Spruce, Tamarack, Leatherleaf

Ecology

Wet areas dominated by black spruce and tamarack are the key to identifying northern forested bogs and fens. Northern forested bogs, sometimes known as swamp conifers, are even more extensive than the white cedar and black ash swamps. Many large areas of intact and often very old (with little disturbance for centuries) northern forested bogs remain in the state, mostly because they are not very "useful" to humans. In the northern part of the state, some of the forested bogs have been disturbed by the harvest of larger tamarack, the removal of sphagnum moss, and the conversion of some areas to farms. Despite these disturbances, many areas of intact forested bogs and fens are still present in northern Wisconsin, and some relicts are found in the south.

The typical forested bogs or fens are on or around glacial lakes, ponds, or water-filled depressions. Rates of succession and peat accumulation are highly variable depending on the length of the growing season. Over time, bogs and fens convert to forest, but at different rates in different places, depending on water depth, acidity, and the size of the watershed. The forest around lakes and ponds forms concentric rings of vegetation. Intertwining roots of shrubs

Bog? Fen?

Bog, fen, swamp, and *marsh* are all terms used to describe wetlands. Confusion ensues when the terms are used interchangeably. Belden swamp is actually a fen. Bogus swamp is a bog (see color insert). Bibon marsh is a swamp. Whether the wetland is a bog, fen, swamp, or marsh depends on the amount of accumulated peat, the water chemistry, and the water flow.

Short growing seasons and cool temperatures do not permit decomposers such as mites and springtails enough time to keep up with the annual volume of plant growth, and organic material accumulates over the course of time in the form of peat, partially decomposed organic matter (primarily sphagnum moss or sedges). Peat builds faster in acidic conditions because the harsh conditions keep the decomposers at low populations.

Water chemistry refers to the pH of the water—acidity or alkalinity—which helps determine which plant species will grow. Water flow refers to the movement and replenishment of waters with minerals absorbed from the soil.

The four communities can be defined as follows:

- Bog: Permanently wet land with accumulations of sphagnum peat and water coming from precipitation and runoff. The water is very acidic and can become more acidic as time goes on. Nutrients for

Northern forested bog (illustration by Jim McEvoy; courtesy of Wisconsin Department of Natural Resources)

plant growth come primarily from rain and snow. Raised bogs are higher in the center and have scattered stunted black spruce growing with sphagnum and leatherleaf. Trees grow slowly on these raised bogs because of the very acidic conditions.

- Fen: An area where peat accumulates and sedges are the primary peat makers. The water is not as acidic as bogs, and nutrients come from the surrounding soil or flowing water. Tamarack is the most numerous tree species in fens, but many times trees are absent. Types of fens in northern Wisconsin are poor fen, sedge fen, boreal rich fen, and tamarack fen.
- Swamp: Forested area (primarily white cedar and black ash) that can develop from a fen if calcium-rich water continues to enter. Abundant sphagnum moss can be part of the swamp, but it does not accumulate. Soil dwellers such as nematodes and fungi break down the organic material as rapidly as it accumulates, and the decomposers reach a balance. Types of swamps in Wisconsin are white cedar swamps and black ash swamps.
- Marsh: An area where the vegetation contains little or no sphagnum and has abundant grass-like plants. Mineral soil is underwater, and peat does not accumulate. Cattails, bulrush, and pickerelweed areas are dominant marsh plants (Crum 1988).

Rose pogonia (photograph by Thomas A. Meyer; used by permission)

Open Bog

The typical open bog described in textbooks is the first step in the process of changing lake to land. An open bog is a long-lived community, dominated by sphagnum and low shrubs that eventually develops into a tree community. These open bog communities extend over water by interlocking roots, and mosses fill the gaps. The rate of the bog's growth is quite variable, from an increase of several feet every 50 years on one body of water to growing only inches on the next. In some areas, the open bog is only a few feet wide, while in other areas it might cover hundreds of acres.

On the surface, many bogs appear the same, but great changes in species composition occur with different water chemistry. In areas where there is hard water flowing into an open bog via drainage or seepage, many rare orchids can occur, such as the rose pogonia. These alkaline bogs have more plant diversity, especially sedges and orchids. Some species are found more often in open bogs than anywhere else.

- Characteristic Groundlayer Species: bog rosemary, leatherleaf, pale laurel, bog birch, wild calla, pitcher plant, buckbean, large cranberry, small cranberry, round-leaved sundew, marsh cinquefoil, spread-leaved peat moss, leafy liverwort
- Butterflies and Moths: arctic skipper, black arches, bog holomelina, cranberry fruitworm, iris borer, pink-edged sulphur, pitcher plant borer, red-fronted emerald, old-maid, two-spot dart, white-streaked looper
- Beetles: bog ground beetle, cranberry rootworm, cranberry weevil, marsh beetle, minute bog beetle
- Insects: American emerald, beaverpond baskettail, chalk-fronted skimmer, cranberry toad bug, curve-tailed bush katydid, dusky-faced meadow katydid, four-spotted skimmer, grannulated pygmy grasshopper, huckleberry planthopper, northern crested pygmy grasshopper, northern grasshopper, pitcher plant mosquito, racket-tailed emerald
- Mollusks: short-ended pea clam

and sedges can grow over water, which forms a ring of vegetation without trees. When the accumulated plant material deepens, small trees can gain a foothold, resulting in another ring of small trees. More build-up of organic material permits the trees to grow much taller. The combined effect is a landscape with rings of similar vegetation around the lake.

Range

See the map in the previous chapter (p. 56) for the range of all large wetlands in the state, including forested bogs and fens.

Pitcher Plant

The pitcher plant, which grows in bogs and fens, is well known as a species that supplements its nutritional needs with insects. An insect is attracted to the pitcher plant by its sweet-smelling nectar and the colorful inside of the water chamber. The hapless bug then tries to feed on the nectar and is directed to the pool by downward-pointing structures. It eventually falls into the sticky liquid where it is dissolved. Amazingly, the pitcher plant mosquito lays its eggs only in pitcher plants, and the larva lives out its youth in this same liquid without being dissolved. The larvae feed on the remains of other insects before emerging as adults.

Thousands of lakes in Wisconsin have some form or another of northern bogs and fens around their shores. The most common type in Wisconsin is lake-edge bogs containing black spruce. Typically, they develop around an open lake without water movement. In other areas, drainage patterns allow shallow sheets of water to move slowly across the land, promoting abundant growth of sedges and tamaracks. These fens have much less sphagnum moss and black spruce than the bogs.

In a few areas of Wisconsin there are large raised bogs, often called muskegs. These bogs form when accumulating peat develops on large, shallow, relatively uniform areas. After a period of time, the peat in the bog raises above the bog–upland edge. Rainwater or snowmelt accumulates at the lowest point, the depression at the edge of the bog. Plant species found at the moat-like pond edge are much different from species found at the center of the bog. Wild calla and blue flag iris are commonly found around the edge, but not in the bog's center. These raised bogs have scattered stunted black spruce growing with sphagnum and leatherleaf. Trees grow slowly on these raised bogs because of the very acidic conditions. Large muskegs can be explored in Lincoln, Iron, and Douglas Counties. Other examples of northern bogs and fens can be found in Lakewood Area, Northern Bayfield County, Ridges Sanctuary, Three Lakes, and Vilas County.

Activities

- Stand on the edge of a large muskeg. Observe the rise toward the center and the ponding of water near the edge. Consider the succession pathways the muskeg could take, such as converting to white cedar swamp, becoming more dominated by black spruce, or becoming even sparser in trees as the site becomes more acidic (Crum 1988).

- In a bog lake, launch a canoe and explore the edges. List the species growing at the mat edge or on a partially submerged log. Describe the succession pathways available to the plants growing on that log.
- Sketch the zones around the bog lake and describe their differences.

Characteristic Species

Plants

bunchberry
Canada blueberry
Canada mayflower
cinnamon fern
creeping snowberry
fancy wood-fern
Labrador tea
northern bog violet
speckled alder
starflower
swamp false Solomon's seal
tussock cottongrass

Mosses and Lichens

blood lichen
common beaked moss
common tree moss
feather moss
mealy goblet lichen
nodding moss
old man's beard
peat moss
pitted cetraria
puffed shield lichen
reddish peat moss
Schreber's cedar moss
square-leaved peat moss
twig lichen
velvet tree apron moss
wood reveler

Mushrooms

blue-staining cup
bog conocybe
chocolate milky
hollow-stalked larch suillus
larch polypore
larch suillus
larch waxy cap
late fall waxy cap
red hot milky
sphagnum bog galerina

Butterflies and Moths

American idia
banded purple
bicolored moth
black zigzag
Canada tiger swallowtail
Columbia silkmoth
Compton's tortoise shell
cranberry spanworm
false hemlock looper
garden tiger moth
Harris' checkerspot
larch casebearer
mustard white
neighbor
northern pine sphinx
rusty tussock moth
spring azure
spruce budworm
two-spot dart
unicorn caterpillar
white streaked looper

Beetles

banded longhorn
eastern larch beetle
eastern spruce bark beetle
ribbed pine borer
strawberry root weevil
white-spotted sawyer

Insects

Acadian pygmy grasshopper
orchid fly
spined soldier bug

Spiders

daring jumping spider
goldenrod spider

Landsnails

brown pupa snail
Lea's pill snail
moss pupa snail
shining zonite shell
tiny star shell
white swamp snail

Reptiles and Amphibians

central newt
green frog
red-bellied snake

Birds

American goldfinch
black and white warbler
chipping sparrow
Nashville warbler
olive-sided flycatcher

pine siskin
purple finch
raven
red-breasted nuthatch
swamp sparrow
veery
white-throated sparrow
winter wren
yellow-rumped warbler

Mammals

masked shrew
red squirrel
short-tailed shrew

Great Lakes Shoreland Forest

Indicators: Balsam Fir, White Spruce, White Pine, Balsam Poplar

Ecology

Excluding mountainous areas, Wisconsin is the southernmost spot in the world touched by circumboreal forest, the spruce–fir forest of the Northern Hemisphere, which can be found along the Lake Michigan shore of Door County, 100 miles farther south than any other known stand. This forest is unique among boreal forests, and some ecologists believe that it should be called the Great Lakes shoreland forest, as a specific forest type. Conservation organizations have ensured partial protection of the forest for the near future by protecting many sites in Door County. Unfortunately, the main portion of the original boreal forest, which formerly grew on the clay plains along the south shore of Lake Superior, was almost totally obliterated by logging and conversion to farmland. After the white spruce, balsam fir, and white pine were logged, the area re-

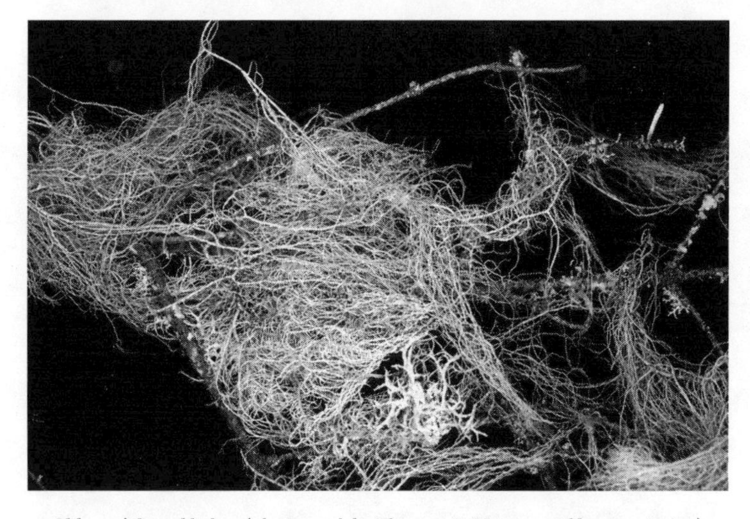

Old man's beard lichen (photograph by Thomas A. Meyer; used by permission)

Northern Parula
(image adapted by author)

Northern Parula

Many species require a specific nesting material in their habitat to reproduce successfully. This requirement is found throughout the natural world, but it's not often that the same species requires specific nesting material in distant parts of a continent. Northern parula warblers commonly nest in the southern United States and use Spanish moss (actually a flowering plant that hangs from many trees branches) for nest material and structure, building their nests within the hanging moss. These same birds, however, are also found in the boreal areas of North America where they use a much different nesting material, Unsea lichens, for building their hanging nests. This lichen, sometimes called old man's beard, grows abundantly in old spruce and fir stands, old tamarack stands, and old-growth white cedar stands.

grew to a forest of aspen, paper birch, and balsam fir. This type of forest regeneration mimics regeneration that occurs after catastrophic events such as fires, wind storms, or spruce budworm infestations—all natural processes of these forests.

In Wisconsin, Great Lakes shoreland forests are a mix of boreal and transitional species, which is different from those found in Ontario and even more different from the boreal forests of Siberia, which don't contain transitional species (Curtis 1959). Lake-effect snowfalls prolong winter, and the cold waters of Lake Superior and Lake Michigan modify summer temperatures. These long winters and cool summers produce ideal conditions for the development of these northern forests.

Range

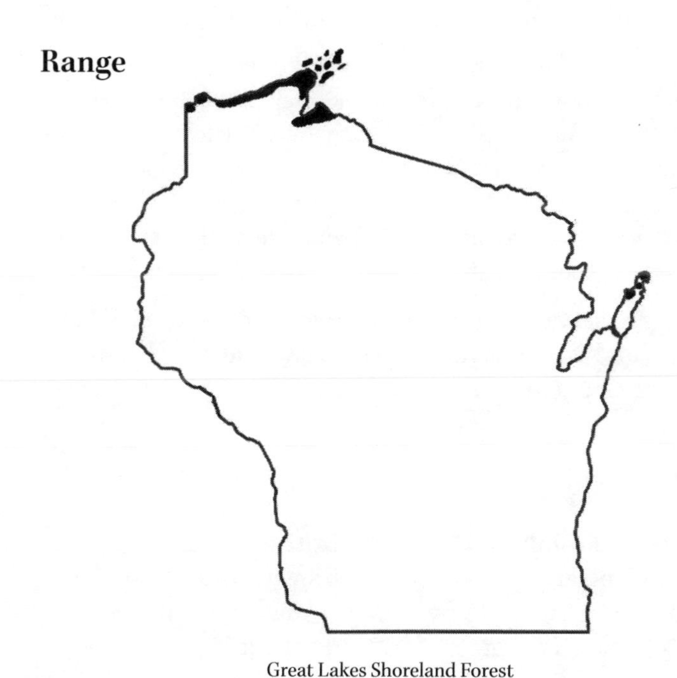

Great Lakes Shoreland Forest

These forests grow on heavier clay soils, with the better stands in the deep drainage ravines adjacent to Lake Superior. Dominant species vary with every locality and can include balsam fir, white spruce, hemlock, yellow birch, white pine, paper birch, or white cedar. Many areas formerly dominated by boreal forest now show an increase in white pine, white spruce, and balsam fir, and perhaps some day we will again be able to have large areas of Great Lakes shoreland forests in Wisconsin.

You can explore the Great Lakes shoreland forest at the Apostle Islands, Brule River, Newport State Park, Northern Bayfield County, and Ridges Sanctuary.

Activities

- Visit the shoreland forests in mid-May and observe the migrating birds. On very cold and wet days, the birds will often be found on or near the ground. Millions of tiny insects, mostly midges, inhabit these areas and are a favorite food of migrating birds. The midges live out most of their lives in the shallow waters of the Great Lakes and the adjacent shoreland forest.
- Visit one of the sites listed above (these are the oldest, most stable examples remaining) and make a plant list. Make a similar list from another younger shoreland forest in the same area. Compare the species of tree and shrub and consider the time it will take the second-growth forest to attain the composition of the old forest (Pratt 1995).
- Near the city of Superior, observe the tree and shrub composition and numbers on the flat clay plains. Observe the extent of standing water in spring and return to the site periodically as it dries. Do different tree species affect the evaporation rates? Does it appear that evergreens can evaporate more water, due to their being green all year as compared to aspen, which have leaves for less than four months?
- Observe how trees and shrubs grow near the shore of Lake Superior. Describe the impact wind and snow have on their shapes (Arno 1984).

Characteristic Species

Plants

American fly honeysuckle
American mountain-ash
beaked hazelnut
big-leaved aster
bluebead
bunchberry
Canada mayflower
dog violet
dwarf raspberry
goldthread
ground-pine
hairy honeysuckle
lady fern
mountain maple
oak fern
one-sided shinleaf
pink shinleaf
red baneberry
rough-leaved ricegrass
round-leaved dogwood
small Bishop's cap
starflower
sweet-scented bedstraw
thimbleberry
twinflower
twisted stalk
wood anemone

Mosses and Lichens

blood lichen
bristly lichen
British soldiers
brown-fruited cup cladonia
dog lichen
frayed lichen
giant byrum
homalia moss
Iceland moss
knight's plume
mountain fern moss
old man's beard
pitted cetraria
puffed shield lichen
sieve lichen
silvery byrum
spoon lichen
tan grape lichen

velvet tree apron moss
wood reveler

Mushrooms

bulbous cort
clustered blue chanterelle
clustered coral
elf cup
fragile russula
golden jelly cone
gray almond waxy cap
hairy black cup
pig's ears
purple club coral
separating trich

Butterflies and Moths

alternate woodling
American angle shades
arched hooktip
arcigera flower moth
Atlantis fritillary
birch casebearer
birch dagger
birch skeltonizer
black zigzag
brown-lined sallow
Canada tiger swallowtail
Compton's tortoise shell
drab-brown wave
dreamy dusky wing
galium sphinx
gray scoopwing
green comma
one-eyed sphinx
rose hooktip
scarlet-winged lichen moth
sleeping baileya
spear-marked black
spruce budworm
sweetheart underwing
unspotted looper
wanton pinion
white-marked tussock moth

Beetles

black-horned pine borer
cedar beetle
eastern spruce bark beetle
fire beetle
northern pine weevil
Pale's weevil
pine engraver
pine-stump borer
poplar borer
red turpentine beetle
shot-hole borer
two-toothed longhorn
white-spotted sawyer

Insects

balsam wooly aphid
big-leaf aster plant bug
black ichneumon
broad-winged katydid
coral-winged grasshopper
island grasshopper
poplar plant bug
spotted camel cricket

Spiders

bowl and doily spider
forest wolf spider
inconspicuous crab spider

Landsnails

fraternal pill snail
handsome vallonia
minute disk snail
open disk snail
tiny harp snail

Reptiles and Amphibians

mink frog
red-backed salamander
red-bellied snake

Birds

black and white warbler
blue-headed vireo
Canada warbler
chipping sparrow
magnolia warbler
Nashville warbler
olive-sided flycatcher
ovenbird
pine siskin
purple finch
raven
red-breasted nuthatch
veery
white-throated sparrow

Mammals

coyote
eastern chipmunk
least chipmunk
red squirrel

Prairies

Wisconsin lies at the northern and eastern edges of the prairie region in North America. It is astonishing that most books about prairies do not even recognize Wisconsin as having prairies. These books must have been written by authors unfamiliar with Wisconsin's natural history.

There are many accounts by explorers, military personnel, and settlers describing vast acres of prairie in Wisconsin. Early settlers recognized the importance of the prairie on Wisconsin's landscape with place names such as Sun Prairie, Prairie du Sac, Prairie du Chien, and Star Prairie. More recently, extensive soil surveys have identified soils that could only have developed in prairie landscapes.

More than 2,000,000 acres of Wisconsin were prairie at one time. Less than 8,000 acres of known prairie still remain, with approximately one-half in some type of protected status. Prairies would disappear in Wisconsin without regular disturbances such as fire, grazing by elk and deer, and digging by soil movers such as badgers.

Remarkably, our prairies, as well as others in the eastern parts of the prairie realm, have the highest species richness even though they are the farthest from the Great Plains. The remaining large tallgrass prairies in Kansas, Oklahoma, Missouri, and the Dakotas are dominated by grass species, while the eastern prairies are dominated by many species of broad-leaved

Prairies

Mississippi River area prairie, mid-1800s (courtesy of State Historical Society of Wisconsin)

flowering plants called forbs. The eastern prairies also have more insect diversity because many insects feed on specific plants for food. Also, the diverse plant community contributes to a diverse fungi community because many fungi are host-specific. Species vary from prairie to prairie, mainly because of differences in moisture and soil nutrients.

Dry Prairie

Indicators: Little Bluestem, Side Oats Grama, Big Bluestem, Long-Stalked Panic Grass

Ecology

There are four basic types of dry prairie: two that develop on very thin soils over bedrock—dry limestone prairie and sandstone bluff prairie—and two that are found on dry, nutrient-poor soil—dry sand prairie and gravelly glacial till prairie. Plants able to survive in these areas are shorter and less robust than individuals of the same species on a deep soil prairie because the bedrock limits root development and therefore the plants' ability to find water.

Dry limestone prairies, sometimes called goat prairies, are the best preserved of the prairie types in Wisconsin. Perhaps 3,500 acres of these limestone prairies still exist. Possibly another 2,000 to 3,000 acres are pastures that still have several dry prairie species left. This is an extremely high percentage of remaining prairie land compared with other prairie–savanna community types. Not surprising, these prairies are found over hundreds of sites and average only a few acres each. Most of these small patches of dry prairie are unprotected and unmanaged.

Limestone prairies occur in the driftless area of Wisconsin. Similar prairies are found along the Mississippi River in Minnesota, Iowa, and Illinois. Farther west, the dry prairies have a climate that promotes more grasses and fewer flowers than our limestone prairies. To the east, dry prairies reside on gravelly glacial till and sands.

Sandstone bluff prairies are much rarer than limestone prairies. They develop on very thin soil over sandstone, and sandstone hills and bluffs have fewer nutrients available to plants than limestone prairies have. Also, sandstone rock does not crack as easily as limestone, further preventing roots from reaching water. The dominant grasses are June grass, side oats

Wing Snaggletooth

The extremely steep slopes along the Mississippi River are one place where land managers question the effects of fire on the prairie. The soils here are very thin, the steepness permits water to run off rapidly, and the ground dries out on the slope facing the hottest sun, making the area unsuitable for most trees and shrubs. So why would regular fire be needed to maintain this dry prairie community?

A small very rare snail called the wing snaggletooth lives only on the steep dry slopes along the Mississippi River. This minute snail forages on decaying matter in very localized areas. It cannot tolerate dense plant growth. Lack of fire not only permits a dense layer of old plant material (called duff) to accumulate, it also changes the microclimate of the soil. The duff retains enough moisture for exotic predatory sowbugs (isopods) to thrive and prey on the rare snail. The conundrum for managers is that fire might kill individual snails, but the lack of fire allows an environment in which the snails are killed by other animals.

Mississippi River bluffs, 1837 (courtesy of State Historical Society of Wisconsin)

grama, and hairy grama. Sandstone bluff prairies are found in central Wisconsin, mostly west and south of Eau Claire. The only publicly owned sandstone bluff prairie of any significant size is described in the Lake Pepin site chapter.

Dry sand prairies occupy the sandy areas and terraces along our major rivers. Sometimes they occur in the floodplains, also. These dry sand prairies are distinctly different from the thin-soiled prairies. The species are approximately the same, but ones that are commonly found on sand prairies occur in low numbers on bluff prairies, including June grass, round-headed bush clover, stiff goldenrod, and horse mint. The sandy soil is ideal for animals that dig, burrow, and tunnel, such as badgers and gophers.

Gravelly glacial till prairies are found primarily in southeastern Wisconsin. Several areas in Waukesha, Walworth, Jefferson, and Rock Counties still have small remnant prairies of

Soil Movers

Badgers, pocket gophers, foxes, and other mammalian soil movers have a significant influence on plant species composition. The effect they have on the prairie mosaic is probably best exemplified on sand prairies.

These animals constantly move soil from one place to another. New fresh soil is always popping up on the surface of the sand prairie. Species that cannot favorably compete for nutrients in a stable prairie are able to become established on the new soil. As badgers and other diggers constantly shift the soil, they provide new habitat for these early colonizing plants, which couldn't persist without the actions of these animals. These colonized plants remain for many years before succumbing to the long-lived grasses (Gibson 1989).

this type. A few very small gravel prairies also can be found in Pierce, St. Croix, and Polk Counties. Side oats grama and little bluestem are the dominant grasses.

Range

Curtis (1959) estimated that 105,000 acres of dry prairies were present when the European settlers arrived. The remaining 5,500 to 6,500 acres of dry prairie are still here because of one reason—the soil was too thin for the plow, so the land was used only for sporadic grazing. While domestic cattle and sheep grazing is highly destructive to tallgrass prairies, it is only moderately destructive to dry prairies. Animals do not graze the slopes as much because slow plant growth on dry bluffs produce poor forage when compared with the lush valleys. The more drought-tolerant species found in dry prairies, especially the grasses, can handle some grazing and remain on many steep slopes.

Another reason these dry prairies are still here is their resistance to the encroachment of woody vegetation, which proceeds very slowly on the thin rock soils of dry prairies. Most mesic (moist) prairies rapidly converted to a tree-dominated landscape when fire suppression became the norm after European settlement.

Most of the preserved dry prairie remains are on steep bluff faces, as well as on sandy or gravelly areas. Areas of pristine dry prairie remnants can be found at Green County Preserves (dry limestone), Lake Pepin (dry limestone and sandstone bluff prairies), Lower Wisconsin River Valley (sand prairie and dry limestone), Nelson Dewey State Park (dry limestone), Perrot State Park (dry limestone), and South Kettle Moraine (glacial till prairie).

Activities

- Map vegetation patches from a high vantage point. Observe patches of slightly different colors reflected by the grasses or the showy flowering plants that occur in clumps. After mapping, go to each patch and list the plants observed in each and compare the lists.
- Return to the same patches as mapped above and collect insects. Collections will need to take place day and night throughout the summer to get a picture of insect diversity. Also, collect the insects from a single plant species, and compare that list to the patch list. Some insect species will

be found only on one species of plant. The observer can separate the species using the whole prairie from those using patches and those using a particular plant regardless of where it's found on the prairie.

- The Mississippi River bluffs have numerous prairies, most with black walnuts on the edges. List the species growing near the walnuts and those far away. Compare the lists and describe what accounts for the differences. Hint: the root system of black walnut exudes a chemical that inhibits growth in many plants (Kricher 1988).
- When a strong storm with very low pressure rumbles out of the lower plains, make plans to visit southwestern Wisconsin prairies in its wake. These storms often transport hapless insects. Caught in the fury, many insects were first recorded in Wisconsin in the southwest after such storms.

Characteristic Species

Plants

beach wormwood
cylindrical blazing star
daisy fleabane
false boneset
flowering spurge
fringed puccoon
heath aster
June grass
lead-plant
old-field goldenrod
pasque flower
prairie tickseed
purple prairie clover
rough pennyroyal
silky aster
smooth sumac
stiff goldenrod
stiff sandwort
whorled milkweed
wild rose

Mosses and Lichens

British soldiers
cornucopia cladonia
xerophytic liverwort

Mushrooms

beaked earthstar
sulfur tuft

Butterflies and Moths

arge moth
asteroid
black-spotted prominent
broken dash
buckeye
chalcedony midget
connected looper
dimorphic gray
dogbane tiger moth
feeble grass moth
figured tiger moth
fleece-winged dart
fringed dart
graphic moth
gray hairstreak
lead-plant flower moth
lynx flower moth
married underwing
mixed dart
painted lady
stalk borer
subflexus straw
three-staff underwing
veiled ear moth

Beetles

American minute lady beetle
Brandel's anthill beetle
dubious checkered beetle
milkweed lady beetle
potato stalk borer

Insects

American grasshopper
antlion
ash-brown grasshopper
clover leafhopper
dusky grasshopper
little black ant
rose leafhopper
straight-lanced meadow grasshopper
tanglewing fly

Spiders
- ant mimic
- black and yellow argiope
- burrowing wolf spider
- crab spider
- goldenrod spider
- jumping spider
- line weaving spider
- shamrock spider

Reptiles and Amphibians
- blue racer
- bullsnake
- eastern milk snake

Birds
- common nighthawk
- eastern kingbird
- grasshopper sparrow
- horned lark
- western meadowlark

Mammals
- striped ground squirrel
- white-footed mouse

Dry-Mesic Prairie

Indicators: Big Bluestem, Little Bluestem, Porcupine Grass, Prairie Dropseed

Ecology

Dry-mesic prairies occur on well-drained, deep, rich soils. The deeper soils retain moisture much longer than dry prairies but less than mesic (moist) prairies. Dry-mesic prairies suffered more losses to the plow than prairies that were drier, more rocky, or more gravelly. Today, the dry-mesic community is nearly nonexistent.

Dry-mesic prairies are usually found in two settings. In steep hilly areas, the lower slopes of dry limestone bluffs tend toward dry-mesic prairies because they have deeper soils and retain moisture better. In a level or rolling landscape, ridge tops and crests of hills harbor dry-mesic prairies. These soils are relatively deep, but the position at the top of hill permits excellent drainage. In combination with midsummer drying, the soils can never become truly moist all the time. Only a handful of very small dry-mesic prairies on this land type remain.

Range

Dry-mesic prairies, which once covered 630,000 acres in the state, are now down to around 1,000 acres. Most of these remaining acres are found in two large prairies, Avoca Prairie and Chiwaukee Prairie, which form the only extensive stands of dry-mesic prairie remaining in the state. Both are on sandy plains: Avoca Prairie formed on a river terrace along the Wisconsin River, and Chiwaukee Prairie is on an old sandy lake dune. Each prairie has higher areas that form slight ridges with drier soil in an otherwise wet site. These high, mostly parallel strands harbor many dry-mesic prairie species. The only hope of perpetuating this community is in preserving and managing these remnants.

Pollinators

Prairies have an abundance of flowering plants, with as many as several hundred species in each community. The grasses and sedges are pollinated by wind. The forbs, however, usually depend on insects for pollination.

The total number of pollinating insect species can be quite high. In studies from other states, more than 300 species have been recorded on prairie flowers (Reed 1994). Nearly one-third of this group is bees, but butterflies, moths, beetles, flies, and other groups also pollinate prairie plants. Most pollinators are attracted to any plant with pollen, but a few only pollinate a specific plant species.

Activities

- Visit a prairie in late summer or fall when seeds and fruits are ripening. Look closely at the seeds—are they being eaten? If you open the pod from a wild indigo plant, you might find that seed weevils have eaten most of the seeds. If most of the seeds are eaten, how do plants maintain populations over time (Putnam 1994)?
- Observe a dry-mesic prairie. Describe what you think would happen if the entire prairie were burned every year, every fourth year, or never. Or if the prairie were mowed and the hay baled every year or every other year. Or if cows were pastured on the site every year for only one month per year (Collins and Wallace 1990).
- Obtain a descriptive account of a prairie area from the original land-survey records. Visit your local library and read the accounts of the first settlers to the area. Early newspaper stories often tell of successful hunts for prairie chickens and other species that are rare or gone from the state now. Compare those early accounts to what you see in the prairie today.
- Most people do not realize that the night crawlers and red worms we use for fishing are not native. When the settlers first arrived, our prairies did not have these earthworms. Describe the ecological consequences of having nonnative earthworms introduced to our prairies. Does the ecology change by having more soil aeration, added organic matter, and a food source for species not normally associated with prairies? Fire tends to dry the prairie and makes it less suitable for earthworms, whereas mowing tends to keep moisture in the prairie soil. Answer the previous question in this new light (James 1988).

Characteristic Species

Plants

common evening-primrose
false boneset
flowering spurge
hoary puccoon
Indian grass
lead-plant
prairie cinquefoil
prairie sunflower
purple prairie clover
rough blazing star
rough pennyroyal
Scribner's panic grass
sky blue aster
stiff goldenrod
thimbleweed
western sunflower
white prairie clover
whorled milkweed
wild bergamot
wild rose
yellow coneflower

Mushrooms

liberty cap

Butterflies and Moths
- aphrodite fritillary
- artichoke plume moth
- asteroid
- black-etched prominent
- chalcedony midget
- connected looper
- dimorphic gray
- dogbane tiger moth
- frothy moth
- glorious flower moth
- gray hairstreak
- lead-plant flower moth
- nais tiger moth
- northern cloudy wing
- painted lady
- primrose moth
- small brown quaker
- smaller pinkish dart
- southern cloudy wing
- stalk borer
- sunflower borer
- three staff underwing

Beetles
- Brandel's anthill beetle
- hairy spider beetle
- metallic flower beetle
- potato stalk borer
- rose chafer
- rose leaf beetle
- striped blister beetle

Insects
- American grasshopper
- bee assassin
- black-horned tree cricket
- colletid bee
- marsh meadow grasshopper
- meadow spittlebug
- rose leafhopper
- straight-lanced meadow grasshopper
- two-striped grasshopper

Spiders
- black and yellow argiope
- brown daddy-long-legs
- crab spider
- jumping spider
- line weaving spider
- shamrock spider

Reptiles and Amphibians
- blue racer
- bullsnake
- eastern milk snake

Birds
- grasshopper sparrow
- horned lark
- western meadowlark

Mammals
- cottontail
- striped ground squirrel
- white-footed mouse

Mesic Prairie

Indicators: Big Bluestem, Porcupine Grass, Prairie Dropseed, Prairie Panic Grass

Ecology

Mesic, or moist, prairies are functionally extinct as an ecosystem in Wisconsin. Mesic prairie and oak savanna, intermixed for thousands of years, now occur only as scattered and highly disturbed remnants. Place names, such as Sun Prairie and Star Prairie, remind us that a tall-grass mesic prairie once existed in Wisconsin, but the landscape bears little resemblance to the original prairie at those places. The complete conversion from a tallgrass mesic prairie to row-crop agriculture came about because of the exceptional richness of the soil.

The deep-blacksoil tallgrass prairie is the mesic prairie type most commonly described in the accounts of the settlers and scientists of the 1800s, but the recent discovery of two very different and equally rare mesic prairie types, the mesic sand prairie and the north-slope limestone prairie, makes us pause to contemplate even more ecosystem losses. Depressions in sandy areas that have adequate moisture permit the development of the mesic sand prairie. The depressions are closer to the water table and hold water much longer than the higher, drier sands. Prairie dropseed is clearly the most dominant species, sometimes forming a nearly pure community of one dominant plant. North-slope prairies are unique. They are usually small and form over very thin soil. The combination of cool wet cavities in the limestone and the north-facing aspect keeps the slopes moist year round. These unusual conditions allow a combination of dry-mesic prairie species, wet-mesic prairie species, and even fen species to merge into the natural community.

Range

Any remnants of the original tallgrass mesic prairie are pitifully small and mostly isolated from other natural communities, let alone other prairies. Most persist for indirect and unexpected reasons. Most remaining mesic prairies are along railroad tracks and in abandoned corners of cemeteries. Perhaps only a few hundred acres of Wisconsin's original 840,000 acres of mesic prairie are still in existence. This number is increasing with restoration projects, but we don't know if these plantings will ever achieve the historical mix.

Known tallgrass mesic prairies can be found at Lafayette, Fond du Lac, St. Croix, Green, Dane, Rock, and Kenosha Counties. Because of their very scattered and isolated position on the land, none are identified in part 2 of the book.

Remaining examples of mesic sand prairies are at Lake Pepin (in the northern part of the Tiffany Wildlife Area) and along the Lower Wisconsin River (in the midst of an oak barren).

The best-known examples of north-slope prairies are on private lands in Dane County. Smaller sites are in Lafayette, Grant, Trempealeau, and Pierce Counties. A very small example of less than one-quarter acre can be observed at Perrot State Park.

Soil Communities

Soil development in prairies is much different than in forested areas. The forest plants have root systems that spread out near the surface in order to capture all the moisture possible. Prairie species have root systems that penetrate deep into the ground. These roots are very finely branched and take up a tremendous amount of subsurface space. Around the rootlets are several species of mycorrhizal fungi that assist in nutrient intake by the plants.

The total numbers and species diversity of the organisms in prairie soils are staggering. The numbers of bacteria, microfungi, molds, and viruses in small samples of soil might contain more than 10,000 individual cells. These living organisms can provide abundant nourishment for other species, such as protozoa, earthworms, beetle larvae, nematodes, and mites. These groups have a tremendous diversity of species and are found in astonishing densities. Some scientists have estimated that the below-ground abundance is ten times that above ground. A six-inch by six-inch soil sample that you can hold in your hands could have 200 to 300 species and 150,000 individual animals. These species may only constitute 1 percent of the soil's weight, but they are essential for developing the soil's fertility because they keep the soil porous and mix the nutrients. This amazing diversity is why prairie soils are so rich and productive (Samson and Knopf 1996).

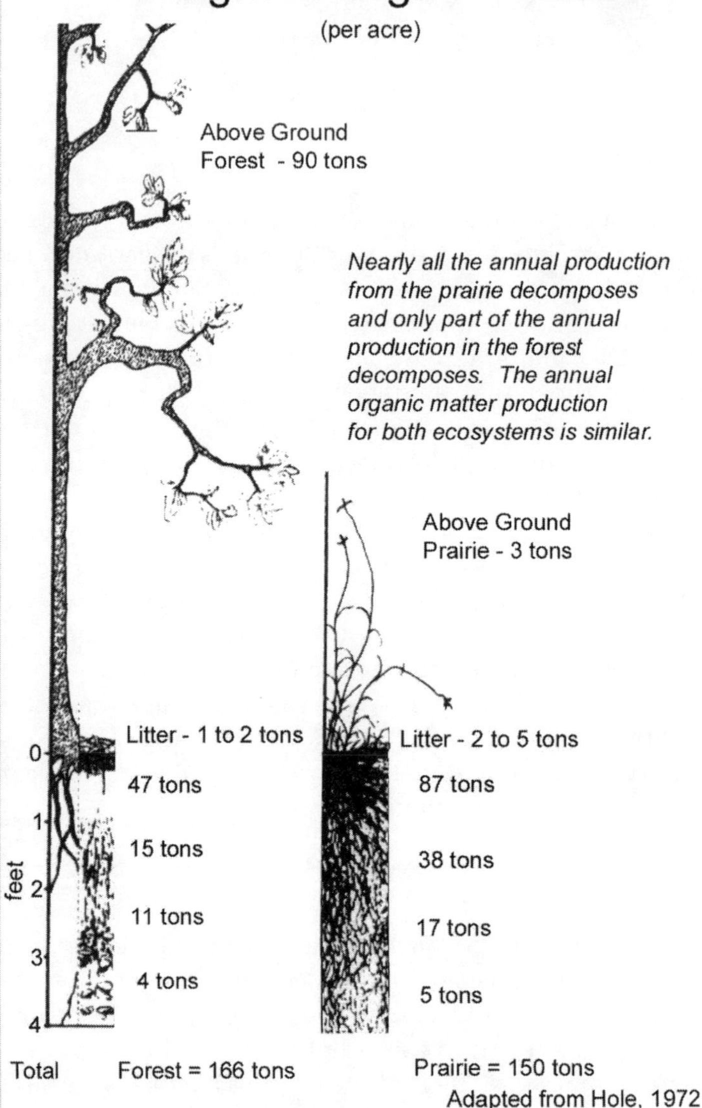

Weight of organic matter per acre in prairie and forest soils (illustration adapted by author)

Activities

- Obtain a copy of the Endangered and Threatened Species list for Wisconsin from any Department of Natural Resources office or your local library. Tabulate the species found in forests and those found in prairies. Consider why the list for prairies is so much larger. Does loss of habitat have any connection to endangerment?
- While standing in a mesic prairie, scoop up enough soil to fill your cupped hands. Now focus on the 100 or more species of microbes, arthropods, nematodes, algae, fungi, and others you have in your hands at one time. Describe the reason for this diversity. Imagine how deep the prairie roots go, and draw your interpretation of the underground ecosystem of a prairie (Samson and Knopf 1996).
- After a prairie burn, map the location of the ant mounds on the prairie. Return after subsequent burns and map the locations again. Do you note any change in numbers or locations of mounds?

Characteristic Species

Plants

common milkweed
compass plant
cream wild indigo
Illinois tick-trefoil
lead-plant
Missouri goldenrod
New Jersey tea
pasture thistle
prairie sunflower
prairie tickseed
purple prairie clover
rattlesnake master
rough blazing star
showy goldenrod
smooth blue aster
stiff goldenrod
wild bergamot
yellow coneflower

Mushrooms

hard agrocybe
tough puffball

Butterflies and Moths

aphrodite fritillary
artichoke plume moth
broad-lined erastria
clouded crimson
doubtful agroperina
eupatorium borer
frothy moth
glorious flower moth
gray half-spot
intermediate cucullia
long dash
nais tiger moth
pale-banded dart
primrose moth
ruby tiger moth
six-spotted gray
small brown quaker
smaller pinkish dart
southern cloudy wing
straight lined seed moth
sulphur moth
wedgeling
white-dotted groundling

Beetles

hairy buprestid
margined blister beetle
milkweed lady beetle
rose chafer
tumbling flower beetle
two-lined soldier beetle

Insects

Allegheny mound ant
bee assassin
black-horned tree cricket
colletid bee
flower thrips
four-lined plant bug
marsh meadow grasshopper
potato leafhopper

rose leafhopper
tanglewing fly

Spiders
brown daddy-long-legs
crab spider
cross spider
jumping spider
line weaving spider
shamrock spider

Reptiles and Amphibians
blue racer
bullsnake
plains garter snake
smooth green snake

Birds
bobolink
eastern meadowlark
savannah sparrow

Mammals
deer mouse
masked shrew
red fox

Wet-Mesic Prairie

Indicators: Big Bluestem, Bluejoint Grass, Canada Wild Rye, Prairie Panic Grass

Our wet-mesic prairies (halfway between moist and wet) fared better than the mesic prairies, but ever so slightly. The soil was nearly as rich and desirable for farmland, but the soil often was too moist for the settlers to farm, so they plowed only during drier years, resulting in an increase in overall dryness. Until the inception of drainage plans, such as ditching and installing drain tiles, many farmers mowed wet-mesic prairies for "marsh hay" rather than plow them. This activity didn't eliminate the wet-mesic prairie community as a whole, but served to diminish the number of species. If mowing took place at a similar time each year, those species blooming at the time of mowing would be affected. Conversely, if mowing took place at different times or only in alternate years, the species were not adversely affected.

The wet-mesic prairies formed best on the beds of glacial lakes or in floodplains between the lowland forest and the upland prairies. They are slow to warm up in the spring. Cool air tends to collect in midsummer, promoting more water retention and higher humidity.

Range

Extensive areas of wet-mesic prairie are hard to find, since only about 2,500 of the original 420,000 acres are still present. You can study stands at Avoca Prairie, Avon Bottoms, Chiwaukee Prairie, Muir Park, South Kettle Moraine, and White River Marsh.

Activities

- Is fire or lack of fire a disturbance in prairies? List the positive effects of a single fire in one column and negative effects in another. List possible effects of burning or not burning for 20 straight years (Collins 1995).
- List some reasons why wet-mesic prairies or prairies in general should be preserved. Evaluate the list. Are your reasons written in terms of benefits to humans? Would there be any benefits if humans were prevented from using the prairie?
- Chart the flowering times of the prairie plants. Visit a site weekly and keep a list of those species blooming. Also, record the insect visitors to any flower.
- In late summer, especially the first two weeks of September, visit a prairie with many flowers. Relax and observe a patch. Occasionally, hummingbirds will visit, but often it's an imitator. Several sphinx and hummingbird moths are daytime feeders and visit the flowers then. Many of these moths are distinctively marked and can be identified using binoculars.

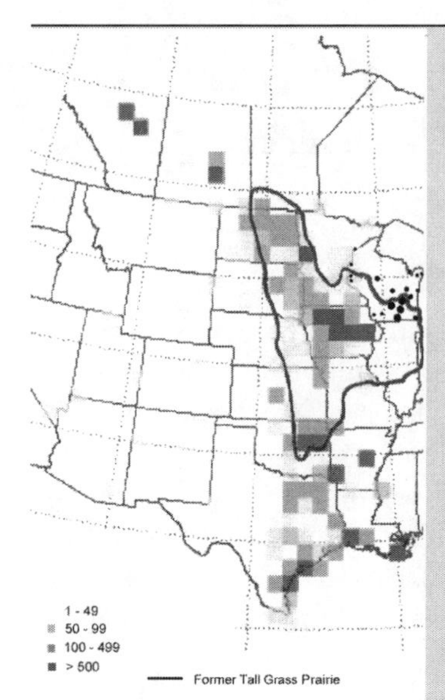

USGS survey of migrating American golden plovers. The American golden plover is a long-distance migrant that nests on the Canadian tundra and winters in southern South America. Its population has been reduced significantly. This species was tracked in migration by volunteers participating in a program sponsored by the U.S. Geological Service. Wisconsin was not included in the survey, but data exist from local bird watchers. A small dot represents 1–9 birds, a medium dot 10–49 birds, and a large dot over 50 birds, including many records of flocks of more than 150 birds. Notice the close alignment with the former tallgrass prairie (adapted by author from USGS data)

Temporary Pond Communities

While it's well known that migrating shorebirds use Wisconsin's lakeshores as habitat, it isn't widely known that they also use many other habitats, including prairies.

The American woodcock prefers alder thicket and shrub-carr along with moist forest floors, common snipes prefer sedge meadows, and killdeer inhabit many types of open areas. Most migrating shorebirds utilize muddy fields and lakeshores of sandy beaches, but some prefer grassy areas.

The upland sandpiper is the most well known grassland shorebird. This species, whose epic migratory flights to Argentina are legendary, prefers short to medium-height grasses. They move unto grassy uplands just greening up from recent burns in the spring. Other migrants, such as the American golden plover, the pectoral sandpiper, Baird's sandpiper, and the marbled godwit, also readily forage on the same habitat. Formerly, long-billed curlews nested on our prairies, and the nearly extinct Eskimo curlew utilized prairies in its migration.

Migrating water birds make heavy use of temporary ponds. Some shorebird species use these places annually, preferring ponds with prairie affinities or lower Mississippi valley connections. Temporary ponds have relatively few plant species, and those that do occur, such as algae and duckweed, are found in low populations. However, temporary ponds are inviting to invertebrates that prefer habitat that alternates between periods of high water and total dryness.

Temporary ponds are also extremely important to amphibians, especially salamanders, which breed in the ponds that linger in floodplain and mesic forests. Unfortunately, temporary ponds in prairies are very rare. To see prairie examples visit Avoca Prairie, Horicon Marsh (Highway A and W Area), and Chiwaukee Prairie.

In addition, many other temporary users add to the diversity of prairie ponds:

- Insects and Crustaceans: bloodworms, fairy shrimp, narrow-winged damselfly, phantom midge, snow melt mosquito, white-faced meadowhawk
- Mollusks: amphibious fossaria, blunt prairie physa, flatly coiled gyraulus, modest fossaria, tadpole snail, swamp fingernail clam
- Reptiles and Amphibians: blue-spotted salamander, spotted salamander, spring peeper, wood frog
- Fish: pirate perch
- Birds: dunlin, greater yellowlegs, least sandpiper, lesser yellowlegs, migratory dabbling ducks, pectoral sandpiper, semipalmated sandpiper, solitary sandpiper

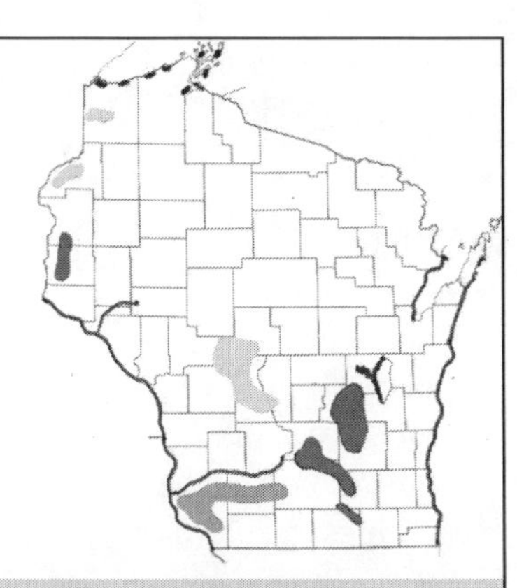

Shorebird map. The light gray shading represents large areas of bog and sedge meadows. These areas are used extensively by migrating greater yellowlegs (listed as an important migrant in Wisconsin), lesser yellowlegs, solitary sandpiper, and common snipe. The medium gray shading is an area of former tallgrass prairie that still attracts "grasspipers" such as Baird's sandpiper, white-rumped sandpiper, pectoral sandpiper, and American golden plover. The dark gray shading represents areas with prairie potholes and shallow water areas in former prairie land. These four areas attract many shorebird species, and they are especially attractive to four species listed as important migrants in Wisconsin—least sandpiper, white-rumped sandpiper, long-billed dowitcher, and Hudsonian godwit. The black areas along Lake Michigan, in and around the wetlands west of Lake Winnebago, and scattered locations along Lake Superior are attractive to sanderling, ruddy turnstone, whimbrel, and red knot. The black area along the Mississippi River, which includes the lower Wisconsin River and the lower Chippewa River, attracts just about any shorebird species, especially in fall during low-water stages (illustration by author)

Characteristic Species

Plants

American vetch
arrow-leaved aster
black-eyed Susan
common mountain mint
Culver's root
false toadflax
meadow anemone
New England aster
northern bedstraw
prairie blazing star
prairie dock
prairie phlox
purple meadow-rue
rosinweed
sawtooth sunflower
shooting star
showy tick-trefoil
sky blue aster
smooth blue aster
switch grass
wild bergamot
wild rose
wild strawberry
yellow coneflower

Mosses and Lichens

pointed mnium
spoon-leaved moss
toothless twisted moss

Butterflies and Moths

agre moth
aphrodite fritillary
bella moth
bog holomelina
doubtful agropernia
eyed brown
frothy moth
glorious flower moth
gray marvel
intermediate cucullia
ironweed borer
long dash
milkweed tussock moth
northern flower moth

Orchid Pollination

Listed as threatened under the U.S. Endangered Species Act, the rare prairie white-fringed orchid lives only on wet-mesic prairie soils. Just a few years ago this species was thought to be widespread. Why, then, has it been listed as a rare and threatened species? The answer lies in an intriguing scientific inquiry. A similar orchid is found on the prairies of Minnesota, North Dakota, and Iowa and for a long time was thought to be the same species as the prairie white-fringed orchid. Even though the flower structure was slightly different, taxonomists believed that this was a geographic variation. Confirmation of species uniqueness came with the use of high-speed photography. Both orchids are pollinated by sphinx moths, and by filming pollination, it was discovered that the pollen sacs were different shapes and were attached to the eyes of the sphinx moths at different locations. One pollen sac attaches to the front of the eye and only fits into a pollen-receiving structure on the prairie white-fringed orchid. The other pollen sac attaches to the top of the eye and doesn't fit into the pollen-receiving structure, making cross-species pollination impossible (Sheviak and Bowles 1986).

olive arches
orange sulphur
red groundling
rigid sunflower borer
ruby tiger moth
ruddy quaker
smaller pinkish dart
southern cloudy wing
spotted straw
sunflower borer
tawny holomelina
Virginia ctenucha

Beetles

black blister beetle
carpet beetle
hairy buprestid
hairy spider beetle
margined blister beetle
metallic flower beetle
narrow sap beetle
Pennsylvania firefly
Pennsylvania soldier beetle
purple tiger beetle
rose leaf beetle
sawtooth sunflower weevil
striped blister beetle
swamp milkweed leaf beetle
triangular rove beetle
tumbling flower beetle
two-lined soldier beetle

Insects

Allegheny mound ant
bee assassin
black-horned tree cricket
clover leafhopper
colletid bee
marsh meadow grasshopper
meadow spittlebug
northern green-striped grasshopper
oblong-winged katydid
phlox plant bug
rapid plant bug
round-winged katydid
spined assassin bug
stilt bug
two-striped grasshopper

Spiders

arrowshaped micrathena
banded argiope
black and yellow argiope
crab spider
cross spider
jumping spider
platform spider
shamrock spider

Landsnails

common disk snail
golden amber snail
prairie pond clam
small zonite shell
southern pill snail
toothless pupa snail

Reptiles and Amphibians

Cope's gray treefrog
eastern hognose snake
painted turtle
tiger salamander

Birds

bobolink
common yellowthroat
eastern meadowlark
savannah sparrow
sedge wren

Mammals

masked shrew
meadow vole
short-tailed shrew

Wet Prairie

Indicators: Bluejoint Grass, Prairie Cord Grass, Marsh Wild Timothy

Ecology

When wet prairies developed, they formed an interesting border between many wetland habitats and the drier areas adjacent. Sometimes they developed around depressions in old lake beds or river meanders. Although they developed in the same areas of the state as the wet-mesic prairies, they covered fewer acres, perhaps 100,000 acres before European settlement. Essentially, they could be found anywhere in the prairie realm where a perched water table was present.

This tallgrass community shares many characteristics with sedge meadows, calcareous fen, and even emergent marshes. Because it is wetter than the other types of prairies, the wet

Borer Moths

Sometimes called the jewels of the prairie by moth enthusiasts, many species of borer moths (genus *Papipema*) are found in prairie situations. Each species of borer moth lays its eggs only on a single host-plant species, such as prairie dock, prairie blazing star, saw-toothed sunflower, meadow-rue, and rattlesnake master. In the early fall the adult females lay their eggs at or near the base of the host plant. The eggs overwinter, then in early spring when a critical temperature is reached, the eggs hatch, and the minute larvae bore into the stalk of the host plant to feed on the pith of the stem and roots for the entire growing season. Finally in late September, when the temperature remains above 50 degrees, the adults emerge for a few nights of frantic mating before the eggs are laid and the adults die. A keen observer can notice the telltale browning on plant leaves that indicates larval presence and return in September to glimpse the adults.

Turtlehead borer moth (photograph by Les Ferge; used by permission)

prairie often has a wetland appearance. Many plant species found in the wet-mesic prairies are not found in the wet prairies.

Dense stands of nearly pure cordgrass dominate many areas. In other locations, bluejoint grass is the most abundant plant. Muhly grasses and several sedge species are common. Flowering plants are much less evident in these situations, probably because of the dense shading that cordgrass creates.

Range

Wet prairies were much too wet for the plow and in many years too wet for "marsh hay" mowing. Farmers mainly used wet prairies as grazing pasture, which is the main reason for the wet prairie's demise. The grasses associated with the wet prairies could not withstand the pressure from grazing. Drainage schemes also permitted excessive silt to deposit on wet prairies. The combined disturbances favored bluegrass, redtop, and reed canary grass, which quickly replaced the native grasses. Because of drainage and conversion to different grass species, the wet prairie is mostly gone. Small areas occur at Avoca Prairie, Chiwaukee Prairie, and White River Marsh, plus a few other areas in the southeast section of the state, the total probably not exceeding 200 acres.

Activities

- Visit a wet prairie several times in the summer and try to identify and locate the insect singers. Similar-sounding insects may actually be different species singing at different times of the year. An example from wet prairies is two meadow grasshoppers. The gladiator meadow grasshopper, sometimes called meadow katydid, begins singing its five-second-long buzz in late June. The buzz does not appear to rise at the end. Around the end of July this species is replaced by the common meadow grasshopper. The species also sings a five-second buzz, but it rises slightly at the end. These two species can be identified by their songs and the time of summer in which they sing (Alexander et al. 1972).
- Place the wet prairie in one of two categories—terrestrial or wetland. Describe the reason for your placement.
- Observe a wet prairie in early May and again in late summer or September, and notice the amount of water stand-

ing in the prairie. Describe the effects of conducting a fire in these two seasons. What would be the effect of fire on shrubs standing in water during the spring versus the late summer?

- Take a slow walk through a wet prairie in early fall. Write down your impression of the insects you flushed. Describe the direction of flight, where they landed on the vegetation, and any predators that pursued them.

Characteristic Species

Plants

black-eyed Susan
common mountain mint
cowbane
Culver's root
field horsetail
golden alexanders
late goldenrod
narrow-leaved loosestrife
New England aster
prairie willow
purple meadow-rue
sawtooth sunflower
wild strawberry

Mosses and Lichens

juniper hair cap
pointed mnium
spoon-leaved moss
toothless twisted moss

Butterflies and Moths

arge tiger moth
bella moth
bog holomelina
bronze copper
common tan wave
Delaware skipper
drab brown wave
eastern black swallowtail
eyed brown
figured tiger moth
gray marvel
Henry's marsh moth
ignorant apamea
lined ruby tiger moth
orange sallow
parthenice tiger moth
pearly wood nymph
rigid sunflower borer
ruby tiger moth
ruddy quaker
silver bordered fritillary
silvery checkerspot
spotted grass moth
tawny holomelina
three-lined flower moth
yellow-collared scape moth

Beetles

argus tortoise beetle
big-headed ground beetle
black-legged tortoise beetle
bumble bee flower beetle
carrot beetle
golden tortoise beetle
minute marsh-loving beetle
mottled tortoise beetle
redbud bruchid

Insects

American grasshopper
black and yellow mud dauber
black-legged meadow grasshopper
blue mud dauber
carrot rust fly
common meadow grasshopper
gladiator meadow grasshopper
northern green-striped grasshopper
round-winged katydid
stilt bug
striped ground cricket

Spiders

banded argiope
black and yellow argiope
lattice spider
pirate wolf spider
shamrock spider

Landsnails

common disk snail
golden amber snail
small zonite shell
southern pill snail
toothless pupa snail
white swamp snail

Reptiles and Amphibians
- central newt
- Cope's gray treefrog
- eastern hognose snake
- painted turtle
- tiger salamander

Birds
- common yellowthroat
- sedge wren
- swamp sparrow

Mammals
- masked shrew
- meadow vole
- mink
- short-tailed shrew

Tall Shrub Communities

MANY PEOPLE ARE familiar with tall shrub communities. An angler carefully maneuvering the tip of a fishing rod through alder brush knows the dense, gnarly nature of the thicket. An early morning paddler knows to look for a doe and her fawns huddled among dew-drenched willows. A duck hunter finds good cover for a blind among the dogwoods at the edge of a beaver pond. A spring worshiper collects the bows of pussy willow to brighten the home.

These communities are recognized and used by many people but often little understood because they are generally part of a larger mosaic of plant communities. Alder, willow, and dogwood species, along with bog birch, constitute a majority of the shrubs. Tall shrub communities mostly occur in corridors and edges between other communities, seldom forming larger patches. Furthermore, tall shrub communities do best and are most persistent on wet mucky soils, which discourages many people from entering the area.

A comparison of the past and present extent of tall shrub communities is not possible. The early land surveyors recorded the trials of making their way through the thicket but did not give a clear picture of how much existed. We do know through various written accounts that thickets of brush recorded by explorers and missionaries still occur along the same rivers today.

The plant and animal composition of tall shrub communities will vary according to species present in the surrounding natural communities. Some species are consistently found more in the tall shrubs than anywhere else.

Shrub-Carr

Indicators: Red-Osier Dogwood, Pussy Willow, Beaked Willow, American Elm

Ecology

Shrub-carr, an unusual name for a plant community, develops best around lakes and meadows and along streams and is dominated by shrubs other than alder, primarily dogwood and willow. At the time of European settlement, shrub-carr formed narrow bands of transition between southern sedge meadows or wet prairies and wet hardwood forests. Since then, conversion to agriculture has eliminated many original shrub-carrs.

Before European settlement, fires helped shape these shrub communities. Springtime fires didn't have much influence on the shrubs because they usually stand in water in the spring. But by late summer and fall the wet areas dry out, and fire could burn the base of the woody vegetation. Trees adapted to wet soils usually have thin bark and cannot handle repeated fires, but shrubs, whose growing parts are insulated by the soil, could, and a relatively stable shrub community developed.

Range

A natural-world observer may see extensive areas of shrub-dominated lowlands in southern Wisconsin (under similar conditions in the north, alder becomes the dominant species). Most of these shrub communities developed after being abandoned as agricultural or pasture land. They are now succession communities, slowly transforming from open disturbed ground, through a shrub stage, on the path to becoming forest. Past attempts at intensive farming transformed some open areas into very simple shrub communities, containing but a few species, most of them exotics. Transition areas next to high-quality meadows are still the richest and most diverse shrub communities.

The total extent of shrub-carr is most likely more now than a hundred years ago. This new acreage varies from very simple to diverse communities. In many remnant natural communities, fire suppression has altered the natural dynamic, and many fens, wet prairies, and sedge meadows have changed to shrub communities. Draining sedge meadows also converted many acres to shrubs. The shrub border effect generated by damming streams and creating waterfowl flowages has added many acres of shrub-carr. Without fire, the narrow shrub fringe between uplands and wetlands is relatively unstable as a long-lived community. The slow process of recovery from excessive disturbance, however, can give longevity to the succession community.

Examples of this community are found at Avoca Prairie, Black River State Forest, Buena Vista Marsh, Cadiz Springs State Recreation Area, Lower Wisconsin River Valley, Muir Park, Necedah National Wildlife Refuge, North Kettle Moraine, South Kettle Moraine, Upper Mukwonago River, and White River Marsh.

Activities

- The shrub-carr community has numerous fruiting shrubs that are used by many species. Visit a shrub community every other week from midsummer through the following spring and chart the fruit development and use. Many fruits have substances that make them bitter in the fall. These chemicals break down in the freezing process over the winter and become palatable in spring (Kricher 1988).
- Map the zone of shrubs around a sedge meadow or wet prairie. Return to the site every few years to map the movement and growth of the trees. Do the same in a disturbed shrub area. Notice how disturbed areas have a scattered shotgun-type succession process with trees developing throughout.
- In midsummer visit similar shrub communities in the north and south. Both places should have the same species of butterfly, *Limenitis arthemis*. The southern one, named the red-spotted purple, is charcoal colored with red spots paralleling blue spots on the outer edges of the wing. The northern one, named the white admiral, is charcoal colored with a large white band near the edge of the wing. Hybrid males have a blue band next to the white band (Opler and Malikul 1992).

Characteristic Species

Plants
- bluejoint grass
- common water-horehound
- fowl meadow-grass
- great water dock
- late goldenrod
- marsh wild timothy
- orange jewelweed
- panicled aster
- purple meadow-rue
- spotted joe-pye weed
- swamp milkweed
- tussock sedge

Mosses and Lichens
- common apple moss
- juniper hair cap
- pointed mnium

Mushrooms
- black jelly roll
- buff crust
- willow milky

Butterflies and Moths
- Acadian hairstreak
- black-rimmed prominent
- blind-eyed sphinx
- cecropia moth
- darling underwing
- dimorphic bomolocha
- dreamy dusky wing
- fragile white carpet
- hag moth
- Henry's marsh moth
- hermit sphinx
- hummingbird clearwing
- lynx flower moth
- one-eyed sphinx
- parthenice tiger moth
- pepper and salt geometer
- puss moth
- red-humped caterpillar
- red twin-spot
- spear-marked black
- tawny-edged skipper
- unicorn caterpillar
- viceroy
- white underwing
- yellow-collared scape moth

Beetles
- banded net-wing
- bumble bee flower beetle
- checkerboard fungus weevil

dogwood twig borer
goldsmith beetle
narrow sap beetle
pale flower beetle
Pennsylvania firefly
poplar and willow borer
red-blue checkered beetle
short-winged leather-wing
spotted blister beetle
two-lined collops
vivid metallic ground beetle
willow leaf beetle

Insects

bee assassin
black and yellow mud dauber
curve-tailed bush katydid
Dawson's grasshopper
diamond-backed spittlebug
differential grasshopper
dogwood spittlebug
gladiator meadow grasshopper
hooded pygmy grasshopper
northern green-striped grasshopper
northern mole cricket
oystershell scale
potter wasp
red plant bug
Say's bush cricket
short-winged grasshopper
slender meadow grasshopper
snowy tree cricket
two-striped grasshopper
Walsh's grasshopper
willow plant bug

Spiders

arrowshaped micrathena
bola spider
crab spider
elongated long-jawed orb weaver
hammock spider
jumping spider
purse-web spider
shamrock spider
star-bellied spider

Landsnails

apple seed snail
brown pupa snail
common disk snail
golden amber snail
handsome vallonia
shining zonite shell
small zonite shell
tiny star shell
white swamp snail

Reptiles and Amphibians

eastern milk snake
northern water snake
pickerel frog

Birds

American woodcock

American woodcock (illustration by Jim McEvoy; courtesy of Wisconsin Department of Natural Resources)

common snipe
common yellowthroat
swamp sparrow
willow flycatcher
yellow warbler

Mammals

ermine
masked shrew
meadow vole
mink
short-tailed shrew

Alder Thicket

Indicators: Speckled Alder, Meadowsweet, Red-Osier Dogwood, American Black Currant

Ecology

In the north, alder thickets occupy places on the landscape similar to shrub-carr in the south. Alders are more common in the north and progressively less common toward their range limit in Illinois. As with shrub-carr, alder thickets develop around lakes and meadows, along streams, and in extensive lake plains, often forming a narrow community only one plant wide. Similar to the case with shrub-carr, fire helped determine the location of alder growth. Also, the persistent and ubiquitous beaver transformed many alder thickets to ponds then sedge meadows. Direct conversion to farmland did not affect the alder thickets as much as shrub-carr.

Many conifer swamps changed to immense alder thickets due to changes in the water table and peat depletion caused by fire. Alder has been considered a succession species, coming in for one generation, then succumbing to dominant trees. While stream-side alder thickets are very stable, changing only with dramatic water changes, alder thickets in burned areas are single-generation communities before changing back to conifer swamps.

Range

The critical factor in all alder thicket development is water movement. If the water source stagnates or dries up, the alder thicket community will succeed into other communities. Where water remains stable, the community can remain stable. Alder thickets can be found

Ancient Alder

For decades managers and scientists considered alder a short-lived succession species. While individual trees and stems may live for a short period of time (30 to 50 years on average), the natural community can persist for centuries. A land surveyor in the 1840s in Waushara County described in his journal descending a slope along the White River, crossing through a few chains of alder, and ascending a slope on the other side. In 1994, the alder was in the same spot and measured within a foot of the same distance recorded in the 1840s survey (Wisconsin Department of Natural Resources Natural Areas Survey, unpublished). The community was in the same spot for at least 150 years. Most surveyors did not bother describing the alder, but historic accounts of the first missionaries on the Bois Brule River described about 10 miles of dense shore brush in the same area that alder dominates today, present as a natural community for over 300 years.

Jewelwings

During the summer months on streams lined with alders, an observant naturalist can view the largest and showiest of the damselflies, which can be found cruising the waterways or perched on streamside vegetation. The ebony jewelwing is metallic green with black wings. The female can be distinguished by a white spot near the tip of the wings and by her copper abdomen. The female submerges for up to one-half hour when depositing her eggs onto underwater plants. After she emerges, she flies to a sunny spot to dry off.

at Brule River, Kohler–Andrae State Park, Mead Wildlife Area, Necedah National Wildlife Refuge, Northern Bayfield County, Point Beach State Forest, Powell Marsh, Three Lakes, and Wisconsin Point.

Activities

- Visit the library or obtain copies of the land-survey records from your county. Compare the description of a streamside area today with that written by the original land surveyors.
- Make a nighttime visit to an alder-lined stream in late summer, particularly one with banks that have patches of exposed mud and burrows. Listen for very low-pitched chips, one to three per second, repeated in a regular fashion. Use a flashlight to find the songster. You might just happen to view the bizarrely shaped northern mole cricket.
- Visit the alder area in early spring to observe marsh marigolds. This common, well-known flowering plant is one of our first blooming species each spring and very important for nectaring insects. Observe the number of insects visiting the flowers. Consider the plant's impact on the survival of these insects when no other flowers are blooming.

Characteristic Species

Plants

arrow-leaved tearthumb
bluejoint grass
crested wood-fern
great water dock
highbush cranberry
marsh bellflower
marsh fern
marsh marigold
panicled aster
rough bedstraw
royal fern

sensitive fern
skunk cabbage
spotted jewelweed

Spotted jewelweed (illustration by Jim McEvoy; courtesy of Wisconsin Department of Natural Resources)

spotted joe-pye weed

Mosses
common apple moss
pointed mnium
spread-leaved peat moss

Mushrooms
brown alder mushroom
scurfy alder cup
witches' butter

Butterflies and Moths
arched hooktip
barred itame
black-banded carpet
black-dotted lithacodia
common tan wave
currant spanworm
dock rustic
double-banded carpet
Doubleday's baileya
double-lined gray
euporatium borer
gray half-spot
Harris' checkerspot
lined ruby tiger moth
many-lined carpet
mourning cloak
mustard white
orange-barred carpet
parthenice tiger moth
puss moth
question mark
scribbler
sensitive fern borer
spotted thyris
sweetfern underwing
white-banded toothed carpet

Beetles
alder flea beetle
carpet beetle
lurid flat-headed borer
red-blue checkered beetle
short-winged leather-wing
vivid metallic ground beetle

Insects
carrot rust fly
currant fruit fly
hooded pygmy grasshopper
lance-tailed meadow grasshopper
northern mole cricket
short-winged meadow grasshopper
slender meadow grasshopper
soldier fly
stilt bug

Spiders
bola spider
crab spider
elongated long-jawed orb weaver
jumping spider
marbled spider
purse-web spider

Landsnails
brown pupa snail
common disk snail
handsome vallonia
Lea's pill snail
shining zonite shell
tiny star shell
white swamp snail

Reptiles and Amphibians
eastern milk snake
northern water snake
pickerel frog

Birds
alder flycatcher
American woodcock
common yellowthroat
gray catbird
swamp sparrow
tree swallow
veery
yellow warbler

Mammals
beaver
masked shrew
meadow vole
mink
muskrat
short-tailed shrew

Sedge Meadows and Fens

Sedge meadows and fens are types of wetlands that develop where water percolates through the soil or collects in low areas. These natural communities are mostly devoid of trees and harbor only scattered shrubs, if any. Tussock sedge, bluejoint grass, and wiregrass sedge are the most common plants. They are quite persistent and may reach remarkable old age.

Species composition is usually less diverse than other open treeless communities. Some species, however, are tightly aligned with sedge meadows and fens. Due to the rigorous conditions of flooding and drying, only species adapted to the stress of growing in standing water persist in these communities.

For decades, sedge meadows and fens were viewed as nothing but a hindrance to progress. Schemes to make these natural communities "useful" seemed to spawn on a regular basis. With each change in drainage technology, drainage was tried anew. Those meadows that weren't drained were often pastured or used as mowing meadows for marsh hay. Eventually, though, most people began to recognize the role of wetlands in maintaining water quality.

When the influx of European settlers entered Wisconsin, approximately 1.2 million acres may have been sedge meadow and fen. We do not know the acreage with great reliability because several types of fens have only recently been classified. Today, about 40,000 acres of high-quality sedge meadows and fens remain in the state. Estimates by type are: 10,000 acres of poor fen (poor fens have less species diversity than other sedge communities); 10,000 acres of northern sedge meadow; 20,000 acres of southern sedge meadow; and a few hundred acres of calcareous fen.

Calcareous Fen

Indicators: Bluejoint Grass, Fowl Meadow-Grass, Fringed Brome, Big Bluestem

Ecology

A fen is a low, wet area where peat accumulates from decomposing grasses, sedges, or reeds. In a calcareous fen, which is an herb or herb–shrub community, the peat develops under the influence of water rich in calcium or magnesium bicarbonates. The chemistry of the calcareous water, the landforms, and the magnitude of flow greatly influence the species composition. Several distinct types of calcareous fens occur in Wisconsin.

The most prevalent type is the prairie fen, which has active seeps or areas where water flows to form many rivulets. The flowing areas have a distinct sedge composition, especially spike-rushes. Between the flow areas, sedges, grasses, and forbs show a strong, moist tallgrass prairie influence.

A mound fen is formed by peat that accumulates where pressure forces water to the surface in flat areas. These mound fens, created by artesian water sources, have different species from the prairie fen, such as mat muhly and sterile sedge.

A third type of fen occurs in low-lying areas with standing calcium-rich water. Ponds form during wet periods, and the soil remains saturated during dry periods. The mineral waters promote a hybrid community of prairie and fen with abundant shrubby cinquefoil and valerian. The term fen meadow may be appropriate

Marl flats, the final type of fen, form on shallow flat areas, usually along the shores of a lake. In Wisconsin, most are found around a few lakes in southeast and east central Wisconsin. Marl flats have an abundance of sedge species.

Other communities accurately called fens are not calcareous fens. They meet the definition of peat accumulation but do not have active flow of calcium- and magnesium-rich waters. Shrubs or trees are often found on these other fens. A cedar swamp, discussed earlier, forms under the influence of mineral-rich water and can correctly be called a treed fen (a fen with trees). A tamarack fen, a sedge-dominated area with tamaracks, grows in water with less mineral content; without tamaracks, this same fen type is called a poor fen.

Range

Due to their unique requirements, fens have always been rare in the state and are now even more so. Compared to other states, however, Wisconsin has a tremendous number of fens, with at least 80 known fens and probably many as yet undiscovered because of their small size. Wisconsin's fens probably cover less than 1,000 acres. Despite the small number of acres, more than half has been acquired for long-term protection. Excellent fens can be seen at Muir Park, South Kettle Moraine, and Upper Mukwonago River.

Swamp thistle and swamp metalmark butterfly (illustration by Jim McEvoy; used by permission)

Metalmarks

Swamp thistle grows abundantly in calcareous fens, but it also grows in sedge meadows, alder thickets, shrub-carr, along streams, and elsewhere in the state. The swamp metalmark is a small butterfly whose sole larval food is the swamp thistle or woodland thistle. Unfortunately, the metalmark is found in only five locations in the state. Why should one species be rare when its host plants are relatively common?

This butterfly is small, less than 1½ inches tip to tip, and delicate, with a small body and very thin wings. One explanation for its rarity is the purely mechanical problem of reaching the larval plant to deposit its eggs. The metalmark needs to lay its eggs directly onto the basal leaves of a swamp or woodland thistle. If the thistle is surrounded by thick sedges or brush, the insect cannot reach the plant. In calcareous fens, however, the saturated soils keep the vegetation low in spots, and the metalmark can reach the thistle rosettes.

Activities

- Walk the perimeter of a fen and note the number of active seeps and flowing rivulets. Then describe the effects of changes in water flow, elimination of all flow, and pollution to the water source. Many species require continuous water flow and would be eliminated if the water stopped. Identifying ground water recharge areas involves sophisticated sampling techniques.
- Map the subtle changes in vegetation from both a top and side view. Note the differences in composition and height of the plants at different locations in the fen.
- Conduct invertebrate inventories in the same habitat patches mapped above. Describe the differences in the ways invertebrates utilize plants.
- A special activity for landowners that may have animals grazing a fen: First determine if you have a fen. Do you have pastured areas that remain wet all year? Do cows and humans avoid these areas in all but the driest years? Do the cows sink in up to their knees when walking through the wet spots? Does the vegetation appear different colored in early spring or late fall than the surrounding vegetation? The sedges along the rivulets will green early, but the remainder of the fen often remains brown well after the surrounding land is flush with spring growth. If the answer is yes to any of these questions, try to fence the area for a year to see the response. A fen should contain many of the characteristic species listed in this chapter. Many superb fens have been located immediately after grazing ends.

Characteristic Species

Plants

common mountain mint
common water-horehound
field horsetail
grass-of-parnassus
marsh bellflower
marsh fern
northern bedstraw
northern water-horehound
panicled aster
purple meadow-rue
red-osier dogwood
shining aster
shrubby cinquefoil
small fringed gentian
southern blue flag
spotted joe-pye weed
swamp betony
swamp milkweed

Mosses

juniper hair cap

Butterflies and Moths

arge tiger moth
bella moth
black arches
bog holomelina
Canadian owlet
cobbler
common tan wave
Delaware skipper
delicate cycnia
eupatorium borer
eyed brown
goldenrod flower moth
gray marvel
hermit sphinx
iris borer
long dash
milkweed tussock moth
orange sallow
parthenice tiger moth
pink-patched looper
silver bordered fritillary
silvery checkerspot
six-spotted gray
straight-lined looper
tawny-edged skipper
three-lined flower moth
white-dotted groundling

Beetles

bee-like flower scarab
big-headed ground beetle
black blister beetle
narrow sap beetle
Pennsylvania soldier beetle
short-winged leather-wing
triangular rove beetle
two-lined collops

Insects

American grasshopper
bee assassin
black and yellow mud dauber
carrot rust fly
colletid bee
crested pygmy grasshopper
diamond backed spittlebug
differential grasshopper
low-ridged pygmy grasshopper
northern green-striped grasshopper
phlox plant bug
red plant bug
spined assassin bug
two-striped grasshopper

Spiders

arrowshaped micrathena
banded argiope
black and yellow argiope
brown daddy-long-legs
cross spider
grass spider
lattice spider
shamrock spider
star-bellied spider

Landsnails

common disk snail
southern pill snail
tiny star shell

Reptiles and Amphibians

central newt
eastern milk snake
northern water snake
painted turtle
tiger salamander

Birds

American woodcock
common snipe
common yellowthroat
sedge wren
swamp sparrow
willow flycatcher

Mammals

masked shrew
short-tailed shrew

Southern Sedge Meadow

Indicators: Tussock Sedge, Bluejoint Grass, Fowl Meadow-Grass

Ecology

Sedge meadows require wet soils and develop best where spring-melt waters drain very slowly, or "pond," so standing water is present until at least midsummer. More than one-half of the plant species are sedges. These meadows dry in late summer, and abundant flowers appear in late summer and early fall. Sedge meadows are always in the zone between too dry or too wet. Wet prairies, shrub-carr, and forests develop when sedge meadows become drier. If persistent standing water is the norm, cattail or bulrush marshes develop.

Sedge meadows are most numerous in old glacial lake beds and around the shores of lakes and streams just above the permanent water table. The common sedges in these meadows grow into round clumps of vegetation (hummocks) that extend above the normal water levels. Examples can readily be seen in the low areas of cattle pastures, although hoof compaction exaggerates these hummocks.

Fires are also important for maintaining sedge dominance, but fires during droughts can burn deep into the peat. These burned areas can become ponds with the onset of rainy periods. Other disturbances such as drought or drainage create dry conditions that can change the composition, for example, to nearly pure stands of stinging nettle. These drastic impacts may last for centuries. In areas where silt is laid down every year, reed canary grass can become almost the sole plant species.

Range

Many ancient ecological communities in Wisconsin are now on the brink of extinction. For most of these communities, the events leading to their breakdown occurred between 1850 and 1950. But the southern sedge meadow has suffered more in recent times and is still currently under siege. At one time this diverse natural community covered nearly 1,000,000 acres. Losses started a little later than in upland communities but have accelerated with time because of technical advances in converting wetlands to agriculture. The amount of sedge wetland loss has been staggering in the mid-1900s, with fewer losses in the late 1990s due to environmental regulations, and only about 200,000 acres remain in the state. Even this acreage is highly misleading because most of these sedge meadows have invasive exotic species, such as reed canary grass, as the dominant plants. As few as 20,000 acres may be all that is left of the community of plants and animals known as southern sedge meadow.

Large expanses of southern sedge meadows are still present at Black River State Forest, Necedah National Wildlife Refuge, and White River Marsh. Smaller patches can be found at Horicon Marsh and Upper Mukwonago River.

Activities

- Walk through a sedge meadow in late August or September. Notice the numerous insects fleeing your footsteps. Many will land in the sticky trap of an argiope spider (see color in-

Meadow Bugs

August and September is a time of great abundance in sedge meadows. While birds, reptiles, amphibians, and mammals are the obvious occupants of the meadows, it's the insect populations that rule the community.

On an early autumn morning, a naturalist can get an indication of the predatory intensity in the meadow by the large number of spider capture nets outlined with droplets of dew. Spiders are the top predators in the meadow, and the largest and most obvious are several species of a group called argiopes, which can be seen hanging upside down, waiting for prey to enter their orbs.

Another top predator, the ambush bug, is a master of disguise and surprise attacks. Ambush bugs hide on flower heads, especially the large umbels of joe-pye weed and goldenrods. These well-camouflaged hunters attack nearly anything landing on the flower. Even huge prey, such as butterflies and bumble bees, can be seen futilely trying to escape.

While life and death struggles occur daily, so does the search for a partner. Many insects meet by using chemical attractants, but other groups of insects use sound or light to attract mates.

An evening in the sedge meadow can reveal as many as 10 species of crickets, katydids, and grasshoppers singing to attract the favors of a mate. Several cricket species look very similar, but find each other by sound. Each species has its own call, and some have different calls depending on how close the mate is. A few experts can identify many of these crickets and katydids by their "songs." It is hypothetically possible to get population counts of these species and track trends over time by counting singing males.

Fireflies (also called lightning bugs or glow bugs) find their mates by light. Fireflies are well known to rural Wisconsinites, but most people do not realize that there are several species, each of which has its own pattern of light flashes. The frequency and intensity of the flashes identify the species for potential mates. However, a predatory lightning bug that feeds on other fireflies mimics the flash pattern of the other species. When the duped bug comes to investigate, the predator captures and eats it.

sert). If a web has many trapped insects, sit and watch the actions of the spider. Note the size of insect the spider will go after first, then the second, and so on.

- Find a large umbel-shaped flower and examine it closely for the presence of an ambush bug, which should not be hard to find because they are relatively common. You can sit quite close to the flower, as most insects will come and go oblivious to your presence. Sooner or later you should be able to see the ambush bug capture its prey. Imagine the strength of those legs when they are grasping a bumble bee.

- Find patches of flowering vegetation. Many sedge meadow species tend to form in loose clones, where stems apparently separated on the surface are actually connected underground. List the plant and insect species found in each patch and compare the species found. Are these patches important for maintaining diversity (Forman and Godron 1986)?
- Visit wet meadows at different places in your region. Find an undisturbed sedge meadow, a meadow that has developed behind a water impoundment, and a wet meadow where ditches have channeled the water. Develop a plant species list for each area and compare them. How have human activities affected species composition?

Characteristic Species

Plants

common cattail
common water-horehound
field horsetail
marsh pea
meadow anemone
northern water-horehound
orange jewelweed
panicled aster
purple meadow-rue
southern blue flag
spotted joe-pye weed
swamp milkweed

Mosses

Schreber's cedar moss

Butterflies and Moths

Baltimore
bronze copper
chalcedony midget
Delaware skipper
eyed brown
Henry's marsh moth
ignorant apamea
Putnam's looper
silvery checkerspot
spotted grass moth
spring azure
Virginia ctenucha

Beetles

pale firefly
Pennsylvania soldier beetle
purple tiger beetle
red-blue checkered beetle
two-lined collops

Insects

black-sided meadow grasshopper
crested pygmy grasshopper
curve-tailed bush katydid
differential grasshopper
gladiator meadow grasshopper
low-ridged pygmy grasshopper
sedge pygmy grasshopper
striped sedge grasshopper

Spiders

banded argiope
black and yellow argiope
elongated long-jawed orb weaver
pirate wolf spider
shamrock spider

Reptiles and Amphibians

smooth green snake
western chorus frog

Birds

American bittern
common yellowthroat
sandhill crane
sedge wren
sora
swamp sparrow

Mammals

mink
muskrat

Northern Sedge Meadow

Indicators: Bluejoint Grass, Wiregrass Sedge, Few-Seeded Sedge, Tussock Sedge, Rattlesnake Grass

Ecology

Northern sedge meadows are similar to southern sedge meadows in outward appearance, hummocks, and water requirement, but there are many differences. Because of the cooler northern climate, many southern species cannot survive the short growing season. Northern sedge meadows have a significant moss influence, which impacts other species. Also, the northern meadows have fewer spring bloomers than their southern counterparts. Most of the bloom comes in late summer. Swamp aster, grass-leaved goldenrod, bog goldenrod, and rattlesnake grass are found more often in this community than any other.

Range

The sedge meadows of the north have fared much better than the southern sedge meadows, primarily because farmers in the north had a difficult enough time clearing and farming the uplands without also attempting to drain and farm the wetlands. Most of the northern sedge meadows border lakes and streams. They often form after a beaver family has moved on or died out, and the dams have failed. The rising water kills the trees and shrubs, but the perimeter of the pond can support sedge meadow species. When the pond recedes, the sedge meadow species invade the newly opened land. Fire, especially in northwest Wisconsin, can also kill trees and encourage sedge meadow species.

A much different type of sedge meadow occurs where fires regularly burned over peatland. Wiregrass sedges dominate these frequently burned sites. There are few forbs in these sedge meadows, which gives the meadows a uniform appearance.

The best wiregrass sedge meadows are at Crex Meadows and Powell Marsh. Larger areas of the more typical northern sedge meadow are also at Crex Meadows, as well as Three Lakes and Mead Wildlife Area.

LeConte's Sparrow

LeConte's sparrow is closely associated with northern sedge meadows. This five-inch bird is distinguished by its buffy face pattern. However, most visitors to these meadows will probably never see one. It is a secretive sprite that seldom emerges from the tall sedges, rarely perching high enough for observation. Its high-pitched insect-like trill can be hard for most people to hear or even recognize as a song. Also, the male is difficult to locate by his song because he sings mostly at night, usually stopping around dawn.

LeConte's sparrow (illustration by Jim McEvoy; courtesy of Wisconsin Department of Natural Resources)

Activities

- Find two sedge meadows in the north: one in the sand barrens land in northwest Wisconsin, the other elsewhere. List the plant species composition for the two sites and compare them. The northwest site will invariably have an abundance of wiregrass sedges and few flowering plants, whereas the other will have wide-leaved sedges and numerous flowers. Does fire frequency and water-level change account for the differences?
- In late August and September, visit a sedge meadow with a forest surrounding it after the passing of a cold front. If you slowly walk the edges, you should be able to observe large numbers of migrating birds. Insects apparently concentrate in these areas and are exploited by the birds.
- Conduct inventories of aquatic and terrestrial invertebrates in the wiregrass sedges and compare those results with the wide-leaved sedge meadows. Fewer species should be found in the wiregrass meadows because the plant diversity is much less.
- Visit the meadows from late spring into June. You should see several northern (boreal) butterflies flying at this time. Many seem to prefer these large open areas.

Characteristic Species

Plants

arrow-leaved tearthumb
common boneset
common cattail
dark-green bulrush
grass-leaved goldenrod
hardhack
marsh bellflower
marsh fern
marsh skullcap
northern blue flag
northern water-horehound
panicled aster
sensitive fern
spotted joe-pye weed
swamp aster

Mosses

Schreber's cedar moss

Butterflies and Moths

American bird's-wing
aphrodite fritillary
Baltimore
bent-line carpet
black-banded carpet
common tan wave
double-banded carpet
drab brown wave
gray half-spot
Henry's marsh moth
ignorant apamea
Leonard's skipper
many-lined carpet
meadow fritillary
orange sallow
Putnam's looper
spring azure
Virginia ctenucha
white-banded toothed carpet

Beetles

clay-colored billbug
long-necked ground beetle
pubescent ground beetle
redbud bruchid
swamp milkweed leaf beetle

Insects
- black and yellow mud dauber
- blue mud dauber
- curve-tailed bush katydid
- northern grasshopper
- sedge pygmy grasshopper

Spiders
- banded argiope
- black and yellow argiope
- elongated long-jawed orb weaver
- pirate wolf spider
- shamrock spider

Reptiles and Amphibians
- smooth green snake
- western chorus frog

Birds
- American bittern
- common snipe
- common yellowthroat
- northern harrier
- sandhill crane
- sedge wren
- sora
- swamp sparrow

Mammals
- mink
- muskrat

Savannas

At the time the first French missionaries arrived in Wisconsin, savannas covered nearly 9,600,000 acres, more than one-quarter of the land. These natural communities also covered huge areas of Minnesota, Iowa, Illinois, Indiana, Michigan, and Ohio. Savannas are ecosystems with extensive grass areas and scattered trees or shrubs. The vast savannas take many forms in North America, but Wisconsin has only four savanna communities: bur oak openings, black oak–Hill's oak barrens, pine barrens, and sand barrens.

Prior to widespread European settlement, Wisconsin's savannas formed a wide transition boundary between the prairies to the west, oak forest to the east, and pine forest to the north. The settlers plowed nearly all the prairie land and most of the savannas. Unplowed savanna remnants survived in hilly and wet areas or on dry, infertile sands, where the "barrens" type of savanna developed. Savannas in the driftless area were grazed and burned regularly up to the 1950s. Aerial photography from that era shows many acres of sparse tree cover (savanna). Today, with fire suppression and changes in farming practices, these lightly wooded savannas have closed to forest, and precious little savanna remains.

We do not know the precise composition of plants and animals in these original savanna communities, although historic accounts from the first explorers and settlers give us indications. Even with these accounts, it is difficult to understand what these savannas were like

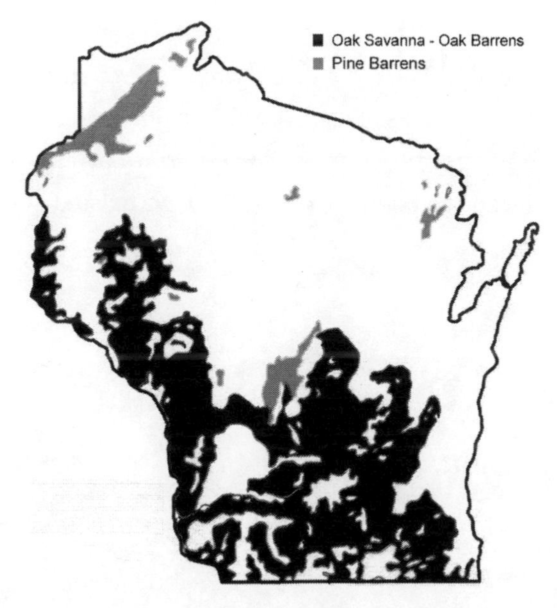

Savannas–Barrens

Savanna (illustration by Jim McEvoy; courtesy of Wisconsin Department of Natural Resources)

because there is so little left. Without the extensive savannas to give us context, we can only visualize how a savanna system might have functioned: Imagine a wildfire sweeping across the flat prairies, with widely scattered oaks, of Rock County. The fire would burn hot and fast, killing any tree except the thickest-barked bur oak. When the fire reached the hilly moraines in the eastern part of the county, it would rush up the first west-facing slopes with great speed and intensity, again killing nearly all the oaks. Upon reaching the crests, the fire would burn with much less intensity. On the down slopes, a few more bur oaks would survive. The fire would race up the next line of hills, but with less intensity. After a few lines of hills, white oaks and red oaks could withstand the fire, as could more shade-tolerant forbs.

Fire shaped all of the savannas. However, different fire intensities, different aspects, and different amounts of moisture permitted great variations over short distances. There might have been many subtypes of savanna. For example, the composition of the bur oak opening on the glacial till of the kettle moraine is certainly different from a ridge top bur oak opening in the driftless area. Other types probably developed on relatively flat prairie areas (dense bur oak groves). Floodplains had swamp white oak openings. Black oak–Hill's oak savannas dominated the sandy areas of central Wisconsin. Sand barrens developed along the sandy terraces of the Wisconsin and Chippewa Rivers. Pine barrens developed in the sandy areas in northern Wisconsin, and red pine savanna may have existed at one time.

Bur Oak Opening

Indicators: Bur Oak, White Oak

Ecology

Bur oak openings were lost to the same forces that transformed our prairies and for the same reasons: the rich soil, which was irresistible to farmers, and fire suppression, which allowed encroachment by shrubs and trees. Today, these natural bur oak openings are as rare as any natural community on earth, which is surprising since they once covered 5,500,000 acres of Wisconsin's landscape.

Fire suppression and tilling the soil, as well as logging and heavy grazing, have transformed these bur oak openings. Their structure and composition has changed so completely that now they are almost unidentifiable as natural communities. Unmanaged bur oak openings are barely recognizable as such because the characteristic "opening" trees are imbedded in a sea of younger trees. Furthermore, the bur oak opening groundlayer plants have been mostly replaced with forest species. These groundlayer species can still be found in open areas or at the edges of woods and trails, though rarely. The only way we will ever have a functioning bur oak opening ecosystem again is to restore degraded and overgrown bur oak openings. Even with the most ambitious program, however, the entire ecosystem cannot be duplicated. The elk and bison are gone, as are the lynx and cougar.

These original bur oak openings varied in the distribution of trees, with many scattered orchard-like trees and occasional dense groves. The understory composition varied according to light intensity, from species of open sunny prairies to those of closed oak forests. Yearly fires burned through these oak openings, either started by Native Americans or generated by lightning strikes. Resultant shrub patches, especially of black oak and New Jersey tea, were common.

Range

This community has been almost completely destroyed by conversion to farmland, pasture, and oak-dominated forests. Most remaining bur oak opening areas are almost exclusively

Swallow-Tailed Kite

Though some farmland that once was savanna still contains scattered oak groves, it's not enough to provide the landscape requirements for certain savanna species, such as the swallow-tailed kite. This bird, which nests in large isolated trees, has undergone a massive shrinking of habitat in the last 200 years. It forages over, through, and under the scattered trees for reptiles, birds, insects, and small mammals. Loss of the savanna landscape was too much for the kite to overcome, and it is now found only in the extreme southern United States in pine savannas (Meyer 1995).

Bur oak opening (photograph by Eric Epstein; used by permission)

used for pasture. Areas with unmodified understory are nearly nonexistent. The best examples of these exquisite bur oak openings can be seen at Avoca Prairie, Avon Bottoms, Chiwaukee Prairie, and South Kettle Moraine.

Activities

- Sketch the tree forms and spatial patterns of trees in an oak opening. Bur oak is the common tree and usually grows in strange-shaped forms. Also, observe the size of the first branches, which can give a clue to the tree's age. Ancient bur oak trees can live 300 to 400 years, and two-foot-diameter branches are common on these monarchs.
- Observe the leaf-litter decay process by digging a cross-section of the leaf litter using a shovel. Oak leaves have large amounts of tannin in them, and this chemical is difficult for many species of fungi to break down. The leaf litter needs a few years to mechanically break down before the specialist fungi are able to utilize the leaves. In a functioning oak savanna, fire would consume the oak leaves every few years. Consider the effects fire might have on the leaf-litter fungi community.
- Observe the ends of oak branches. Often these branches will be whitened. This feature, called white rot, is caused by decomposing fungi. A remarkable fact regarding white rot is the number of species involved and how they partition the branch. As many as 12 species can live along a two- to three-foot section of branch.

- Visit an oak savanna throughout the growing season to document the singing insects. As many as 90 species of crickets, cicadas, katydids, and grasshoppers make sounds. These insects can be identified and tracked throughout the year by the different sounds they make. These singing insects can be found in four general habitats: on the ground, in herbaceous vegetation, in shrubs, and in trees. Oak savannas provide the most diverse combination of those habitats and thus have the highest proportion of singing insects when compared to other communities (Alexander et al. 1972).

Characteristic Species

Plants

American hazelnut
big bluestem
bird's-foot violet

Bird's-foot violet (photograph by Thomas A. Meyer; used by permission)

false Solomon's seal
false toadflax
flowering spurge
gray dogwood
hoary puccoon
hog peanut
lead-plant
little bluestem
New Jersey tea
prairie tickseed
purple prairie clover
red raspberry
riverbank grape
spreading dogbane
wild bergamot
wild geranium
wild rose
wild strawberry

Mosses and Lichens

blister lichen
matted byrum
reddish dog lichen
sod lichen
urn moss

Mushrooms

black jelly drops
burn site ochre cup
gemmed amanita
green stropharia
grisette
oak-loving collybia
parasol
purple-gilled laccaria
skull-shaped puffball
spindle-shaped yellow coral
white worm coral
yellow blusher

Butterflies and Moths

common wood nymph
dotted graylet
fawn sallow
greater wax moth
hobomok skipper
least skipper
little wood saytr
oak skeltonizer
painted lady
southern cloudy wing

Beetles

black firefly
eyed click beetle
fiery hunter
hermit flower beetle
rose chafer
rose leaf beetle
tile-horned prionus
white oak borer

Insects

black-horned tree cricket
colletid bee
common walkingstick
narrow-winged tree cricket
oak treehopper
partridge scolops
short-legged shield-back
sulphur-winged grasshopper

Spiders

arrowshaped micrathena
bola spider
brown daddy-long-legs
crab spider
eastern wood tick
jumping spider
purse-web spider

Reptiles and Amphibians

blue racer
bullsnake
Cope's gray treefrog
eastern hognose snake
eastern milk snake
tiger salamander

Birds

American kestrel
brown thrasher
chimney swift
eastern kingbird
eastern meadowlark
gray catbird
indigo bunting
song sparrow
vesper sparrow

Mammals

coyote
fox squirrel
red fox
woodchuck

Black Oak–Hill's Oak Barrens

Indicators: Black Oak, Hill's Oak

Ecology

The black oak–Hill's oak barrens is a savanna community that differs both in appearance and in species composition from the bur oak openings savanna community. Species more commonly found in the black oak–Hill's oak barrens include lupine, goat's-rue, frostweed, and butterfly weed.

Frequent fires did not consume the massive thick-barked bur oaks, but black and Hill's oaks were often burned. Survival after a fire for black oaks and Hill's oaks depends on underground shoots developing into new trees. This savanna was sometimes referred to as "oak grubs" because the only way to eliminate them was to dig deep around the roots and "grub" them out. Under a regime of regular fire at short intervals, grubs could persist for hundreds of years. This phenomenon was so pervasive in certain areas that the original surveyors referred to the areas as brush prairies. Black oak and Hill's oak dominate these areas, and bur oak is a minor component.

Range

Oak barrens develop to their greatest extent on plains, moraines, and bluffs where sandy soil or sandstone bedrock is present. The effects of poor nutrient availability and excessive drying help determine which species thrive. Intermediate frequency of intense fire (every 40 to 60 years), drought, grazing, and oak wilt result in the most diverse assemblage of plants and animals. Less fire permits oak forests to develop, and more fire pushes the community toward brush prairie. The community is most widely distributed in southwestern, central, and west central Wisconsin.

Black oak–Hill's oak barrens are concentrated in the sandy areas in Waushara, Marquette, Adams, Juneau, and Jackson Counties. Other significant acreages occur along the Chippewa, Wisconsin, and Sugar Rivers on sandy outwash terraces; in a large area of Dunn, Chippewa, and Barron Counties; and in the more southerly parts of the pine barrens areas in Polk and Washburn Counties.

Large areas of black oak–Hill's oak barrens or scrubby oak forests remain and can be found at Black River State Forest, Blue River, Crex Meadows, Lower Wisconsin River Valley, and Necedah National Wildlife Refuge.

Activities

- Visit a sandy oak barrens throughout the growing season. Concentrate the visits on open sunny areas with very little vegetation. These spots are often the haunts of robber flies. Watch carefully as they attack flying prey. Robber flies need to lay their eggs in a host to provide food for their larvae. Beetles are excellent hosts, but they have an impenetrable armor at rest. When in flight, however, the soft abdomen is exposed, leaving an opening for the robber flies to attack.

Pigeon By-Products

The passenger pigeon once inhabited Wisconsin sometimes in the millions. The species was nomadic, nesting one year and migrating through the next. Consider the profound effect this bird had on the oak savannas and woodlands of Wisconsin. Late in their existence, an estimated 136,000,000 birds nested in an 850-square-mile area (Schorger 1982). The amount of oak mast and other food gathered could be nearly 10,000,000 pounds per day. After nesting, some roost sites had even more birds, and they were known to pack into even smaller areas.

Far more important, consider the effects of bird droppings at the nest and roost sites. As much as 70 pounds of fertilizer could be deposited around a single nest tree during the brief nesting period. At roost sites, 35 tons of droppings per acre per day could be deposited on the ground. These waste products added vast amounts of nitrogen and uric acid to the savannas. What did this nutrient loading do to the composition of the savannas, and how long did this concentrated fertilization affect any one site? We do know concentrated fertilization can kill many plants, but we do not know how long the effects lasted because extinct birds' habits cannot be studied.

Passenger pigeon flight (courtesy of Wisconsin Department of Natural Resources)

- Map the patterns of oak wilt in managed and unmanaged situations. Note: the managed areas will have the dead trees cut, and you may notice a trench around the perimeter of the affected area. Come back for several years to document the spread of the oak wilt and dead trees. Describe the differences in spread between the site where management is trying to stop the oak-wilt spread and the unmanaged site.
- Visit the barrens in early spring when the first flowers are beginning to bloom. Notice the insects visiting those flowers. Several species of bees and bee-mimic flies pollinate these early bloomers. One native fly looks very similar to the honey bee, but the honey bee is an exotic brought in by the pioneers. Also consider the abundance of deer flies and horseflies occupying the barrens in midsummer. The numbers of individuals and species can be staggering. When you swat and perhaps kill that biting fly, do you ever consider the diversity of the deer fly fauna?

Characteristic Species

Plants

American hazelnut
black-eyed Susan
common rockrose
common spiderwort
false Solomon's seal
false toadflax
field pussytoes
flowering spurge
gray goldenrod
hoary puccoon
June grass
lance-leaved ground cherry
lead-plant
little bluestem

prairie tickseed
rough blazing star
round-headed bush clover
starry false Solomon's seal
wild rose
wild strawberry
yellow coneflower

Mosses and Lichens
awned hair cap
British soldiers
common hair cap
common pygmy moss
cornucopia cladonia
juniper hair cap
ladder lichen
matted byrum
purple-horned tooth moss
reindeer lichen

Mushrooms
collared earthstar
eastern cauliflower mushroom
ochre jelly club
rounded earthstar
saltshaker earthstar
torn fiberhead
umbrella polypore

Butterflies and Moths
black and yellow lichen moth
common ptichodis
Edward's hairstreak
Ernestine's moth
feeble grass moth
lynx flower moth
Melsheimer's sack-bearer
nais tiger moth
oak skeletonizer
orange-tipped oakworm
pearly-winged lichen moth
pink-striped oakworm
placentia tiger moth
rubbed dart
sheathed quaker
subflexus straw
tephrosia moth
three-lined angle

Beetles
black oak weevil
caterpillar hunter
cloudy flower beetle
dingy ground beetle
green June beetle
oak branch borer
rose chafer
sand tiger beetle
six-spotted tiger beetle

Insects
American grasshopper
antlion
dusky grasshopper
oak apple gall
oak treehopper
sand wasp

Spiders
burrowing wolf spider
crab spider
eastern wood tick
jumping spider
line weaving spider

Reptiles and Amphibians
blue racer
bullsnake
Cope's gray treefrog
eastern hognose snake
eastern milk snake
tiger salamander

Birds
brown thrasher
eastern towhee
gray catbird
indigo bunting
vesper sparrow

Mammals
coyote
fox squirrel
red fox
striped ground squirrel
woodchuck

Pine Barrens

Indicators: Jack Pine, Hill's Oak, Red Pine

Ecology

Pine barrens dominated by jack pine are the northern savanna counterpart of the black oak–Hill's oak barrens found in southern Wisconsin. Today, these pine barrens are extensive, but small stands are separated from each other by adjoining forest land. At the onset of European settlement, pine barrens were found in northern Marinette County; Oneida, Vilas, and northern Lincoln Counties; portions of Adams, Juneau, Wood, and Portage Counties; and in a broad band going from Burnett County, southern Douglas, and Northern Washburn County through the heart of Bayfield County nearly to Lake Superior.

Like the oak barrens, the pine barrens developed on very sandy soil. In areas that burned regularly, jack pine, which depends on fire for seed dispersal, and stunted Hill's oak became the dominant trees. The size of the Hill's oak and the number of jack pine depended on the frequency of the fires. The landscape was a mosaic of open lands, scattered trees, dense shrubby Hill's oak thickets, and groves of jack pine and aspen. Sharp-tailed grouse prefer this shifting landscape mosaic created and shaped by fire. With fire suppression and the development of nurseries and tree-planting programs, the pine barrens were converted to jack pine–Hill's oak forests and jack pine or red pine plantations. The sharp-tailed grouse populations are now mostly restricted to the areas managed with fire.

Range

The pine barrens community, which originally covered 2,300,000 acres, is an important part of our northern landscape. The understory is mostly sandy grassland or bracken grassland, but in some areas blueberries and huckleberries so completely cover the ground that they form a heath.

Landscape Effects of Fire

The former pine barrens landscape had large, recurring fires that burned tens of thousands of acres. These fires did not burn everything and produced a patchy landscape similar to the Yellowstone fires of 1988, with large and small unburned patches. Certain insects, such as the jack pine budworm, adapted to the regular fires, and their populations ebbed and flowed with the availability of jack pine as a food source. Because of the regular fires and patchy landscape, the jack pine budworm population never reached levels that could eat all the jack pine buds in the forest. However, with today's emphasis on logging for paper products, large contiguous areas of dense jack pine covers the landscape, which provides the right ingredients for population explosions of jack pine budworm, which can kill millions of jack pine trees.

Sharp-tailed grouse (illustration by Jim McEvoy; used by permission)

The closer the barrens are to the prairies, the more prairie species are found in them. The areas richest in prairie species are in Polk and Burnett Counties, with progressively fewer prairie species found the farther north you go. The barrens in the northeast have few prairie species associated with them, and the barrens in Vilas, Oneida, and Lincoln Counties, which are even farther from the prairies, are mostly heath dominated. Prime areas for studying pine barrens are Crex Meadows, Douglas County Sharptail Barrens, Dunbar Barrens, Moquah Barrens, Necedah National Wildlife Refuge, and Vilas County.

Activities

- Visit two of the pine barrens noted above. List the plant species found on hikes through the properties and compare the lists. The list for the barrens farthest south and west should contain more species. As the barrens move to-

Bracken Grassland

Bracken grasslands are found in the north on sandy and gravelly soils. After timber was logged from the northern forests, vast areas burned, transforming the forests into bracken grasslands. Scattered pine stumps still survive as an indicator of the former forest community. Mid-summer frost keeps the encroaching forest from reestablishing. Bracken grasslands form mostly in basins with grasses and sedge at the bottom and bracken fern on the slopes. Good examples can be found at Johnson Lake in Vilas County (see the chapter in part 2 on Vilas County) and at Dunbar Barrens. Plant species found more commonly in bracken grasslands than anywhere else include: bracken fern, prairie brome, poverty oatgrass, slender wheatgrass, Canada hawkweed, northern heart-leaved aster, rough-leaved ricegrass, and upland wild timothy.

ward the northeast, a slow but perceptible loss of species occurs, with the least diverse barrens occurring at Moquah.

- Compare the groundlayer species in a fire-managed barrens (usually these areas have very few trees), a wildfire-shaped barrens (two large wildfires occurred in the northwest in the 1970s), a jack pine plantation, and a red pine plantation. Note: a general trend emerges with species requiring full sunlight in the fire-managed barrens to very few species under densely planted red pines
- Map the tree spacing in the same barrens as above. Describe the animal species, such as flycatchers, northern harrier, and Karner blue butterfly, that might use such spacing.

Characteristic Species

Plants

American hazelnut
bracken fern
Canada blueberry
Canada mayflower
early low blueberry
false Solomon's seal
flowering spurge
huckleberry
little bluestem
narrow-leaved loosestrife
prairie redroot
sand cherry
spreading dogbane
starry false Solomon's seal
sweet fern
wild bergamot
wild rose
wintergreen

Mosses and Lichens

awned hair cap
British soldiers
brown-fruited cup cladonia
common hair cap
common pygmy moss
cornucopia cladonia
flabby lichen
Iceland moss
juniper hair cap
ladder lichen
matted byrum
miniature tree lichen
pitted cetraria
purple-horned tooth moss
reindeer lichen
starry lichen
urn moss
water measuring cord moss

Mushrooms

bitter bolete
black morel
common fiber vase
crown-tipped coral
dye-maker's false puffball
Russell's bolete

Butterflies and Moths

black and yellow lichen moth
Compton's tortoise shell
Esther moth
Formosa looper
jack pine budworm
lead-colored lichen moth
manto tussock
neighbor
red-fronted emerald
snowberry clearwing
sordid underwing
stormy arches
sulphur moth

Beetles

cloudy flower beetle
green June beetle
pine engraver
pine heartwood borer
pine root collar weevil
pine stump borer
red pine flat-headed borer
ribbed pine borer
spotted savage beetle

Insects

antlion
coral-winged grasshopper
mining bee

ornate pygmy grasshopper
sanded pygmy grasshopper
sand wasp

Spiders
bowl and doily spider
burrowing wolf spider
crab spider
eastern wood tick
jumping spider
line weaving spider

Reptiles and Amphibians
bullsnake
Cope's gray treefrog
eastern hognose snake
smooth green snake

Birds
eastern towhee
hermit thrush
indigo bunting
mourning dove
northern flicker
yellow-rumped warbler

Mammals
coyote
least chipmunk
red fox
red squirrel
striped ground squirrel

Sand Barrens

Indicators: Little Bluestem, June Grass, Sand Sedge, Hairy Panic Grass

Ecology

An appropriate introduction to the sand barrens community could be "welcome to the Wisconsin desert." Sand barrens are another natural community that have increased in acreage with human activity. These barrens can range from random sand blows on dry sand prairies to large expanses of wind-shaped sand landscapes. They are different from lake dunes, both in formation and vegetation. These barrens were formerly dry sand prairies, but through tilling and over grazing and reduction of vegetation, the prairies lost their capacity to hold the soil. As a result, sand blows form, along with a compliment of many desert-like features, such as dunes formed by wind-blown shifting sand, small desert pavements formed by pebbles concentrated in the hollows, and surface soil temperatures approaching 160 degrees. Along with these physical changes, the vegetation has desert-dwelling adaptations: succulent leaves, deep tap roots, and many short-lived annuals.

Sand barrens, which owe their existence to disturbances, can tolerate some small disturbances. This community, however, cannot, tolerate off-road vehicles. One pass by an off-road vehicle is equivalent to thousands of footsteps. Continued use can impede vegetation

Lark sparrow (illustration by Jim McEvoy; courtesy of Wisconsin Department of Natural Resources)

Box Turtle

Animal species more commonly found in the southern Great Plains can be observed in the Wisconsin desert. Ornate box turtles are mostly limited in the state to the sandy areas along or near the Wisconsin River (see color insert). This turtle may be more appropriately called a tortoise, and it rarely goes to water. It lives out its life wandering the barrens for its favorite succulent foods—prickly pear cactus, spiderwort, and berries. At one nature preserve the box turtles have a fondness for mullberries; a turtle's entire head can be stained purple after days of gorging.

The lark sparrow, a bird of dry places in the western United States, is also found in Wisconsin's sand barrens. These birds prefer to forage for weed seeds in very sparse grassland. If the vegetation becomes too thick, they will move on. A resource management conundrum occurs in sand barrens. Lark sparrows like to nest near the base of small trees; in Wisconsin they prefer nesting near red cedar. But if the red cedars are not controlled, they tend to fill in an open area, eliminating lark sparrow habitat. If the red cedars are entirely removed, however, the lark sparrow population crashes. One answer is to provide the preferred habitat by cutting the larger red cedars, although this is a continual and labor-intensive management technique.

growth, increase sand movement, and turn a desert into a sand wasteland.

Range

The dune-like formations occur along sandy river terraces of the Wisconsin and Mississippi Rivers. These have increased in area and are quite stable; they are only gradually being taken over by oaks. Many plants able to survive extremely dry conditions are important because they stabilize the wind-blown sands.

Wonderful examples of sand barrens can be found in Blue River Sand Barrens and Lower Wisconsin River Valley.

Activities

- Explore the flat area between dunes. Sketch the composition of this area in comparison with the moving sands nearby. This feature is called desert pavement. The wind blows the lighter sands, which form the many and various-shaped dunes. The wind cannot move pebbles and rocks, and a flat hard surface develops. Carefully observe these areas for species.
- A continuation of the above activity is to describe how plants and animals have adapted to survive in this harsh environment. Consider the mid-day temperature in summer and the extreme cold of winter. Rains that fall on these sand barrens quickly disappear deep into the ground. All these stress factors limit the species that survive to those having special adaptations.
- Map pocket gopher mounds and how they increase over the course of a year. Track these mounds for a few years to document the plants growing on them. These mounds can provide primary habitat that is unavailable elsewhere on a sand prairie or barrens and consequently harbor species that only grow on them (Gibson 1988).
- If you happen to come upon a crow, blue jay, or other bird stretched out on the ground, acting intoxicated with wings spread and uncoordinated movements, do not assume the bird is injured. Carefully observe the bird before approaching. The animal may be "anting." This activity involves a bird, often a crow, lying on an ant hill. The ants defend their home by spraying formic acid as a repellent. The benefit for the crows is that the formic acid repels troublesome parasites, mites, and ticks (Terres 1980).

Characteristic Species

Plants

- common ragweed
- common rockrose
- common spiderwort
- fall witchgrass
- few-flowered panic grass
- flowering spurge
- hairy puccoon
- horse mint
- rough blazing star
- round-headed bush clover
- sand cress

Mosses and Lichens

- awned hair cap
- British soldiers
- common hair cap
- common pygmy moss
- green-felt moss
- juniper hair cap
- ladder lichen
- matted byrum
- miniature tree lichen
- purple-horned tooth moss
- reindeer lichen
- soft lichen

Mushrooms

- barometer earthstar
- buried-stalk puffball
- collared earthstar
- rounded earthstar

Butterflies and Moths

- black and yellow lichen moth
- broken dash
- checkered skipper
- feeble grass moth
- lead-colored lichen moth
- olive-shaded bird-dropping moth
- pearly wood nymph
- ragweed flower moth
- small bird-dropping moth
- southern cloud wing

Beetles

- American bembid
- bronze tiger beetle
- dingy ground beetle
- noble tiger beetle
- sand tiger beetle
- six-spotted tiger beetle

Insects

- antlion
- dusky grasshopper
- sand wasp

Spiders

- ant mimic
- burrowing wolf spider
- crab spider
- jumping spider
- line weaving spider

Birds

- common nighthawk
- horned lark

Mammals

- striped ground squirrel

Aquatic Communities

Aquatic communities have standing water more than six inches deep or flowing water throughout most of the year. This book uses a simple classification that separates natural communities into shallow marsh, deep marsh and lakes, Great Lakes reefs, rivers, and springs. With a primary focus on vascular plants, this simplified classification can be useful to the budding naturalist.

Advanced nature observers will recognize many more differences in aquatic communities than presented here. Lake communities can vary dramatically in species composition based on limnological characteristics. Differences in lake depth, pH, water chemistry, drainage patterns, and shoreline characteristics can all affect species composition. Similar parameters can affect river-community composition, which generally becomes more diverse as the river gets larger. Only a handful of people may have the knowledge and means to study the deep Great Lakes reefs, but they are intriguing nonetheless. An endemic species, the Wisconsin well amphipod, lives in ground-water aquifers and may never have its life history studied.

Wisconsin aquatic communities include 1.7 million acres of Lake Superior, 4.7 million acres of Lake Michigan, and nearly 14,000 inland lakes encompassing more than 1,000,000 acres. Inland lakes have dramatically increased since European settlement with the creation of more than 650,000 lake acres by impounding rivers. Wisconsin still has approximately 33,000 miles of river communities available for exploration.

Shallow Marsh

Indicators: Bulrushes, Arrowheads, Bur-Reeds, Pickerelweed, Wild Rice

Ecology

The shallow marsh is characterized by the presence of standing water (normally six inches to three feet deep) all or most of the year and by perennial plants that extend their growing parts well beyond the surface of the water. These wetland communities are easily separated from forested wetlands by the absence of woody plants.

Cattail marshes are the most recognizable type of marsh, but in some marshes, bulrushes are the dominant species. Some marshes may have an abundance of sedges, grasses, or, unfortunately, nearly pure stands of purple loosestrife, an invasive exotic. Other shallow-water areas have dominant plants appearing only late in the growing season. Usually these places are very attractive to animals, especially migrating waterfowl. These areas can be dominated by nearly pure stands of wild rice, pickerelweed, bur-reed, or American lotus (in backwater areas along the Mississippi).

Other types of shallow marshes develop in places where landforms or rainfall and soils limit their distribution. Small areas of shallow-water, sedge-dominated communities occur along the Great Lakes. Shore features are formed by the action of currents and waves along the shore. These currents carry sand and deposit it as sand bars or spits that protect the bays from the pounding of waves. This protection from the Great Lakes' influence is not complete because these areas are subjected to small tides with water flowing in and out. This water movement and the natural system it creates can technically be called an estuary.

Coastal-plain marsh communities develop along lakes with greatly fluctuating water levels (as much as 15 feet). On these sandy soils, rainfall determines the water level in any particular year. These lakes are found in Marquette, Adams, Waushara, and Portage Counties. The high-water cycle keeps these coastal-plain marshes free of shrubs. In low water, the marshes exhibit a remarkable flora with several species found only in these marshes, in Indiana and Michigan marshes, and along the Atlantic coast.

Pickerelweed (illustration by Jim McEvoy; courtesy of Wisconsin Department of Natural Resources)

An unusual emergent community, the interdunal wetland, is found in depressions of open dunes. These rush, sedge, and shrub wetlands also fluctuate in wetness, depending on the level of the nearby Great Lake. The lake's influence is not direct but comes under the dunes through saturated sands. The best examples in the state are in Kohler–Andrae State Park.

Shallow marsh plants grow in permanently or seasonally shallow water, which needs to be at least six inches deep for part of the year. Dense areas of bulrushes, lake sedges, lotus, arrowhead, bur-reed, or pickerelweed can dominate the shallow marsh in various combinations, depending on water depth and water hardness.

Different combinations of species can profoundly change the use by animals. Bulrushes are favorite nesting places for yellow-headed blackbirds and the rare red-necked grebe. Lake sedge may harbor dense colonies of sedge wren. Lotus and pickerelweed marshes are preferred by American coot and common moorhen.

Range

The best developed shallow marshes are in the managed wetlands of the state. Good examples are found in Apostle Islands, Black River State Forest, Brule River, Cadiz Springs State Recreation Area, Crex Meadows, Lakewood Area, Mead Wildlife Area, Muir Park, Necedah National Wildlife Refuge, Northern Bayfield County, North Kettle Moraine, Powell Marsh, Ridges Sanctuary, South Kettle Moraine, Three Lakes, Upper Mukwonago River, White River Marsh, and Wisconsin Point.

Activities

- Find a shallow marsh with no fish (usually they are isolated basins) and a shallow marsh with fish (usually they are connected to streams of deeper water areas). Compare the dragonfly diversity. Visit the sites a few times during the summer, using the dragonfly guides in the reference section to list the species found. Note: fish love to eat dragonfly nymphs, which may explain the differences.
- List the plant species found in a shallow marsh with few or no cattails and a marsh with abundant cattails. Compare the lists and note the differences in diversity.
- Visit the same shallow marshes from the first activity in early April, mid-May, and late June. List the frogs heard and the approximate numbers (one to 10, dozens, too many to count). Fish also eat frogs and their eggs.

Characteristic Species

Plants
bur-reed
chairmaker's rush
common arrowhead
great bladderwort
great bulrush
hard-stemmed bulrush
needle spike-rush
Small's spike-rush
wool grass

Mosses
common apple moss

Butterflies and Moths
bent-line carpet
bronze copper
buckbean munroessa
golden looper
obscure paraponyx
pickerel weed borer
pondweed synclita
Putnam's looper
white-tailed diver

Beetles
four-spotted bembid
giant diving beetle
giant scavenger beetle
knotweed weevil
narrow scavenger beetle
patruus bembid
rice water weevil
river tiger beetle
striped diving beetle
water scavenger beetle
water weevil

Insects
American emerald
black-mantled glider
black-tipped darner
blue dasher
calico pennant
Canada darner
common baskettail
common bluet
common whitetail
curve-tailed bush katydid
dot-tailed whiteface
four-spotted skimmer
green darner
hooded pygmy grasshopper
pied skimmer
prince baskettail
racket-tailed emerald
rice leaf miner
Say's bush cricket
sedge pygmy grasshopper
spiny baskettail
spread-winged damselfly
twelve-spotted skimmer
variable darner
yellow-legged meadowhawk

Spiders
elongated long-jawed orb weaver
pirate wolf spider
red freshwater mite
six spotted fishing spider

Aquatic Insects
anopheles mosquito
backswimmer
burrowing mayfly
common whirligig
crawling water beetle
dobsonfly
fishfly
giant water bug
lake fly
marsh beetle
marsh fly
phantom crane fly
predaceous water beetle
small mayfly
velvet water bug
water boatmen
water penny beetle
water scavenger beetle
water scorpion
water treader

Mollusks and Snails
Adam's pea clam
American ear snail
amphibious fossaria
blunt prairie physa
common stagnicola
flatly coiled gyraulus
graceful fossaria
greater corinate ramshorn
great pond snail
irregular gyraulus
modest fossaria
oval lake limpet
riverbank looping snail
Say's toothed planorbid
slender pond snail
solid lake physa
striped stagnicola
swamp fingernail clam
tadpole snail
ubiquitous pea clam

Fish
bowfin
brassy minnow
central mudminnow

Reptiles and Amphibians
central newt
green frog
mudpuppy
northern water snake
snapping turtle

Birds

- American black duck
- American coot
- black tern
- pied-billed grebe
- purple martin
- sora
- yellow-headed blackbird

Mammals

- mink
- muskrat
- otter

Cattail Marsh

Indicators: Broad-Leaved Cattail, Narrow-Leaved Cattail

Ecology

This shallow marsh community is obviously dominated by cattails and is highly recognizable, with features well known to most naturalists. People maintain or restore this community because they value some of its residents, namely ducks and muskrats. Species diversity is poor, but those present can have high production, especially the species that are hunted and trapped.

Range

Some of the best cattail marshes are found on areas heavily managed for sport hunting, such as Horicon Marsh and Mead Wildlife Area.

Muskrat House

Muskrats can have a profound influence on how plants and animals use a shallow marsh. Their lodge-building activities open extensive areas of dense cattails, thus permitting sunlight to reach the water. Under these open conditions, one-celled plants grow abundantly, and this encourages population growth of the species that feed on those plants, such as the mallard, the blue-winged teal, and the northern shoveller. If left unchecked by predators, muskrat populations soar to the point at which they eat all the available food. Once this happens, their population crashes, and the cattails reestablish dominance.

Characteristic Species

Plants
bulblet water hemlock
common spike-rush
duckweed
great water dock
marsh skullcap
water sedge

Mosses and Lichens
common apple moss

Butterflies and Moths
cattail moth
Henry's marsh moth
pickerelweed borer
sky cosmet

Mollusks and Snails
Lilljeborg's pea clam

Reptiles and Amphibians
central newt
Cope's gray treefrog
painted turtle
snapping turtle
spring peeper
western chorus frog

Birds
marsh wren
sora
swamp sparrow

Mammals
mink
muskrat

Deep Marsh and Lake Communities

Indicators: Floating-Leaved and Submerged Plants

Ecology

Deep marshes have standing water between six inches and three feet deep during the growing season. The plant species of the deep marsh community are those that emerge above the water, those that float, those attached to the bottom with floating leaves, those attached to the bottom that stay submerged, and submerged plankton. Water chemistry, water movement, depth zones, and substrate control the species composition.

Deep marshes contain some of our most recognizable aquatic plants, including several species of water lily, water shield, and floating-leaved pondweed. Deep marshes also provide important feeding areas for many waterfowl species and several well-known fish species, such as largemouth bass and northern pike.

Lakes are permanent bodies of water with little or no current. The plant species present in any lake depends on the depth, bottom materials, water chemistry, and water movement. The characteristic species section lists plants indicative of the hardness of the water. Shallow lakes are usually less than six or seven feet deep and contain floating plants, floating-leaved plants, and submerged plants. Emerging vegetation is usually sparse or nonexistent because the lake bottom has not been exposed to air, which prevents emergent species from becoming established. Plants that require a foothold on the lake bottom cannot grow below a certain light penetration zone. In deep lakes, these submerged plants are absent from portions of the lake. Deep lakes can also vary tremendously in species composition with different bottom material (gravel, boulders, sand, silt, and muck), water chemistry, water source (seeps, springs, rivers), and water basin (the Mississippi River basin, for example, has many species different from the Great Lakes basin).

Range

Dam and dike construction has increased the water acreage beyond what it was 150 years ago. Today, there are nearly 970,000 acres of inland lakes in Wisconsin. However, very few are in a

Rosette Lakes

In a few areas of northern Wisconsin where the sands are sterile and the glacial kettles formed numerous lakes, a special type of soft-water lake can develop. These lakes are very clear and have extremely poor nutrients and production of one-celled organisms. The fish are always stunted, and the large plants are sparse.

One group of plants, called rosette plants, has adapted to live in these stressful conditions. These rosette-shaped species spend most of their lives on the sandy bottom of these lakes. When lake levels drop and the plants are exposed, however, they can flower and produce seeds. Vilas County has excellent sterile rosette lakes for observation.

Deep-water marsh (photograph by Eric Epstein; used by permission)

wilderness setting and even those are subject to human impacts. Changes in chemistry, acidity, nutrients, pollution, shore development, and noise have left very few lakes untouched. Most of the remaining "wild" lakes are shallow, subject to winter kill, and have wetlands around them. There are very few undeveloped deep lakes. The rarest lakes are those that are undeveloped with complete fish communities, from top predators on down. Some of these can be found at Devil's Lake State Park, Lakewood Area, Muir Park, North Kettle Moraine, South Kettle Moraine, and Vilas County.

Activities

- Sketch a cross section of a lake. Take a boat or canoe into the water and observe the different plants at different depths as you proceed away from the shore. You will have to estimate the deepest parts beyond view or use a lake map. Note the most abundant plant species in each zone. Consider the reasons for the development of zones (Pratt 1995).
- Describe the effects purple loosestrife, an exotic plant, has on a shallow marsh. This plant can so overwhelm the native species that only purple loosestrife remains. Could there be a change in animal use as a result?
- While on the water in a boat or canoe, observe the floating-leaved plants. Collect a leaf and feel the slimy surface. Observe the animal life on a single leaf.

- Observe the surface dwellers in shallow water, especially on a sunny day when the full effects of their presence is revealed. View the shadows they cast on the lake or marsh bottom. Sketch the angle of the sun, the position of the insect on the surface, and the position of the shadow on the bottom.

Characteristic Species

Plants

VERY SOFT WATERS

brown-fruited rush
golden pert
least waterwort
pipewort
ribbon-leaved pondweed

SOFT WATERS

alternate-leaved water milfoil
fern pondweed
grass-leaved arrowhead
grass-leaved pondweed
large-leaved pondweed
lake quillwort
needle spike-rush
white stem pondweed

HARD WATERS

common pondweed
common waterweed
eel grass
northern water-nymph
slender pondweed
water stargrass
water beggar's-tick

VERY HARD WATERS

comb pondweed
coontail
Eurasian water milfoil
flatstem pondweed
Fries' pondweed
Illinois pondweed
Richardson's pondweed
white water crowfoot

Mosses

giant fountain moss
water moss

Butterflies and Moths

gyrating munroessa
obscure paraponyx
pondweed synclita

Beetles

false longhorn beetle
minute water scavenger beetle
minute whirligig
water lily beetle

Spiders

six spotted fishing spider

Aquatic Insects

alderfly
backswimmer
blue dasher
burrowing mayfly
common bluet
common whitetail
dobsonfly
eastern pondhawk
fishfly
giant water bug
green darner
large caddisfly
long-horned caddisfly
marsh fly
narrow-winged damselfly
northern caddisfly
phantom crane fly
pied skimmer
prince baskettail
ripple bug
small mayfly
spread-winged damselfly
stream mayfly
toe-biter
twelve-spotted skimmer
variable darner
water boatmen
water treader
widow dragonfly

Mollusks and Snails

Adam's pea clam
American ear snail
bell-mouthed ramshorn
blunt prairie physa
campeloma spire snail
deer toe
dusky lily-pad limpet
eastern floater
eastern ramshorn
fat mucket
fat pea clam
flat-ended spire snail
giant northern pea clam
globular pea clam
graceful fossaria
Great Lakes horn snail

great pond snail
grooved fingernail clam
keeled promenetus
lake fingernail clam
lake stagnicola
lilliput
long fingernail clam
modest gyraulus
mucket
ordinary spire snail
paper floater
perforated pea clam
pigtoe
pink heelsplitter
pocketbook
pond fingernail clam
quadrangular pea clam
rhomboid fingernail clam
ribbed valve snail
ridge-beak pea clam
rusty pea clam
shiny pea clam
short-ended pea clam
shouldered northern fossaria
showy pond snail
small spire snail
spike
strange floater
striated fingernail clam
striped stagnicola
swamp fingernail clam
three-keeled valve snail
tiny nautilus snail
triangular pea clam
two-ridged ramshorn
Walker's pea clam
white heelsplitter

Aquatic Invertebrates

jelly ball

Fish

banded killifish
big mouth buffalo
black bullhead
black crappie
blackside darter
blackstripe topminnow
bluegill
bluntnose minnow
brassy minnow
brook silverside
brook stickleback
brown bullhead
bullhead minnow
burbot
carp
chain pickerel
channel cat
chestnut lamprey
common shiner
creek chub
emerald shiner
fathead minnow
golden shiner
grass pickerel
green sunfish
Johnny darter
lake sturgeon
lake trout
largemouth bass
log perch
longnose gar
mimic shiner
muskellunge
northern pike
orange spotted sunfish
pumpkinseed
quillback
rock bass
silver lamprey
smallmouth bass
smallmouth buffalo
spottail shiner
spotted sucker
stonecat
tadpole madtom
trout perch
walleye
warmouth
white bass
white crappie
white sucker
yellow bass
yellow bullhead
yellow perch

Reptiles and Amphibians

central newt
mudpuppy
painted turtle
snapping turtle
spiny softshell
stinkpot

Birds

black tern
pied-billed grebe
ruddy duck

Mammals

beaver

Lake Superior and Lake Michigan Reefs and Benthos

Ecology

The reefs and bottoms of Lakes Superior and Michigan contain significant communities that contribute greatly to the total diversity of these giant lakes. Deep-water algae and bacteria that can survive the extremes of low light and cold temperatures are the base of the food chain. Additional food for the bottom dwellers, such as dead plants, invertebrates, and fish, fall out of the more productive zones above. Several fish species depend on these reefs as primary spawning areas.

Range

For all practical purposes these communities are out of reach of even the most dedicated observer. Information regarding these communities comes from researchers and specially equipped diving expeditions, and future observations will come from only a few sources. The rest of us must be content just knowing of them. The best developed of these reefs are off the Apostle Islands and Door County.

Characteristic Species

Aquatic Insects
- alderfly

Aquatic Invertebrates
- hydra
- tubifex worm

Fish
- bloater
- burbot
- chinook salmon
- coho salmon
- deep water sculpin
- lake chub
- lake trout
- nine-spined stickleback
- pink salmon
- rainbow smelt
- round whitefish
- sea lamprey
- siscowet
- slimy sculpin
- spoonhead sculpin

River Communities

Indicators: Flowing Water

Ecology

Wisconsin's large rivers have changed dramatically in the last 150 years as river habitat has been lost or degraded. Today, the mighty Mississippi River is not much more than a very long lake. Dams slow many rivers, which warms the water, eliminating habitat for trout and other fish. Mill ponds are common in many of Wisconsin's communities. River habitat changes when silt, including polluted sediments, is deposited above the dam. Fish are physically prevented from swimming upstream. Spawning beds and waterfalls are inundated.

With all these negative factors, it's amazing that we do have clean water and robust fish populations. There are miles of clear, fast trout waters in the state. Many rare dragonfly species, such as St. Croix snaketail and the pygmy snaketail, occur only in mid-sized northern rivers. Even small sections of some larger rivers are free flowing and relatively clean. Pollution abatement programs have significantly improved many rivers.

To understand the dynamics of a river community, a river system needs to be viewed from source to mouth. Species assemble into different natural communities in the headwaters and the middle reaches of a river. The most diverse communities develop in our largest rivers, such as the Mississippi, Chippewa, Wisconsin, St. Croix, and Black Rivers. The species composition also depends on whether the banks are forested or open to full sunlight and on how much human modification has taken place.

A forest headwater river (or stream) has most of its organic material supplied from outside the stream. Leaves falling into the stream provide food for several groups of animals. Some animals, such as stonefly larvae, crustaceans, or snails, shred the organic material. Others, such as some caddisfly and beetle larvae, graze on the bacteria and algae. Still others collect or gather the accumulated organic material on the bottom, including net-spinning caddisfly larvae and the larvae of a few fly species. Some predators, such as dragonfly larvae, feed on the shredders, grazers, and collectors.

When rivers approach 10 meters wide (about 30 feet) sunlight hits the water, which allows a complex community to develop. This community, called the periphyton, adheres to rocks, logs, plants, or other solid material. Species living in the periphyton usually have suction cups, excrete a glue-like substance, or develop a jelly-like slime. In the middle reaches of a river, the

Pygmy snaketail dragonfly (illustration by Jim McEvoy; courtesy of Wisconsin Department of Natural Resources)

Mussel Reproduction

Science continues to reveal amazing interactions between species and their environment. One such interaction is between freshwater mussels, large rivers, and their fish hosts. Freshwater mussels, which many people refer to as clams, can be found throughout the large rivers of the state. Today, however, the populations of many species have plummeted to the point of extinction. Several stresses, especially pollution and competition from exotic zebra mussels, have led to the decline, but unusual life cycles may also doom several species.

Female freshwater mussels expel millions of larvae into the waters of a large river. These larval animals attach themselves to the gills of a host fish or salamander, live on the gills of the fish for awhile, then drop off and sink to the bottom. If they land on the correct bottom material, which varies according to species, they will begin their slow growth to adulthood. Some species can live to the ripe old age of 100 years.

This elaborate relationship is further complicated because some mussel species can only attach to the gills of a single fish species. If the fish becomes rare, so too does the mussel. In the case of the ebony shell, which attaches only to the skipjack herring, the species is doomed in Wisconsin. The skipjack herring is a migratory fish that formerly came to Wisconsin every year in large numbers. With the completion of the Mississippi River lock and dam system, however, this species can no longer migrate into the state. Unfortunately, the ebony shell does not have its host any more, and only very old individuals remain.

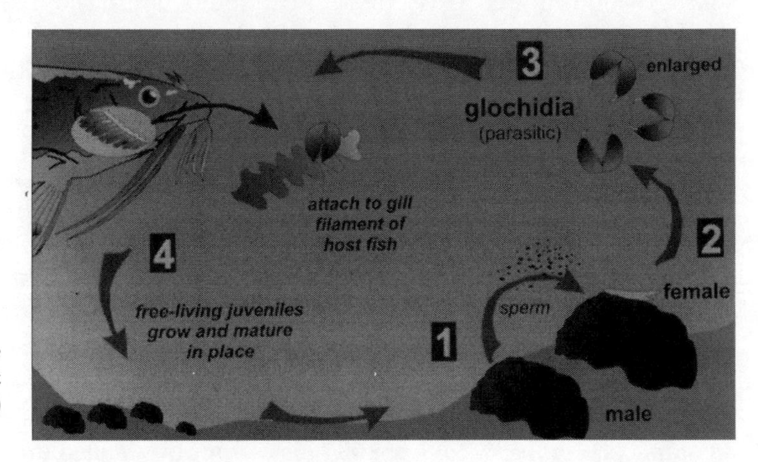

Freshwater mussel life cycle (courtesy of Wisconsin Department of Natural Resources)

community changes from one with many shredders to one with few shredders but more grazers and collectors and their predators. The main channels of large rivers have mostly collectors and predators because the rivers' depth and flow prevent the establishment of a periphyton community.

Range

River sections available for study are found at Avon Bottoms, Black River State Forest, Blue Hills, Brule River, Copper Falls State Park, Interstate State Park, Lake Pepin, Lakewood Area, Lower Wisconsin River Valley, Marathon County, Maribel

Caves, Moose Junction, Nelson Dewey State Park, Northern Bayfield County, North Kettle Moraine, South Kettle Moraine, Three Lakes, Upper Mukwonago River, White River Marsh, Wildcat Mountain State Park, and Wyalusing State Park.

Activities

- Map the features of a stream, such as riffles, meanders, and pools, and describe different uses by the insects (Allan 1995).
- Compare the species diversity in headwater, midstream, and large rivers. Consider why large rivers can have more diversity than a mid-sized stream (Allan 1995).
- Describe the effects floods would have on the features of a river. Would the pools be in the same place after a flood? Would they be as deep after the flood as before? What would the effects of a large log lodged in a pool have?
- Observe the territorial actions of dragonflies on the streamside vegetation. Visit a mid-sized stream in northern Wisconsin and observe the dragonflies and damselflies. Note their actions and aggressive nature toward competitors.

Characteristic Species

Submerged Plants

blunt-leaved pondweed
comb pondweed
common water starwort
common waterweed
eel grass
flatstem pondweed
red pondweed
ribbon-leaved pondweed
Richardson's pondweed
sessile-fruited arrowhead
white water crowfoot

Mosses

common apple moss
green hair moss
plume moss
rivulet cedar moss
water moss
wavy catharinea

Mushrooms

water club

Beetles

American bembid
bronze tiger beetle
crawling water beetle
fasciated diving beetle
marsh beetle
minute moss beetle
mud-loving ground beetle
riffle beetle
riverbank ground beetle
round sand beetle
spiny-legged rove beetle
water penny beetle

Insects

backswimmer
black fly
black-tipped darner
black-winged damselfly
blue dasher
blue mud dauber
broad-winged damselfly
burrowing mayfly
common bluet
common stonefly
common whirligig
crane fly
dobsonfly
dot-tailed whiteface
fawn darner

fishfly
giant stonefly
green stonefly
lake fly
large caddisfly
marsh fly
micro-caddisfly
net-spinning caddisfly
net-winged midge
northern caddisfly
northern mole cricket
phantom crane fly
pygmy mole cricket
ripple bug
rolled-wing stonefly
shore bug
small mayfly
small water strider
small winter stonefly
snail case caddisfly
snipe fly
soldier fly
stream cruiser
stream mayfly
ten-spotted dragonfly
two-lined stonefly
velvet water bug
water boatmen
water scorpion
winter stonefly
yellow-legged meadowhawk

Spiders

pirate wolf spider
ray spider
red fresh water mite

Aquatic Invertebrates

freshwater sponge
jelly ball
tubifex worm

Mollusks and Snails

Asiatic clam†
black sandshell
buckhorn
compeloma spire snail
cylindrical papershell
deer toe
eastern floater
eastern ramshorn
elk toe
fat mucket
fawn foot
flat-sided horn snail
fluted shell
fragile papershell
hickory nut
lilliput
mucket
Ohio River pigtoe
ordinary spire snail
paper floater
perforated pea clam
pigtoe
pimpleback
pink heelsplitter
pocketbook
purple wartyback
ridge-beak pea clam
river bank looping snail
river pea clam
rusty pea clam
small spire snail
spike
strange floater
sturdy river limpet
swamp fingernail clam
tadpole snail
threehorn
three-keeled valve snail
three ridge
tiny nautilus snail
tiny pea clam
white heelsplitter
zebra mussel†

Fish

American brook lamprey
banded darter
bigmouth buffalo
bigmouth shiner
blacknose dace
bluntnose minnow
bowfin
brook lamprey
brook silverside
brook trout
brown bullhead
brown trout
central stoneroller
channel catfish
chestnut lamprey
common shiner
creek chub
fathead minnow
freshwater drum
gizzard shad
golden redhorse
greater redhorse
highfin carpsucker
hornyhead chub
Johnny darter
largescale stoneroller
least darter
log perch
longnose dace
longnose gar
longnose sucker
mooneye
northern hog sucker
northern redbelly dace
pearl dace
quillback
rainbow darter
rainbow smelt
rainbow trout
river shiner
rosyface shiner
sand shiner
sauger

shorthead redhorse
shortnose gar
shovelnose sturgeon
silver lamprey
silver redhorse
slenderhead darter
smallmouth bass
smallmouth buffalo
southern redbelly dace
spotfin shiner
stonecat
tadpole madtom
western sand darter
white sucker

Reptiles and Amphibians
mudpuppy
snapping turtle
spiny softshell

Birds
belted kingfisher
spotted sandpiper
wood duck

Mammals
beaver
mink
muskrat

Springs

Ecology

Springs are active water sources for rivers remarkably different from headwater rivers. Nutrients available to animals are directly related to the chemistry of the rock, not the surrounding vegetation. The springs community usually has a few shredders, such as caddisfly larvae, but can have a distinctive algae and microbe community that leads to species specialized to graze on them. These groups, along with the predators that prey on them, form the springs community.

Range

Spring communities can be studied at Baraboo Hills, Black River State Forest, Cadiz Springs State Recreation Area, Devil's Lake State Park, Governor Dodge State Park, Lakewood Area, North Kettle Moraine, South Kettle Moraine, Vilas County, and Wildcat Mountain State Park.

Characteristic Species

Plants
- common waterweed
- watercress

Mosses and Liverworts
- giant fountain moss
- rivulet cedar moss
- thallose liverwort
- torn veil moss

Insects
- black fly
- common stonefly
- finger-net caddisfly
- green stonefly
- marsh beetle
- net-winged midge
- ripple bug
- rolled-winged stonefly
- skiff beetle
- snipe fly
- spring stonefly
- tube-making caddisfly
- water scavenger beetle
- water strider

Aquatic Invertebrates
- freshwater sponge
- green hydra
- scud
- water mite

Mollusks
- black sandshell
- brook snail
- ellipse
- fat mucket
- lilliput
- slippershell
- snuffbox
- threeridge
- water penny snail

Fish
- blacknose dace
- brook stickleback
- brook trout
- mottled sculpin
- pearl dace
- rainbow darter
- slimy sculpin

Small Communities

Several distinct natural communities form where special conditions limit the community size. These small communities occupy only a few acres on Wisconsin's landscape but harbor many species unique to these communities.

Different beach communities form on each of Wisconsin's three beach types: sandy beaches, cobble beaches, and beaches that form on flat bedrock where the land and water meet. Well-formed beach and lake dune communities occur along Lake Michigan and Lake Superior shores.

Other small communities are limited to places where rock is at the surface with little or no soil. Most of these rock communities are cliffs with different plants and animals living on different cliffs. The vertical rock and exposure to the sun determines species survival. Another rock community is the acid bedrock glade, which develops on ridge crests of igneous bedrock. These glades have patches of oaks, pines, aspen, and shrubs intermixed with prairie-like areas and extensive areas of lichen-encrusted bedrock exposures. An example can be seen at The Dalles of the St. Croix within Interstate State Park.

Underground communities or communities controlled by an underground feature are the most specialized of the small communities. The cave community is the most obvious, but underground air and water flow can influence species living in small areas, such as algific slopes.

A virtually unknown community lies deep in the ground water of east central Wisconsin. This underground community provides habitat for the Wisconsin well amphipod. No one knows how it lives or what it eats, but a few startled homeowners have seen this critter emerging from their taps. This animal is known only from Wisconsin.

Lake Dunes

Ecology

Dunes develop only near our largest lakes and are created by sands deposited by water and shaped by wind. They develop best where there are constant on-shore winds. Along the Great Lakes, the most magnificent dunes are in Michigan because of the prevailing westerly wind. Wisconsin's dunes along the shores of Lake Michigan are also very well developed, but those along Lake Superior are small.

Dune plant communities form when the wind-shaped material becomes stabilized with beach grass. Then succession gradually takes over, adding species such as wormwood, silverweed, and trailing juniper, which offers enough protection for tree saplings to grow. On many occasions, breaks in succession are caused by storms, which expose bare sand that can easily move across the land by wind until it is again stabilized.

The dunes have a distinct plant community that is similar to plant communities on the Atlantic coast dunes. The upper Great Lakes region has several dune specialists, such as dune thistle and dune goldenrod.

Range

Dune areas can be found at Apostle Islands, Chiwaukee Prairie, Kohler–Andrae State Park, Northern Bayfield County, Point Beach, Ridges Sanctuary, Whitefish Dunes State Park, and Wisconsin Point.

Activities

- Visit a stabilized dune and walk across it from the open beach to the forest. Describe the differences in vegetation and the process of dune succession (Watts 1975).
- Do the same activity, but this time record the fungi component. Make several visits and record the fungi found at different locations along the dune. Does dune fungi succession mimic the plant succession?
- View the open sandy areas of a dune. Describe why they are open (Watts 1975).

Clustered Broomrape

An interesting plant parasite called the clustered broomrape lives on Wisconsin's open dunes. It grows among the roots of the beach wormwood plant, a member of the sage family that is scattered throughout the sparse dune vegetation. The clustered broomrape takes its nourishment from the roots of the beach wormwood and does not need sunlight. Occasionally, a bundle of yellowish-orange broomrape flower stalks emerge from the bare sand. After a brief appearance, the flowers wither, and all that remains is a blackened, dried stalk.

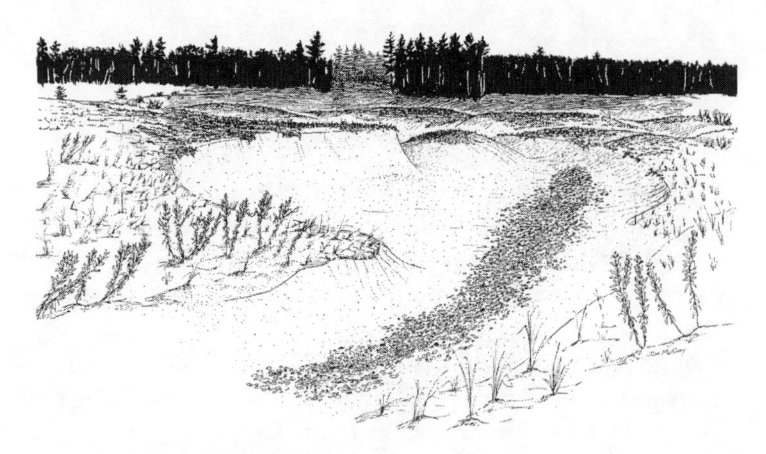

Lake dune (illustration by Jim McEvoy, courtesy of Wisconsin Department of Natural Resources)

- Visit a swale where water stands most of the year. Record the species found and the water depth. Return annually for several years and record the same information. Continue through one cycle of high and low Lake Michigan water levels and then compare the differences.

Characteristic Species

Plants

beach grass
beach wormwood
Canada bluegrass
Canada wild rye
common evening-primrose
common milkweed
silverweed
starry false Solomon's seal
trailing juniper

Lichens

British soldiers

Mushrooms

sandy laccaria
white matsutake

Butterflies and Moths

dimorphic gray
iris borer
stormy arches

Beetles

dingy ground beetle
noble tiger beetle

Spiders

goldenrod spider

Birds

chipping sparrow

Lake Beaches

Ecology

On sand and cobble beaches, plant growth is limited to the upper beach, where plants are safe from normal wave action. The vegetation is in a constant state of flux because of blowing sand, lake levels, storms, and disturbances caused by people. Cobble beaches have little area for the roots of plants to reach soil. Not surprisingly, cobble beaches have few plants, mostly smartweeds. Sandy beaches are the most common and recognizable of our beaches and commonly have sea rocket as the predominate plant.

Most beach plants are annuals and limited to a specific type of beach. One beach community develops on bedrock, where plants grow in the faults and cracks. If the water level rises and falls, the beach is a very tenuous place for plants to live.

The piping plover is one of the few Wisconsin animal species that has an affinity for beach environments. Its habitat is limited to sandy and gravelly beaches, and the loss of nearly all of our beaches to other uses has endangered its existence. Only one or two pairs remain on remote beaches in the Apostle Islands National Lakeshore.

Range

Natural beach communities can be found at the Apostle Islands, Kohler–Andrae State Park, Newport State Park, Northern Bayfield County, Point Beach, Ridges Sanctuary, Whitefish Dunes State Park, and Wisconsin Point.

Dune Tiger Beetle

With the loss of the piping plover from most of Wisconsin, the largest predator on the beach is now an insect. Dune tiger beetles are the undisputed predatory kings of the beach. They have excellent vision and are very fast runners and fliers. These dune tiger beetles will tackle just about anything they can consume. They are most active on bright sunny days. At night and during cloudy, rainy periods they hole up in their burrows. These structures sometimes extend for a foot under the surface and are often mistaken for wormholes.

Mating dune tiger beetles (photograph by Kathy Kirk; used by permission)

Activities

- Map the zones of vegetation on the beach at different times of the year. Describe why most of the beach plants are annuals (Watts 1975).
- Find a secluded part of the beach and watch for tiger beetles. These active creatures' habits can be observed with binoculars. Observe their reaction to other tiger beetles when they approach.
- Visit a beach after a major storm, especially one with strong on-shore winds. Windrows of vegetation and animal carcasses will line the beach. Notice the actions of numerous species scavenging the flotsam.
- Visit an interior beach. These features are found around lakes that fluctuate in depth over a short period of time. Record the changes in vegetation when the water level drops.

Characteristic Species

Plants

beach pea
Canada wild rye
cocklebur
common water-horehound
sandbar willow
sea rocket
silverweed
wild mint

Lichens

brown-fruited cup cladonia

Butterflies and Moths

artichoke plume moth
bent-lined carpet
pearly eye

Beetles

cloudy flower beetle
dainty tiger beetle
dead fish beetle

Spiders

goldenrod spider

Birds

herring gull
purple martin
spotted sandpiper

Cliffs

Indicators: Vertical Rocks, Dry or Wet

Ecology

Cliff communities can develop only where bedrock is at the surface. Limestone, sandstone, and granite rocks all have different species growing on them. Exposed cliffs have virtually no soil, no ground water, and abnormal temperatures because of the sun heating the rocks. Shaded cliffs are almost always moist and are sometimes permanently wet. They still have little soil, and direct sunlight hardly ever reaches the plants. The shaded cliffs of the driftless area have a high incidence of endemic and specialized species, such as birds-eye primrose, muskroot, and cliff cudweed.

Range

In Wisconsin, cliff areas are limited. The Niagara escarpment in eastern Wisconsin, the western uplands, the bedrock areas of the Baraboo and Blue Hills, the Gogebic and Penokee ranges, outcrops in Florence, Marinette, Oconto, Menominee, and Marathon Counties, the Apostle Islands, and scattered small areas elsewhere in the state are places to look for cliffs.

Cliffs are hard to study without ropes and ladders, but they are exciting to study because each is different. Areas to visit with cliff communities are the Apostle Islands, Baraboo Hills, Black River State Forest, Copper Falls State Park, Devil's Lake State Park, Governor Dodge State Park, Interstate State Park, Lower Wisconsin River Valley, Marathon County, Maribel Caves, Moose Junction, Whitefish Dunes State Park, Wildcat Mountain State Park, and Wyalusing State Park.

Activities

- Observe a cliff and focus very closely on the lichen community. Describe the aspects of primary succession, such as organic matter developing on inorganic rock and an increase in species diversity, including animals that eat lichens.
- Visit cliffs in the west or north, especially those facing north and remaining wet throughout the summer. Observe the plants growing on those cliffs. Many species at the limit of their range inhabit these cliffs.
- Diagram a small-scale cliff system. Choose a dry or wet cliff and draw the parts of the ecosystem that are necessary to maintain the communities. Include animals that may use the site for foraging or nesting (Pratt 1995).

Characteristic Species

EXPOSED CLIFF

Plants

American pennyroyal
columbine
harebell
marginal wood-fern
rock-cap fern
rock spikemoss
sand cress
sleepy catchfly
smooth cliff-brake

Mosses and Lichens

blistered rock tripe
boulder lichen
fleecy rock tripe
fringe leaf moss
reindeer lichen
smooth rock tripe
torn veil moss
wiry fern moss

Butterflies and Moths

black and yellow lichen moth
lead-colored lichen moth
Olympian marble
painted lichen moth
red twin-spot
sharp-angled carpet

Spiders

barn spider
furrow spider
zebra spider

Birds

bank swallow
cliff swallow
eastern phoebe
northern rough-winged swallow

SHADED ROCK CLIFFS

Plants

bulblet fragile fern
bush honeysuckle
columbine
creeping fragile fern
lion's foot
orange jewelweed
rock-cap fern
rock cress
slender knotweed
tall forked-chickweed
wild sarsaparilla

Mosses and Lichens

boulder lichen
common apple moss
common beard moss
common tree apron moss
feathered neckera
fringe leaf moss
homalia moss
knight's plume
leafy liverwort
mountain fern moss
pear-shaped thread moss
plume moss
smooth rock tripe
thallose liverwort
torn veil moss
wiry fern moss
yellow wall lichen

Butterflies and Moths

American bird-wing
bent-lined carpet
black-dotted lithacodia
painted lichen moth
pink-shaded fern moth
silver-spotted fern moth

Beetles

minute marsh-loving beetle

Insects

columbine leaf miner

Spiders

barn spider
cave orb weaver
long-bodied cellar spider
ray spider
short-bodied cellar spider

Landsnails

humble pupa snail
moss pupa snail

Birds

barn swallow
chimney swift
cliff swallow
eastern phoebe
northern rough-winged swallow

Mammals

Keen's myotis
little brown bat

Caves

Ecology

The fauna of a cave depends almost entirely on the presence of water and bats (many species use bat guano for their nourishment). Cave fauna can be divided into four categories. Troglobites are small invertebrates without common names that are entirely restricted to caves. The upper Midwest has sixteen species of troglobites. Wisconsin's caves harbor some of these species, but not all. Troglophiles are species, such as luminous moss and cave spiders, that are mostly restricted to caves. Trogloxenes include species such as bats that live part of their lives in caves but can survive outside. The fourth category, accidentals, are species that wander into caves.

Algific talus slopes are communities of unusual boreal and glacial relict species. They form on steep exposed dolomite and talus slopes only in the driftless area. A handful of dedicated scientists have recently searched wet dripping cliffs and talus slopes in eastern and southwestern Wisconsin and documented several very rare snail species. Some of these snails are almost microscopic, while others reach one-half inch in size. The persistence of these animals is amazing. Many of these populations are found on just several square meters of land, and several of these species were thought to have become extinct soon after the glaciers receded from the state. The discovery of several species of this relict fauna and their ability to survive for thousands of years in a restricted space should give us pause when considering land management activities.

Range

Most of Wisconsin's caves are found in limestone areas in counties bordering the lower Wisconsin River and in Pierce and Pepin Counties. The larger caves became commercialized, such as Crystal Cave near Spring Valley and Eagle Cave near Blue River. There are also many smaller caves, including some that haven't been explored, and there may be caves that haven't yet been discovered.

Algific talus slope (illustration by Jim McEvoy; courtesy of Wisconsin Department of Natural Resources)

Characteristic Species

Mosses
- luminous moss

Insects
- house mosquito

Spiders
- cave orb weaver
- long-bodied cellar spider
- short-bodied cellar spider

Birds
- chimney swift
- eastern phoebe
- northern rough-winged swallow

Mammals
- big brown bat
- eastern pipistrelle
- Keen's myotis
- little brown bat

Old Fields and Agricultural and Urban Areas

THE EASIEST LOCATION for naturalists to study nature is near the place they live. In today's landscape, the natural communities near habitation have been almost completely transformed from the landscape the settlers found.

Former agricultural areas usually have smooth brome, Kentucky bluegrass, old-field goldenrod, and young trees. Productive agricultural areas have fields with corn, soybeans, small grains, and hay or alfalfa. Agricultural habitats include pasture, fencerows with brush and trees, sometimes ditches with grasses and shrubs, woodlots, and occasionally streams, lakes, or wetlands. The woodlots are usually small (less than 80 acres) and isolated. Urban habitats include buildings, tree-lined streets, yards, many exotic plants, and closely cropped play areas. Most urban parks have some native vegetation, but they again are very isolated.

Though not our native communities in the historic sense, these areas are the start of new ecosystems dominated by introduced European and Asian species. Some native species persist under heavy disturbance because life itself is tenacious. With so much of the old gone forever, new equilibriums are evolving, along with new interactions among species. Some habitat generalists that lost their original habitat have been able to adapt to these new situations. These new communities have become surrogate habitat for some species, such as grassland birds. These species once lived on prairies; now they survive in fallow fields filled with European and Asian grasses.

Most of the species found in the heavily developed areas are innocuous. However, there are some species that are very serious competitors or predators on Wisconsin's native biota. (As previously, in the characteristic species list, a dagger [†] marks those invasive exotics that can significantly alter natural ecosystems.) Nonnative species compete for resources better than native species, and nonnative predators, for which native species have no defense, can severely reduce the numbers of native species. Landowners should learn these invasive species and remove or control them before they completely dominate the landscape.

Many communities and species listed in this book face an uncertain and tenuous future. The species listed in this section, however, will be with us for a long time. They are easy to find and study in the agricultural and urban areas of Wisconsin.

Old fields and agricultural and urban land (courtesy of Wisconsin Department of Natural Resources)

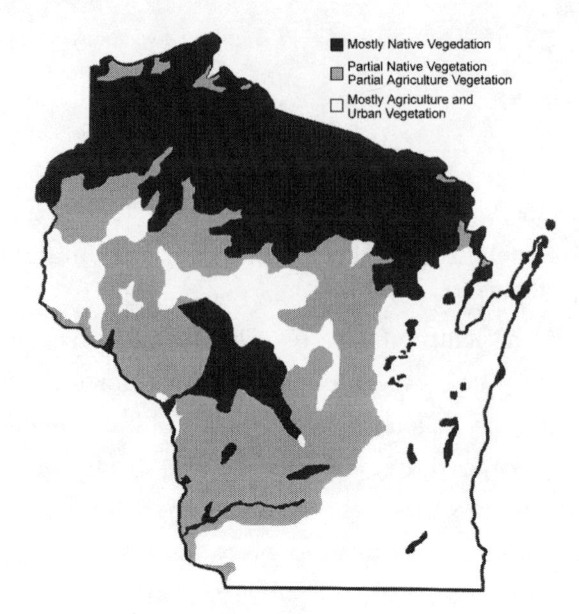

Old Fields, Agricultural and Urban Areas

Range

Agribusinesses, abandoned agricultural land, transportation corridors, and urban areas make up nearly two-thirds of Wisconsin's landscape. While some of this acreage may succeed into native communities, most will be altered for centuries, if not forever. Nearly all of this landscape's natural communities are dominated or directly influenced by Eurasian plants and animals. Only those native species that can cope with persistent disturbance or those that are very adaptable have thrived.

Activities

- Draw a map of the primary natural features of your property. Whether you live in a city apartment, suburban lot, rural area, farm, or recreational land, a map of the primary features of your property can help you make decisions about your land.
- List the species occurring on your land. It's amazing how little we know about the place we live. Most people know if they have white-tailed deer, rabbits, squirrels, grouse, turkeys, or a few selected birds, but very few know what plants they have on their land other than trees, bluegrass, corn, oats, soybeans, and alfalfa.
- Conduct a yearlong recording of phenology on your property. Phenology is the study of when things happen. Make

weekly excursions onto your property and record what you see and hear. Items may include the first flower of spring, the day the robins arrived, the first apple flower, peak cicada singing, the last fall migrant, and the last chipmunk observation. The list is never-ending, and the practice of regular visits can make the natural world a part of your life (Benyus 1989).

Characteristic Species

Plants

alfalfa
alsike clover
apple
barnyard grass
bird's-foot trefoil
black bindweed
black locust†
black medick
black nightshade
bladder campion
blue vervain
Canada thistle†
carpet weed
catnip
clammy weed
cocklebur
common blackberry
common buckthorn†
common burdock†
common buttercup
common evening-primrose
common goldenrod
common St. John's wort
common tansy
common teasel†
cow parsnip
crown vetch†
curly dock
cypress spurge†
daisy fleabane
dame's rocket†
downy brome
English plantain
field horsetail
foxtail
garlic mustard†
giant ragweed†
glossy buckthorn†
goat's beard†
gray dogwood
heal-all
herb-sophia
hoary alyssum
horseweed
Japanese barberry†
late goldenrod
leafy spurge†
Missouri gooseberry
mourning glory
multiflora rose†
nodding thistle†
Norway maple†
orange hawkweed†
orchard grass
ox eye daisy
prickly ash
prickly wild gooseberry
purple loosestrife†
red cedar
red clover
red raspberry
redtop
Russian thistle
sandbur
Siberian elm†
spotted knapweed†
staghorn sumac
summer grape
tower mustard
trembling aspen
turk's cap lily
velvetleaf
white campion
white mulberry
wild asparagus
wild parsnip†
wild plum
yarrow

Lichens

blue-isidate lichen
candlestick lichen
cobra-head lichen
dotted lichen
forked lichen
furrowed-shield lichen
lateral-fruited lichen
little-dog lichen
match stick lichen
narrow crown lichen
orange-dot lichen
orange tree-trunk lichen
oyster shell lichen
rosette lichen
scalloped lichen
slender lichen
wavy-edge lichen
white cup lichen
yellow eyes
yellow-wax lichen

Mushrooms

bladder cup
bristly parchment
destroying angel

fairy ring
fried-chicken mushroom
fringed tubaria
gem-studded puffball
giant puffball
granular puffball
ground pholiota
horse mushroom
lawn mower's mushroom
orange peel
parrot mushroom
shaggy mane
shaggy parasol
spring agaricus
tumbling puffball

Butterflies and Moths

American copper
artichoke plume moth
barberry geometer
catalpa sphinx
checkered skipper
common sooty wing
dark marathyssa
dark-sided cutworm
dun skipper
eastern tent caterpillar
European skipper
faint spotted angle
Ferguson's scallop shell
Garman's quaker
goldenrod gall moth
grape leaf folder
grape skeletonizer
joyful holomelina
little virgin moth
locust underwing
lost sallow
Milbert's tortoise shell
one-lined zale
orange wing
painted beauty
pink-streak
Scirpus wainscot
showy emerald
somber carpet
tawny holomelina
three-lined balsa
verbena moth
vetch looper
yellow bear moth

Beetles

bluegrass billbug
cabbage curculio
clover stem borer
Colorado potato beetle†
community wireworm
dusky sap beetle
garden carrion beetle
goldenrod beetle
goldenrod soldier beetle
green dock beetle
hairy soldier beetle
lesser dung beetle
locust borer
margined blister beetle
pale firefly
Pennsylvania ground beetle
picnic beetle
red milkweed beetle
seed corn beetle
spotted blister beetle
spotted cucumber beetle
spotted June beetle
sumac flea beetle
tobacco wireworm
yellow-scaled scarab

Insects

bald-faced hornet
buffalo treehopper
clear-winged grasshopper
corn planthopper
dogday cicada
fork-tailed bush katydid
four-lined plant bug
garden fleahopper
grape leaf folder
grape leaf skeletonizer
green stink bug
insidious flower bug
little black ant
little carpenter bee
little pasture grasshopper
long-necked seed bug
Lugen's stink bug
mason bee
meadow planthopper
meadow spittlebug
oblong-winged katydid
painted leafhopper
pasture grasshopper
pavement ant
pharaoh ant
pigeon horntail
potato leafhopper
red-legged locust
squash bug
stable fly
thief ant
thread-legged bug
three-ridge planthopper

Spiders

golden lynx spider
grass spider
nursery web spider
platform spider
six-spotted orb weaver
thin-legged wolf spider

Birds

American kestrel
bobolink (north)
brown thrasher
cardinal
chimney swift
common nighthawk
eastern kingbird
eastern meadowlark
eastern screech owl
eastern towhee

field sparrow
gray catbird
gray partridge
horned lark
house finch
house wren
indigo bunting
mourning dove
red-tailed hawk
rock dove†
savannah sparrow
song sparrow
vesper sparrow
western meadowlark

Mammals

giant mole shrew
striped ground squirrel
woodchuck

Worm-eating warbler (Endangered Resources file photograph; courtesy of Wisconsin Department of Natural Resources)

Frog Lake pines, an example of a dry pine forest community (photograph by Thomas A. Meyer; used by permission)

Lungwort lichen (photograph by Thomas A. Meyer; used by permission)

Bogus swamp (photograph by Thomas A. Meyer; used by permission)

Columbia silk moth larva (photograph by Les Ferge; used by permission)

Columbia silk moth (photograph by Les Ferge; used by permission)

Prairie (photograph by Thomas A. Meyer; used by permission)

Blue racer (Endangered Resources file photograph; courtesy of Wisconsin Department of Natural Resources)

Box turtle and cactus (photograph by Thomas A. Meyer; used by permission)

Brule River alder thicket (photograph by Eric Epstein; used by permission)

Argiope spider (courtesy of Wisconsin Department of Natural Resources)

Lulu Lake savanna (photograph by Thomes A. Meyer; used by permission)

Jewelwing damselfly (photograph by Thomas A. Meyer; used by permission)

Roch-a-Cri mound, an example of an open cliff community (photograph by Thomas A. Meyer; used by permission)

Bedrock beach (Endangered Resources file photograph; courtesy of Wisconsin Department of Natural Resources)

Butterwort (Endangered Resources file photograph; courtesy of Wisconsin Department of Natural Resources)

Part 2

Sites

Western Wisconsin

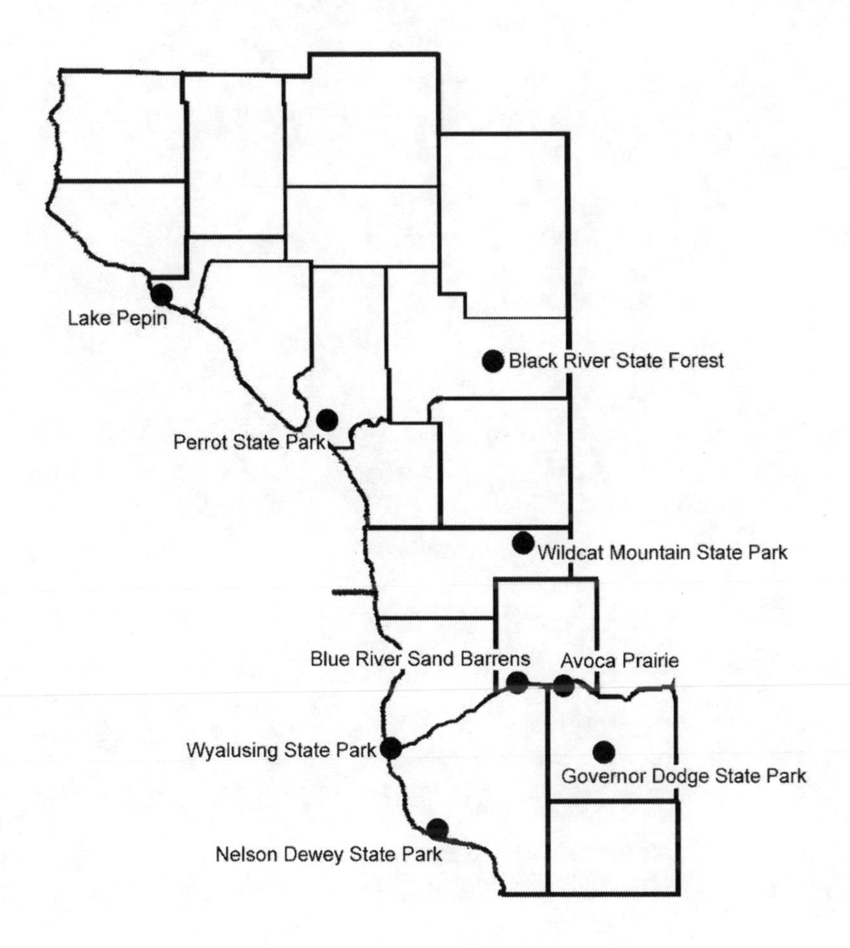

Avoca Prairie

Communities: Dry-Mesic Prairie, Wet-Mesic Prairie, Wet Prairie, Shrub-Carr, Bur Oak Opening

Avoca Prairie is about 1½ miles east of Avoca in Iowa County. The entire prairie is state owned and part of the Lower Wisconsin Riverway, a large state-owned project running from Sauk City to Prairie du Chein. To reach the prairie from Avoca, take State Highway 133 east to Hay Lane Road and follow this road to the parking areas. The first parking lot is where the road descends the hill and is most often used in early spring because of high water in Marsh Creek. If water levels allow fording of the creek, you can drive through the shallow creek over crushed rock to another parking area at the edge of the prairie one-third mile beyond Marsh Creek.

For those seeking the rare, Avoca Prairie may not be the place to visit; the specialties list is relatively small for this prairie. However, with 900 acres of prairie interspersed in a matrix of floodplain savanna, Avoca Prairie is the largest contiguous native prairie east of the Mississippi River. A truly realistic sense of a natural prairie is possible here because this is the only known eastern prairie with vistas void of buildings. The floodplain woods and the prairies appear as they could have hundreds of years ago.

The site is not pristine, though. There have been several attempts to tame the prairie, including limited cattle grazing and, more recently, mowing marsh hay. Mowing, however, did not harm the prairie because it did not occur every year, and the timing of mowing changed with the wetness or dryness of the seasons. Mowing can mimic the effects of fire in controlling shrub growth. Here, the irregular mowing prevented conversion of prairie to shrub-carr or forest. Since fire was reintroduced, shrub expansion is now held in check through burning and selective cutting, although there is a significant shrub component, especially along waterways.

Avoca Prairie developed on an old stream terrace. The sand ridges were originally sandbars in the old streambed, now dry, and the swales, which are low wet spots, were areas of water movement. Sand ridges of dry-mesic prairie occur on high ground next to swales of sedge meadow and wet-mesic prairie. There are a few scattered oaks on the prairie, which give it a definite savanna aspect. The swales and ridges vary in elevation only about four feet. Species composition changes every few feet. The higher areas have big bluestem and white wild indigo as dominates, while the low spots have river sedge and cordgrass. Near the center and along the western edge of the prairie, you will see floodplain savanna. A partially open to nearly closed canopy of swamp white oak dominates these areas. There is virtually no shrub layer, and the characteristic groundlayer species are Virginia wild rye, woodland brome, prairie wedgegrass, cardinal flower, sensitive fern, hairy heart-leaved aster, side-flowering aster, purple joe-pye weed, turtlehead, broad-leaved goldenrod, and wild golden glow.

Butterflies, moths, and other insects use the prairie and savanna vegetation extensively. Because the prairie is so large, it is a natural magnet for migrating and nomadic insects and birds. It provides habitat for a few rare species, such as the Blanding's turtle, which you can recognize by its long neck with yellow underneath and its distinctive shell with yellow flecks on a bluish background. Listed as a threatened species in Wisconsin, this turtle is found in deep marshes and moist prairie areas and is scattered throughout the state in low numbers.

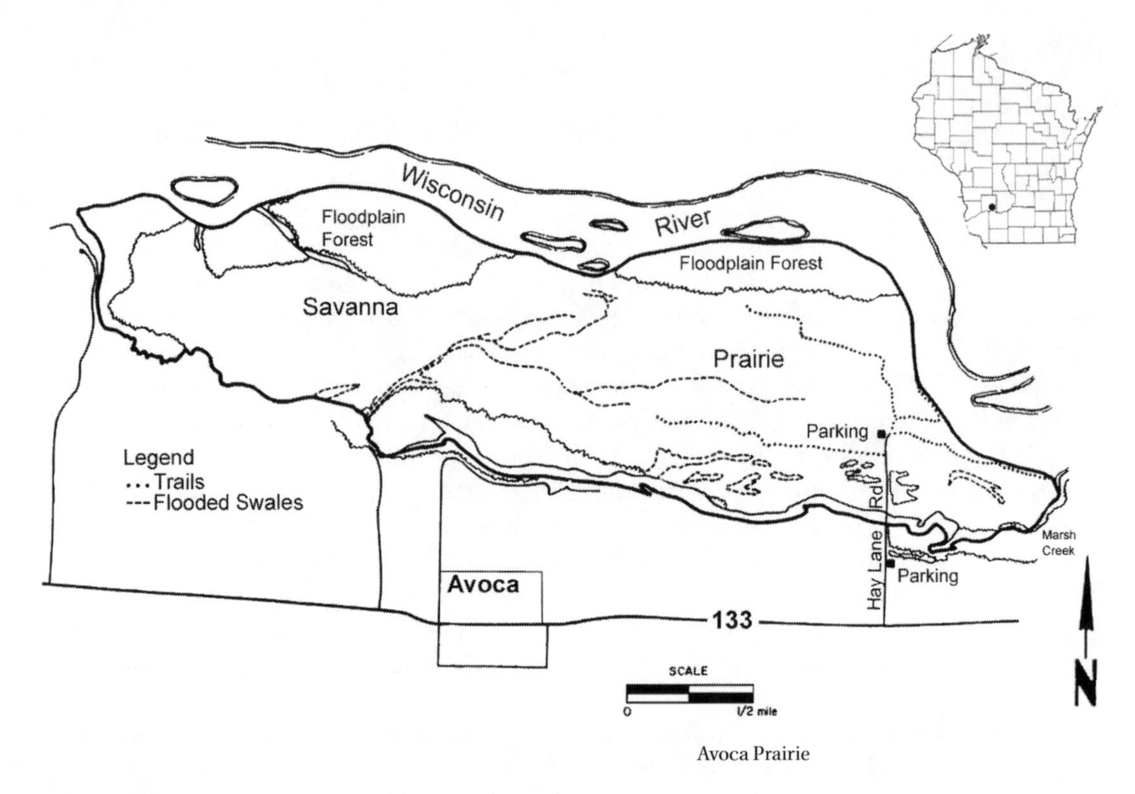

Avoca Prairie

Specialties

Plants

prairie Indian plantain
tall nut-rush
woodland boneset

Animals

baptisia dusky wing (butterfly)
Blanding's turtle
dion skipper (butterfly)
gorgone checkerspot (butterfly)
short-eared owl
six-lined racerunner (lizard)

Uncommon Species

Plants

arrow-leaved violet
bulbous cress
cardinal flower
cup plant
cursed buttercup
downy gentian
grass pink
green dragon
groundnut
hairy heart-leaved aster
indigo bush
low nut-rush
nodding ladies' tresses
nodding wild onion
prairie blazing star
rattlesnake master
Seneca snakeroot
slender beard-tongue
sweet black-eyed susan
sweet Indian plantain
tall green milkweed

Insects

arcigera flower moth
baptisia borer moth
bride underwing
Culver's root borer moth
darling underwing
lined ruby tiger moth
loosestrife borer moth
once married underwing
Oregon cycnia
ottoe skipper
red groundling
sensitive fern borer moth
sneezeweed borer moth
turtlehead borer moth
wingless prairie grasshopper

Reptiles and Amphibians

false map turtle
Ouachita map turtle
smooth softshell

Birds (Summer)
- field sparrow
- red-shouldered hawk
- Virginia rail

Birds (Migrants)
- LeConte's sparrow
- Nelson's sharp-tailed sparrow
- water pipit

Mammals
- badger
- prairie deer mouse
- prairie vole

Black River State Forest

Communities: Dry Pine Forest, White Pine–Hardwood Forest, Forested Swamps, Shrub-Carr, Southern Sedge Meadow, Oak Barrens, Cliffs, River

Black River State Forest covers 65,000 acres in two units. It is owned by the state and managed as a multiple-use forest, combining recreational needs and timber production. An excellent place to begin exploring is the smaller western unit south of Black River Falls. One mile south of town along US Highway 12 is Castle Mound, a butte that was previously an island in glacial Lake Wisconsin. Several buttes and mesas can be seen in the vicinity. A butte is pointed at the top, and a mesa is flat-topped. In the winter you can more easily detect the ancient shoreline of the glacial lake because the fallen leaves allow you to view the water-sculptured sides of the bluffs. Castle Mound does not have any rare plants, but it does have several northern species at the southern edge of their range. One mile farther south is the scenic Perry Creek gorge. You can reach the gorge by driving west from State Highway 27 on Perry Creek Road for 1¼ miles. If you visit in May, keep an eye on the ground for the sand violet in and around the gorge area.

Another excellent site is the area of stunted jack pine barrens near Wazee Lake Park. To observe these barrens, continue east from Castle Mound on W. Castle Mound Road for four miles, then north on Brockway Road to the barrens, where you will see the most diverse plant and animal species composition of any barrens in the state.

A different way to explore Black River State Forest is to canoe down the East Fork of the Black River. Before embarking on such a journey, be sure to consult one of the numerous canoe guides available at your local bookstore or library. The East Fork Black River is an easy stream to canoe, offering a solitude that is unusual this far south in the state. As you canoe down the river, you will see quality pine and hardwood forests along the shoreline.

If you drive the many roads south of the river and north of State Highway 54, you will go through an extensive jack pine forest. Most of this area is used for pulpwood production. The trees are harvested by cutting blocks of forest so that different places are in different stages of

Glacial Lake Wisconsin butte (Endangered Resource file photograph; courtesy of Wisconsin Department of Natural Resources)

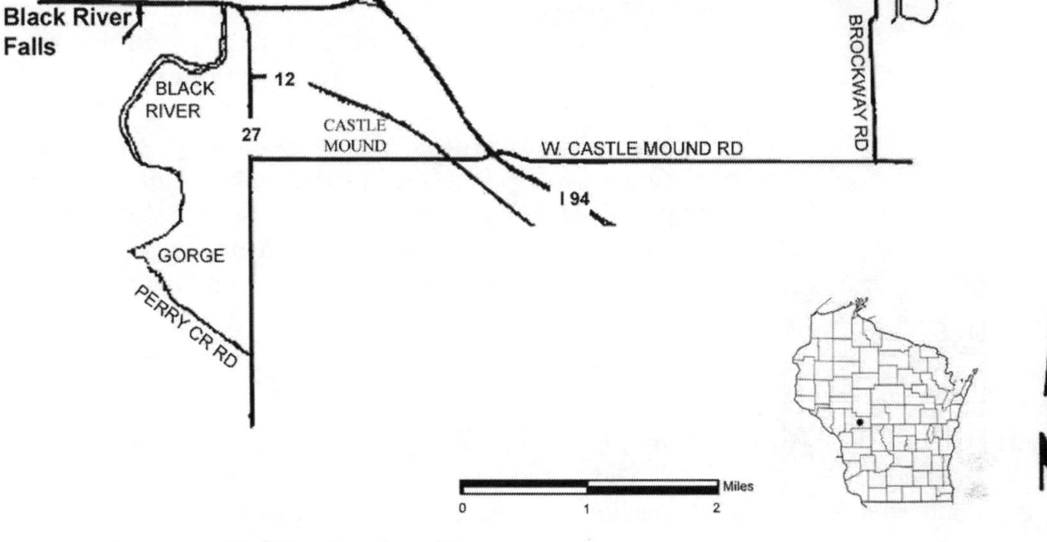

Black River State Forest–West

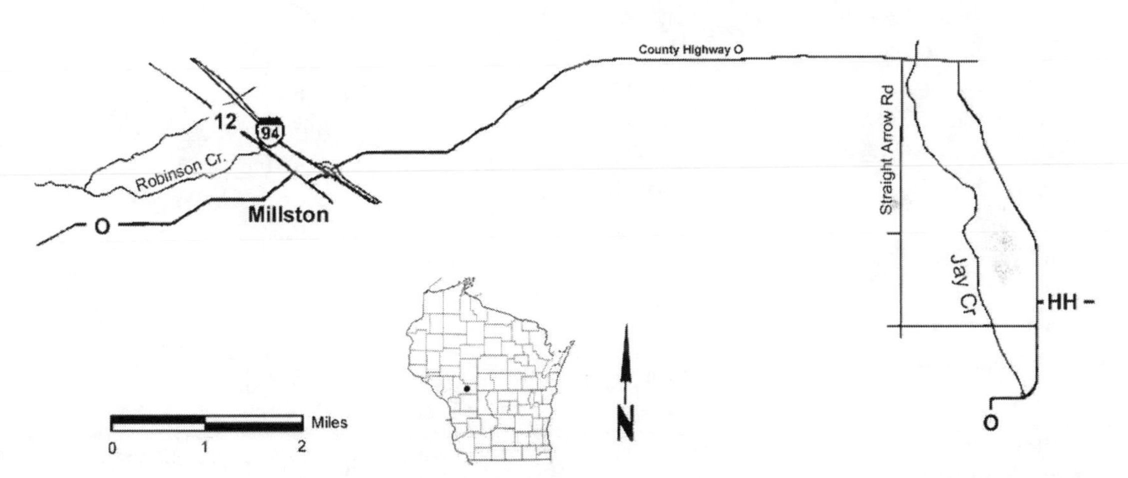

Black River State Forest–East

growth throughout the forest, giving the area a patchwork appearance. With many acres of small, crowded jack pines, not unlike central Michigan, the forest is a magnet for the wayward jack-pine-loving Kirtland's warbler. This federally endangered species made appearances several years in a row in the late 1970s and may do so again in the future.

To explore different natural communities, drive south on I-94 to Millston. Upon reaching the town, you can either go west on County Highway O less than a mile to an excellent area of pines along Robinson Creek or go east on County Highway O toward the Necedah area. Be sure to stop at the state-owned Jay Creek State Natural Area just west of the County O and HH intersection. Jay Creek offers the finest example of a flatwoods in Wisconsin. White pine is the most common tree species here, along with a few red pines and red maples. This natural community forms on wet, sandy soils with small amounts of peat. The groundlayer has abundant fern growth, with dense stands of ostrich fern and Massachusetts fern, a species that normally grows along the eastern seaboard of the United States. The Jay Creek site is the best area in western Wisconsin to observe Massachusetts fern and long sedge. The site is easily identified by the yellow State Natural Area signs.

Specialties

Plants

- clustered sedge
- cross milkwort
- Farwell's water milfoil
- greenish-white sedge
- Hooker's orchid
- long sedge
- marsh willow herb
- Massachusetts fern
- sand violet
- screwstem
- silky willow
- straw sedge
- twin-stemmed bladderwort

Animals

- buck moth
- clamp-tailed emerald (dragonfly)
- dusted skipper (butterfly)
- eastern massasauga (snake)
- five-lined skink (lizard)
- golden-winged warbler
- Henslow's sparrow
- jutta arctic (butterfly)
- Karner blue butterfly
- LeConte's sparrow
- mottled dusky wing (butterfly)
- persius dusky wing (butterfly)
- phlox flower moth
- sharp-tailed grouse
- warpaint emerald (dragonfly)
- western ribbon snake

Other Features

- buttes
- garnet
- mesas
- slate
- wavelite

Uncommon Species

Plants

- clammy hedge-hyssop
- clustered poppy mallow
- dune three-awn
- great St. John's wort
- meadow beauty
- netted nut-rush
- oval milkweed
- purple milkwort
- purple spring cress
- purple fringed orchid
- ragged fringed orchid
- sand milkweed
- short green milkweed
- water pennywort
- yellow false foxglove

Insects

- blue-legged grasshopper
- mottled sand grasshopper
- two-spotted skipper

Reptiles and Amphibians

- Blanding's turtle
- wood turtle

Birds (Summer)

- Kirtland's warbler (very rare)
- LeConte's sparrow
- northern goshawk
- pine warbler
- raven
- sharp-tailed grouse

Mammals

- bog lemming
- otter
- pygmy shrew

Blue River Sand Barrens

Communities: Oak Barrens, Sand Barrens

The sand barrens and dunes of this small area formed after European settlers pastured their cattle on the sand prairie, which so disturbed the vegetation that the sand was free to move at the will of the wind. It piled up in dunes, with small pebbles collecting at the bottom to form a miniature desert-pavement landscape. After the cows were removed, the dunes stabilized as pioneering desert-like plants grew on them, sending down soil-holding roots. Portions of the site had less disturbance, and an oak barrens community developed on the dunes.

The flora of the sand barrens is specifically adapted to harsh conditions, able to thrive with moderate disturbance and to tolerate surface soil temperatures approaching 190 degrees.

The fame flower, which looks very much like a miniature aloe plant, is typical of a desert plant. These small succulents grow throughout the summer, then bloom in July and August. The single flower stalk rises from the rosette of fat leaves. Around 4:00 to 5:00 P.M. on the hottest afternoons, the flowers open. They remain open into the evening but wither and fall by morning.

Because of their proximity to the Wisconsin River, the Blue River Barrens are excellent habitat for several species of turtles that use the area as a nesting site. Unfortunately, several "collectors" annually visit places like the Blue River Barrens to collect reptiles for the purpose of selling them as pets. Collecting here is not only illegal, it's unethical. A quandary arises when promoting places like this because it could increase collecting pressures. I am optimistic, however, that increased awareness and use by people that revere the natural world will deter these unscrupulous individuals.

Very few of these barrens habitats remain. Most sand barrens now grow planted red pines. The remaining open areas are viewed more as giant sandbox playgrounds than as complex ecosystems with wonderfully adapted plants and animals. Many people assume the sand barrens are wastelands in which they can do what they want. This, most often, translates into an area where overgrown kids have a place to play with their toys, such as dirt bikes and off-road vehicles. Though the barrens are publicly owned and off-road vehicle use is prohibited, some people do not respect the law. I hope naturalists quietly studying these unique habitats will some day outnumber the many noisy destroyers of them.

Farther east, near Muscoda, there are several areas where Iceland moss grows, only one of two locations in southern Wisconsin where this species is found. Drive east from Muscoda on Highway 133, and between one-half and one mile after crossing into Iowa County, you will find it growing south of the road. Park safely off the road and search the road right-of-way for a tall, chestnut-brown lichen with curled tips.

Specialties

Plants
- clustered poppy mallow
- fame flower
- Iceland moss

Animals
- ash-brown grasshopper
- Blanding's turtle
- lark sparrow
- long-horned grasshopper
- oithona tiger moth
- Olympia marble (butterfly)
- pawnee skipper (butterfly)
- phyllira tiger moth
- six-lined racerunner (lizard)

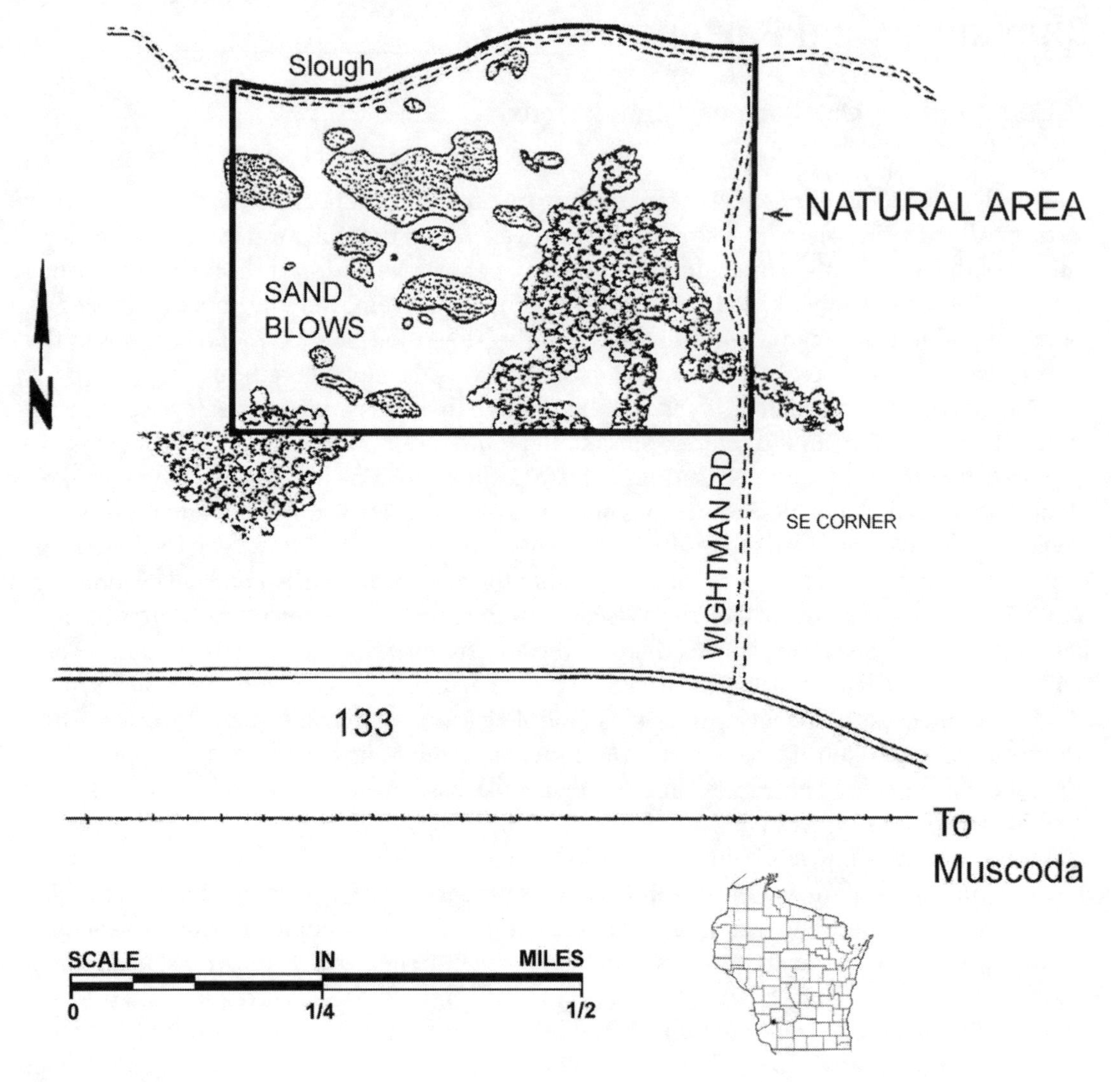

Blue River Sand Barrens

Uncommon Species

Plants
- bearberry
- dune three-awn
- false dandelion
- false heather
- flax-leaved aster
- fork-tip three-awn
- fragrant sumac
- golden aster
- large cottonweed
- long-leaved bluets
- partridge pea
- prickly pear cactus
- rock spikemoss
- sand milkweed
- sand primrose
- short green milkweed
- stiff sandwort
- venus looking glass

Insects
- arcigera flower moth
- bee fly
- broad-necked grasshopper
- gorgone checkerspot
- mottled sand grasshopper
- olive bird-dropping moth
- sand grasshopper
- two-striped mermiria
- white-blotched heterocampa

Birds (Breeding)
- field sparrow

Mammals
- badger
- harvest mouse

Governor Dodge State Park

Southern Oak Forest, Southern Red Oak–Mixed Forest, Dry Pine Forest, White Pine–Hardwood Forest, Cliffs

Governor Dodge State Park covers more than 5,000 acres. Most of the park's lands were heavily farmed by European settlers, and much of the land is disturbed, but a few of the original features of the area remain. The best remaining natural features are found on the bluffs, which still harbor relicts of the original pine stands and the original prairie–savanna species. At a place called Pine Cliff, all three pine species (jack, white, and red) grow together, relicts of our post-glacial past when pine forests covered the southern Wisconsin landscape. It is phenomenal to have all three species still naturally growing together this far south. There are many other bluff areas in this part of the state that have jack or red pine growing on them, but most of those trees have been planted. Other bluff areas have red cedar growing on them, but these cedar stands are not pine relicts; they are former prairie areas overgrown with these invasive conifers.

Farming changed the prairie in Wisconsin forever, but an extraordinary event has occurred in this park. When the state purchased the land for a state park, primarily in the 1950s, farming ceased, and the area reverted to a prairie-like landscape that resembles the original

Long-Lasting Impacts

Most natural communities have the capacity to recover after a single major disturbance, such as a hurricane or catastrophic wildfire. But communities subjected to continuous and prolonged disturbance do not recover well. For example, the predominately forested land of New England was settled by Europeans in the late 1700s, and by the early 1880s most of it had been cleared. Approximately 85 percent of the forests were leveled and converted into farmland, and most of the remaining land was cleared for its timber. The land could only support the marginal farms for a few decades, until they became unprofitable and were replaced by trees. Today's New England landscape is again mostly forested, but the forests' composition is radically different from that of the forests before clearing. Wisconsin's forests have to develop for another 100 years before we can do a similar historic ecology study here.

Even though forest canopy in these areas contains most of the tree species known from the area, a few tree species dominate the forest, and in some areas a nonnative tree, the Norway maple, dominates. The diversity of the forest canopy has been greatly simplified when compared to areas that were not disturbed as much. There are few understory species, and the canopy trees and saplings of those dominant trees form the bulk of the woody growth. Groundlayer simplification has been even more dramatic. Many species of representative forest flowers, such as violets, trilliums, and wild ginger, are missing from most tracts. Some ecologists estimate that it may take several centuries if not a full millennium for the forest to completely recover (Jordan et al. 1987).

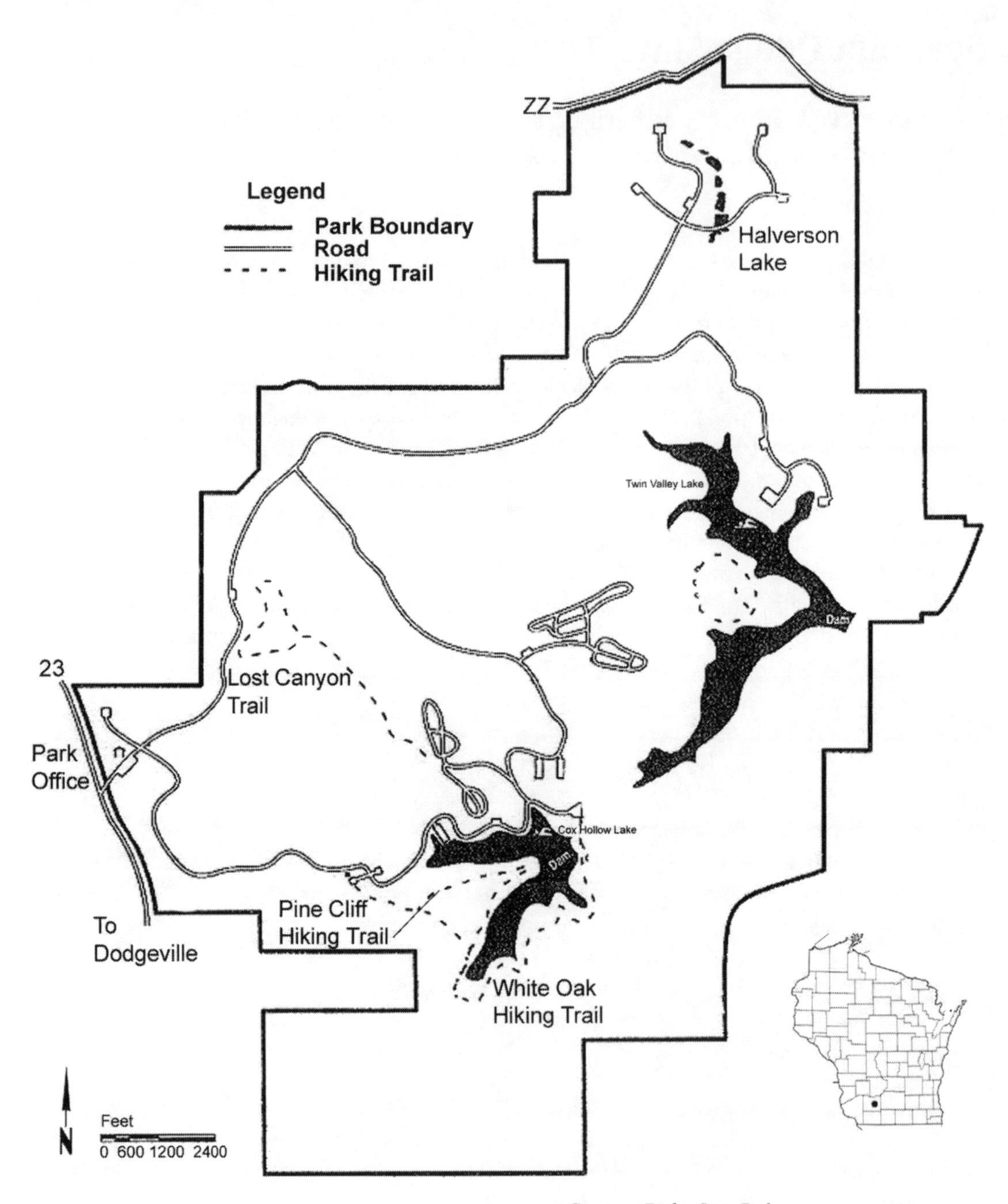

Governor Dodge State Park

landscape enough so that some prairie birds now use it. Of course, the plants are totally different from the original prairie species that grew here, and there are more woods now than then, but the landscape is now aesthetically a grassland–savanna. Formerly, there were prairies on the uplands; these are now grasslands of primarily European species such as quackgrass and smooth brome. Savannas and forests grew on these hillsides 150 years ago. Very hot fires would burn the prairies growing on open southwest-facing slopes, and oak–maple forests with savannas between would grow on the protected northern or eastern slopes. Today, the recovering farmland has shrubby edges resembling the continuum from savanna to woods.

This grassland apparently fits the needs of the Henslow's sparrow, a species threatened throughout its range. Formerly abundant with more than 80 singing males recorded in June 1986, the Henslow's sparrow population is significantly smaller in the park today, due to shrubs and trees colonizing the grassy areas. Planting prairie grasses has helped stem the population decline by providing preferred habitat for the Henslow's sparrow. The shrubs and trees, however, help another rare bird. The Bell's vireo, a threatened species in the state with a steady population of 10 to 12 territorial males within the park, requires shrubby edges embedded with a prairie–savanna landscape.

To view the wildlife and interesting plants, you'll need to avoid the heavy-use areas. Parking lots at Cox Hollow and Twin Lake can be overflowing, and swimming beaches often are wall-to-wall people. The remaining 90 percent of the park is available to you for exploration. The trail system gets you close to the pine relicts and grassland birds. If you are interested in aquatic study, the small Halverson Lake near the group camp is available for such pursuits.

Blanchard's Cricket Frog

Wisconsin's Blanchard's cricket frog is a state-endangered species whose population continues to decline. This little frog was formerly found throughout the southern half of the state. In the 1990s the frog's range in Wisconsin was limited to a few counties along the Illinois border. Surveys indicate the presence or absence of the frog but do not tell us why it is disappearing.

A possible reason for the declining population of Blanchard's cricket frogs was discovered at Governor Dodge State Park's Halverson Lake. These frogs were found in excellent numbers here shortly after the small lake was created. However, after the shoreline and submerged vegetation became dense, they disappeared from the area. Researchers believe that the denser vegetation provided better habitat for other frogs that outcompeted the cricket frog. This explanation of its disappearance is supported by records of several remaining cricket frog populations thriving at stock ponds where the shoreline vegetation is not dense.

Blanchard's cricket frog (illustration by Jim McEvoy; courtesy of Wisconsin Department of Natural Resources)

Specialties

Plants

- rough-stemmed false foxglove
- sullivantia
- violet bush clover

Animals

- Bell's vireo
- blackburnian warbler
- black-throated green warbler
- cricket frog
- Henslow's sparrow
- orchard oriole

Uncommon Species

Plants

- American gromwell
- bare-stemmed tick-trefoil
- black walnut
- bladdernut
- blue cohosh
- bottlebrush grass
- bracted orchid
- clustered black snakeroot
- creeping rattlesnake plantain
- cup plant
- declining trillium
- Dutchman's breeches
- dwarf serviceberry
- false rue anemone
- ginseng
- hooked crowfoot
- Jacob's ladder
- large-leaved shinleaf
- large yellow lady's slipper
- leafcup
- pink corydalis
- pointed tick-trefoil
- poke milkweed
- purple giant hyssop
- Robin's fleabane
- Schweinitz's sedge
- Short's aster
- showy orchis
- sicklepod
- silky wild rye
- slender cliff-brake
- spikenard
- squawroot
- Virginia waterleaf
- white trout lily
- yellow pimpernel

Insects

- red spotted purple

Birds (Breeding)

- Acadian flycatcher
- cerulean warbler
- dickcissel
- field sparrow
- northern bobwhite
- whip-poor-will
- worm-eating warbler

Mammals

- badger
- big brown bat
- gray fox
- southern flying squirrel

Lake Pepin

Communities: Floodplain Forest, Dry Prairie, River

The Lake Pepin sites are the publicly owned natural areas next to Lake Pepin, including the floodplain forests of the huge Tiffany and Nelson-Trevino Bottoms area, the bluffs and prairies near Maiden Rock, and Trenton Bluff.

Since the recession of the glaciers, the heavy load of silt and sand carried by the Chippewa River has formed a natural dam in the Mississippi River because the deposited material is more than can be carried away by the Mississippi. This forms a natural dam that backs up waters, forming Lake Pepin. The main channel of the Mississippi and its accelerated current are next to the Minnesota bluffs.

A huge river bottom forest lies on the delta formed by the Chippewa River deposits. As with all deltas, water continuously carves new routes. There are many bayous, with some starting miles upstream. These floodplain forests, which are called the Nelson-Trevino Bottoms south of Highway 35 and the Tiffany Bottoms north of that road, include nearly 20,000 acres. This wild area is unlike any other in the entire Midwest. It is the largest contiguous block of floodplain forest in the entire upper Midwest and contains numerous backwater channels reminiscent of the bayous of Louisiana. It can be penetrated by foot or boat at various locations along Highways 35 and 25. For the average naturalist, the flavor of the area can be sampled at several locations along these highways, but the interior may be available to experienced trekkers only.

To see other natural areas continue west from the Chippewa River delta on Highway 35, the Great River Road that runs along the Wisconsin side of the Mississippi River. Just east of Stockholm is a roadside park on County JJ that contains a few acres of dry prairie. Upon entering the village of Stockholm, proceed toward the lake and a small village park. It's hard to miss; there are two roads going toward the lake, and they both end in the park. You will see a rock crib used by anglers. If you walk upstream, you should see rattlebox, a plant which is distinctly out of range, usually found much farther south, but which seems to be sustaining itself on the

Migratory Bird Areas

Migratory birds encounter many hardships and hazards on their travels, including exhausting flights across great oceans and land masses, attacks by hungry raptors, and collisions with TV towers. Stopover spots must provide abundant food if the birds are to continue on their way. Birds recognize these areas and return to them year after year. One such area is the floodplain forest of the Chippewa River delta, which plays a vital role in producing food for these birds. Its countless side channels, oxbow lakes, ephemeral ponds, and shallow marshes are a dependable source of abundant insects every year. Tens of thousands of migratory birds utilize the forest along the Chippewa River every May.

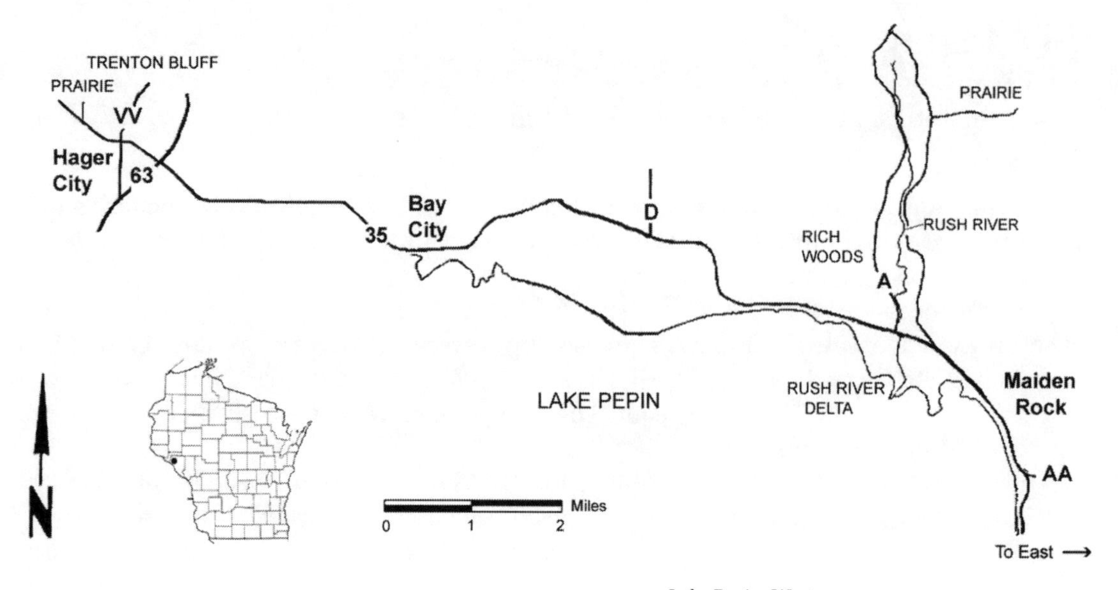

Lake Pepin–West

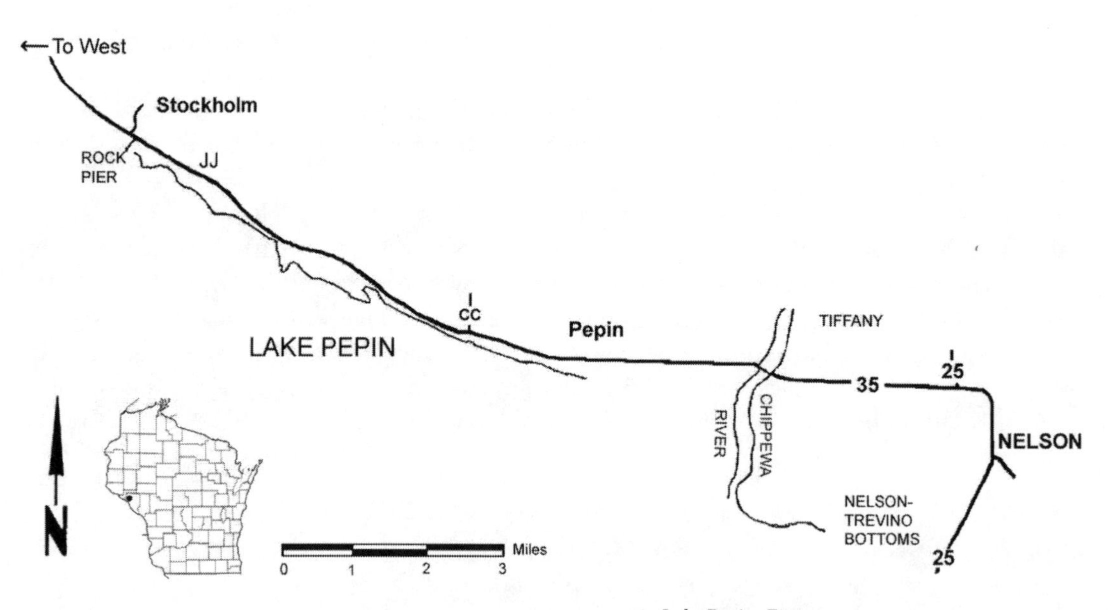

Lake Pepin–East

upstream shoreline. This plant has large black pods with seeds that rattle when shaken. Also, just downstream from the crib is an area used as a swimming beach where you can find many species of freshwater mussels in shallow water.

Upstream between Stockholm and Maiden Rock is an area of alternating ravines and bluffs. The most well-known of these bluffs is Maiden Rock, not to be confused with the village of Maiden Rock. Each bluff has at least a small patch of prairie near its summit; however, all of the bluffs are extremely difficult to explore. Near the village of Maiden Rock are many abandoned mines that formerly produced high-quality sand used in glassmaking. These mines provide excellent habitat for hundreds of thousands of bats. Some bat species use the mines throughout the year, and others use them only for winter hibernation.

Snow trillium (illustration by Jim McEvoy; courtesy of Wisconsin Department of Natural Resources)

Just west of Maiden Rock you can take a scenic drive on County Road A for several miles along the Rush River. If you drive north along the road, you will see older sugar maple–basswood forests that contain rich, diverse groundlayer species, including snow trillium.

Highways 35 and 63 intersect to the west of Bay City, close to Hager City. To reach the Trenton Bluff area, drive west from this intersection on 35 for 1½ miles just northwest of Hager City to where the prairie is next to the road. This series of prairies is owned by the state and has many species associated with the Great Plains, especially the mid-grass plains, including an excellent timber rattlesnake population. The specialties section identifies many plants and animals living here whose main distributions are hundreds of miles to the west. There are several other bluff prairies in the Diamond Bluff area, each with its own appeal, but nearly all are on private lands. Most of these are under consideration for bluff-top second homes. These developments will ultimately destroy the prairies because it will be impossible to allow managed fires with houses sitting on top of the bluffs. I wonder how many owners of these homes have experienced a prairie fire rushing up the slopes at incredible speed. It's a disaster waiting to happen.

Specialties

Plants

- bladderpod
- brittle prickly pear
- Carolina anemone
- cliff goldenrod
- clustered poppy mallow
- dotted blazing star
- downy prairie clover
- dragon sagewort
- fame flower
- glade mallow
- James' polanasia
- pomme-de-prairie
- prairie dandelion
- prairie plum
- prairie sagewort
- prairie thistle
- rattlebox
- small skullcap
- snow trillium
- snowy campion
- toothed evening-primrose
- wild licorice

Animals

- Acadian flycatcher
- arogos skipper (butterfly)
- bald eagle
- cerulean warbler
- dusted skipper (butterfly)
- eastern massasauga (snake)
- golden eagle
- hooded warbler
- Kentucky warbler
- lark sparrow
- northern prairie leafhopper
- ottoe skipper (butterfly)
- plains pocket mouse
- prairie ring-necked snake
- red-shouldered hawk
- six-lined racerunner (lizard)

spotted skunk
velvet-striped grasshopper

Uncommon Species

Plants

PRAIRIES

aromatic aster
bird's-foot violet
broom sedge
butterfly weed
clammy ground cherry
common spiderwort
cream wild indigo
cup plant
downy painted brush
Drummond's rock cress
false pennyroyal
field milkwort
fire-on-the-mountain
fork-tip three-awn
Geyer's spurge
goat's rue
Great Plains ladies' tresses
hairy beard-tongue
hairy grama
hairy lens-grass
hairy umbrellawort
heart-leaved alexanders
large cottonweed
long-leaved bluets
oval milkweed
partridge pea
pink corydalis
prairie blue-eyed grass
prairie dropseed
prairie larkspur
prairie panic grass
prairie satingrass
prairie smoke
prairie sunflower
prairie violet
prairie willow
rough false foxglove
sand cherry
sand cress
sand milkweed
sand-reed
Seneca snakeroot
short green milkweed
showy tick-trefoil
sky blue aster
small skullcap
smooth blue aster
starry campion
stiff gentian
stiff sandwort
tall green milkweed
upland wild timothy
Venus looking glass
violet wood sorrel
western sunflower
white prairie clover
wolfberry

BOTTOM LAND HARDWOODS

American lotus
cardinal flower
early meadow-rue
false aster
false dragonhead
false pimpernel
marsh speedwell
moonseed
mosquito fern
northern bugleweed
northern willow-herb
purple giant hyssop
purple meadow-rue
rock jasmine
water forget-me-not
water purslane
wild bean
winterberry

Insects

dainty sulphur
Georgia prominent
gorgone checkerspot
long-horned grasshopper
pawnee skipper
prairie cicada
rattlebox moth
sand-bar grasshopper
sand grasshopper
western red-winged grasshopper

Freshwater Mussels

black sandshell
deer toe
fat mucket
fawnfoot
fragile papershell
giant floater
hickory nut
lilliput
mapleleaf
mucket
Ohio River pigtoe
paper floater
pigtoe
pimpleback
pink papershell
pocketbook
spike
threehorn
washboard
yellow sandshell

Shells of Potentially Extirpated Mussels

buckhorn
butterfly
ebony shell
fluted shell
monkey face
pink heelsplitter

Fish

American eel
banded darter
bigmouth buffalo
blacknose dace
blacknose shiner

blue sucker
bullhead minnow
crystal darter
emerald shiner
finescale dace
flathead catfish
greater redhorse
highfin carpsucker
least darter
longnose gar
mimic shiner
Mississippi silvery minnow
mooneye
Ozark minnow
quillback
redfin shiner
river carpsucker
river darter
river shiner
sand shiner
shortnose gar
shovelnose sturgeon
slimy sculpin
smallmouth buffalo
trout perch

Reptiles and Amphibians

false map turtle
smooth softshell
timber rattlesnake

Birds (Breeding)

field sparrow
least bittern
nothern bobwhite
orchard oriole
prothonotary warbler
tufted titmouse
whip-poor-will

Mammals

badger
ermine
least weasel
long-tailed weasel
pocket gopher

Nelson Dewey State Park

Communities: Southern Oak Forest, Southern Red Oak–Mixed Forest, Dry Prairie, River

Nelson Dewey State Park is an area of ridges, valleys, and Mississippi River bottom lands with small patches of natural communities and openings that still contain relicts of the original natural communities, including dry prairie, cliff communities, and algific slopes. Nearly all of these areas are still in private hands, and there is a real danger of losing these sites if they aren't managed properly.

Oak forest covers most of the park, but there are relatively large areas of prairie on the limestone bluffs towering 300 feet above the Mississippi. Park roads and trails reach the steep dry prairies. The prairies within the park and surrounding lands are separate entities but are close enough to ensure pollen transfer by insects.

Bottom-land forest can be viewed at various places in the Cassville area. The best way to get the feel for the habitat is by boat. These bottoms are part of the Upper Mississippi National Wildlife Refuge. The public boat landing in Cassville (follow the signs) is a good place to

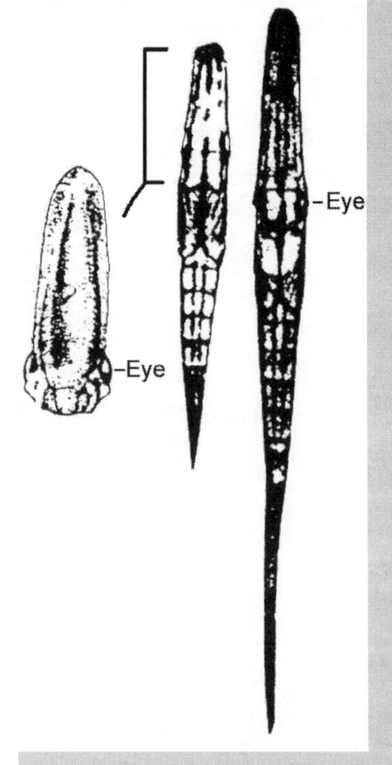

Duck-billed leafhopper (illustration adapted by author)

Duck-Bills

Although Nelson Dewey State Park isn't well known to many nature lovers, it is becoming well known to insect observers. Butterfly aficionados have discovered the richly varied butterfly and moth fauna, and many of Wisconsin's first sightings of rare butterfly species have been recorded here. In fact, rare butterfly observations seem to be the norm at the park.

More recently, surveys for other insect groups have revealed an equally diverse and interesting leafhopper fauna. One species in particular is worthy of special note. This rare leafhopper is large enough to be seen easily and is shaped much differently than other leafhoppers. It has a long structure protruding forward from its head and mouth, giving it a bizarre appearance. The structure reminds me of a duck's bill. Since this leafhopper is not known by many people, a common name does not exist, so I have taken the liberty of calling it the duck-billed leafhopper.

I know leafhopper taxonomists will cringe at the name because the structure is not a bill and it does not reflect leafhopper anatomy. Still, the name will be mine for the critter because it describes the leafhopper in a way that cannot be conveyed using the scientific name. Conservation of obscure groups such as this leafhopper will come more easily if more people understand and recognize the species.

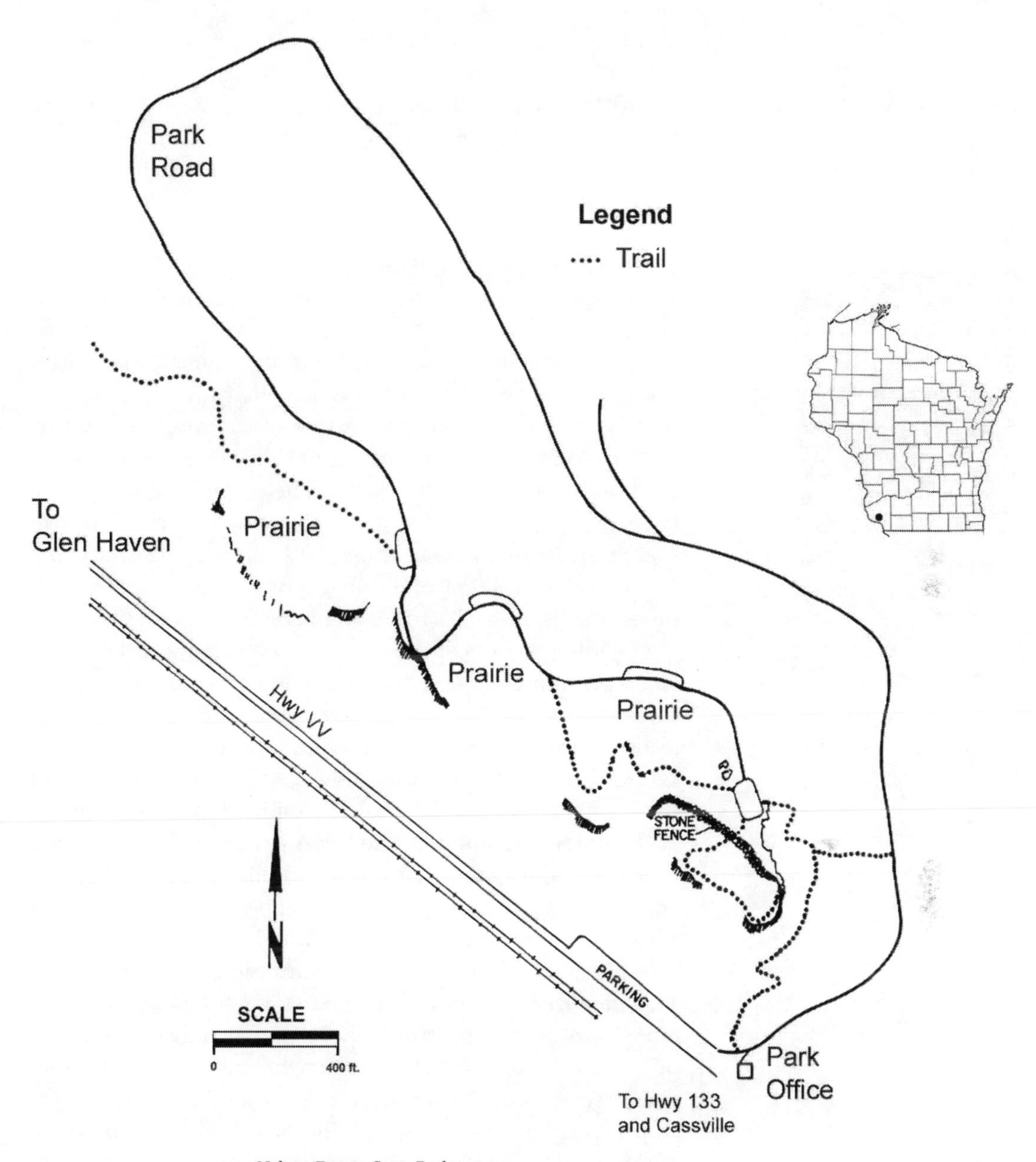

Nelson Dewey State Park

launch and explore the Mississippi River. Many islands are in a wild condition, just ripe for exploration. The silver maples that have fallen into the river are excellent places to observe nearly all of Wisconsin's turtles.

The Bertom Lake area is excellent for studying bottomland forest. Its open sloughs and open flowing river have some of the best remaining freshwater mussel beds in the state. Take State Highway 133 east of Cassville about one mile to Sand Lake Lane, which ends a short way into the bottoms.

You'll need to park your car in the refuge parking lot and explore the Bertom Lake bottom lands on foot.

North of Cassville below lock and dam no. 10 is an area known nationally for its concentration of bald eagles. The wintering bald eagles love the combination of open feeding waters along with protected roosting sites. "Eagle Valley" at one time was the project of a conservation organization. It is now privately owned and not open to the public. The site contains some fine southern oak forests with their associated species.

At Nelson Dewey State Park, stay on County VV and drive north until you intersect with County Highway V leading west into Glenhaven. Algific slopes, one of the most interesting natural communities remaining in Wisconsin, are found on the hills surrounding Glenhaven. These algific slopes, which have only recently been identified and described by researchers, are rocky areas where the ground is constantly cool because of the geology of the area, which allows many out-of-range northern species to survive here. Some of these species are definitely relicts of the ice age, such as a species of landsnails previously thought to be extinct. These slopes keep cool through a system of air circulation. A sinkhole at the surface leads to an area of soft underlying limestone that has dissolved through the years. Southwest Wisconsin has many of these sinkholes that lead to caves. Algific slopes are different. Water enters the sinkhole and freezes, forming ice in fissures. Cool air settles in the sinkhole, descends to the icy cave, becomes very cold, then reappears at the slope surface. As the cold air seeks an outlet, it filters through the talus on the hillsides, keeping the "feet" of the plants in arctic conditions. Thus far, this is the only known area in Wisconsin where algific slopes occur. Unfortunately for the naturalist, they are on private lands, and some landowners are intolerant to visitors viewing these sites. Therefore, this information is included for educational purposes only and in hopes that someday some of these unique communities can be preserved.

Specialties

Plants

intermediate sedge
lance-leaved buckthorn
limestone oak fern
marbleseed
northern monkshood
pin oak
prairie dandelion
prairie Indian plantain
twinleaf
violet bush clover
yellow giant hyssop
yerba de Tajo

Animals

abbreviated underwing
bald eagle
byssus skipper
duck-billed leafhopper
Kentucky warbler
migrant raptors

olive hairstreak
ottoe skipper
pleistocene landsnails
western worm snake
Whitney's underwing

Uncommon Species

Plants

bristle-leaved sedge
cliff goldenrod
downy gentian
downy paintbrush
fire-on-the-mountain
fragrant sumac
hairy beard-tongue
honey locust
prairie brome
prairie crowfoot
purple giant hyssop
purple milkwort
seneca snakeroot
short green milkweed
showy goldenrod
Virginia bluebells
white camas
yellow false foxglove

Insects

Anna's tiger moth
bicolored grasshopper
bisected honey locust moth
catalpa sphinx
columbine dusky wing
common sooty wing
coral hairstreak
cross-line skipper
dejected underwing
double-lined gray
giant leopard moth
green-legged grasshopper
gulf fritillary
handsome grasshopper
honey locust moth
ilia catacola
imperial moth
magdalene catacola
pink-legged tiger moth
Reakirt's blue
Ripley's lappet moth
roadside skipper
rose wing
serene underwing
striped hairstreak
tawny emperor
yellow-banded catacola

Mussels

black sandshell
butterfly
deer toe
fawn foot
fragile papershell
giant floater
hickory nut
Higgins' eye
lilliput
monkey face
paper floater
pink heelsplitter
pocketbook
rock shell
spike
strange floater
threehorn
wartyback
washboard
white heelsplitter

Mussel Shells

buckhorn
ebony shell
elephant ear
fat mucket
Ohio River pigtoe
pink papershell
spectacle case
yellow sandshell

Reptiles and Amphibians

black rat snake
Ouachita map turtle
timber rattlesnake

Fish

bluntnose darter
brook silverside
brook stickleback
golden redhorse
goldeye
log perch
longnose gar
Mississippi silvery minnow
mooneye
Ozark minnow
silver chub
silver redhorse

Birds (Breeding)

cerulean warbler
common moorhen
dickcissel
least bittern
prothonotary warbler
red-shouldered hawk
tufted titmouse
yellow-crowned night heron

Birds (Migrants)

golden eagle
merlin
peregrine falcon
Swainson's hawk

Mammals

big brown bat
eastern pipistrelle
Indiana bat
prairie deer mouse
southern flying squirrel

Perrot State Park

Communities: Southern Oak Forest, Southern Red Oak–Mixed Forest, Dry Prairie, Shallow Marsh

This section can easily be divided into three units: Perrot State Park, Trempealeau National Wildlife Refuge, and Tamarack Creek Wildlife Area. The park and refuge are adjacent, and the wildlife area lies several miles upstream.

Perrot State Park is an area of sharp contrasts. This park, which is bordered by the Mississippi River on one side and a broad sandy terrace on the other, contains high steep hills of refuge for many plants and animals. The steep topography, orientation to the sun, and extent of past human disturbance have determined what species will be found in any particular area. The remnant bluff prairies, foremost of which is Brady's Bluff, harbor many species normally found much farther west. The many park trails provide access to the prairies. The remaining parts of Perrot State Park are overgrown savanna. The only true forests are in areas of steep ravines and north-facing coves.

Trempealeau National Wildlife Refuge is an abrupt change from the bluffs of Perrot. Three dikes hold back the waters of the Trempealeau River in its very flat delta. This is an area of lowland forest and marsh land. The plants here are diverse, with an abundance of wild rice and American lotus. The animal life is even more diverse. A valuable waterfowl area, Trempealeau is home to nesting ospreys, which are quite rare this far south in Wisconsin, and king rail and common moorhen, which are quite rare this far north.

An area of river-deposited sand is found near the headquarters. This feature is easily seen by taking the auto tour of the refuge. At several places along the tour, visitors can see prairie restoration projects and their associated fauna. Near the end of the tour is a parking area that holds a healthy population of downy prairie clover. This western species was never part of any prairie plantings; it survived in isolated patches until the refuge personnel removed trees in the area.

The final area is Tamarack Creek Wildlife Area. As its name suggests, this is one of the largest areas of tamarack bog left in the driftless area. Unfortunately, storms in the summer of 1998 blew many tamaracks down. Time will tell if they regenerate at the site. The bog forms a

Bell's vireo (illustration by Jim McEvoy; courtesy of Wisconsin Department of Natural Resources)

Bell's Vireo

The western side of the Tamarack Creek Wildlife Area along County Highway F has many shrubby hillsides and ravines. This habitat is ideal for the Bell's vireo. This state-threatened species has seen population declines throughout most of its range. A bird of the southern plains, the Bell's vireo prefers brushy areas in an open landscape. Wisconsin has a tremendous amount of brushy abandoned agricultural fields and edges, but this vireo does not inhabit most of these areas. It lives in only a few locations in the southwestern third of the state, probably limited by its finicky habitat selection habits. You can locate a male in a shrubby patch by keying in on its loud, long, penetrating song.

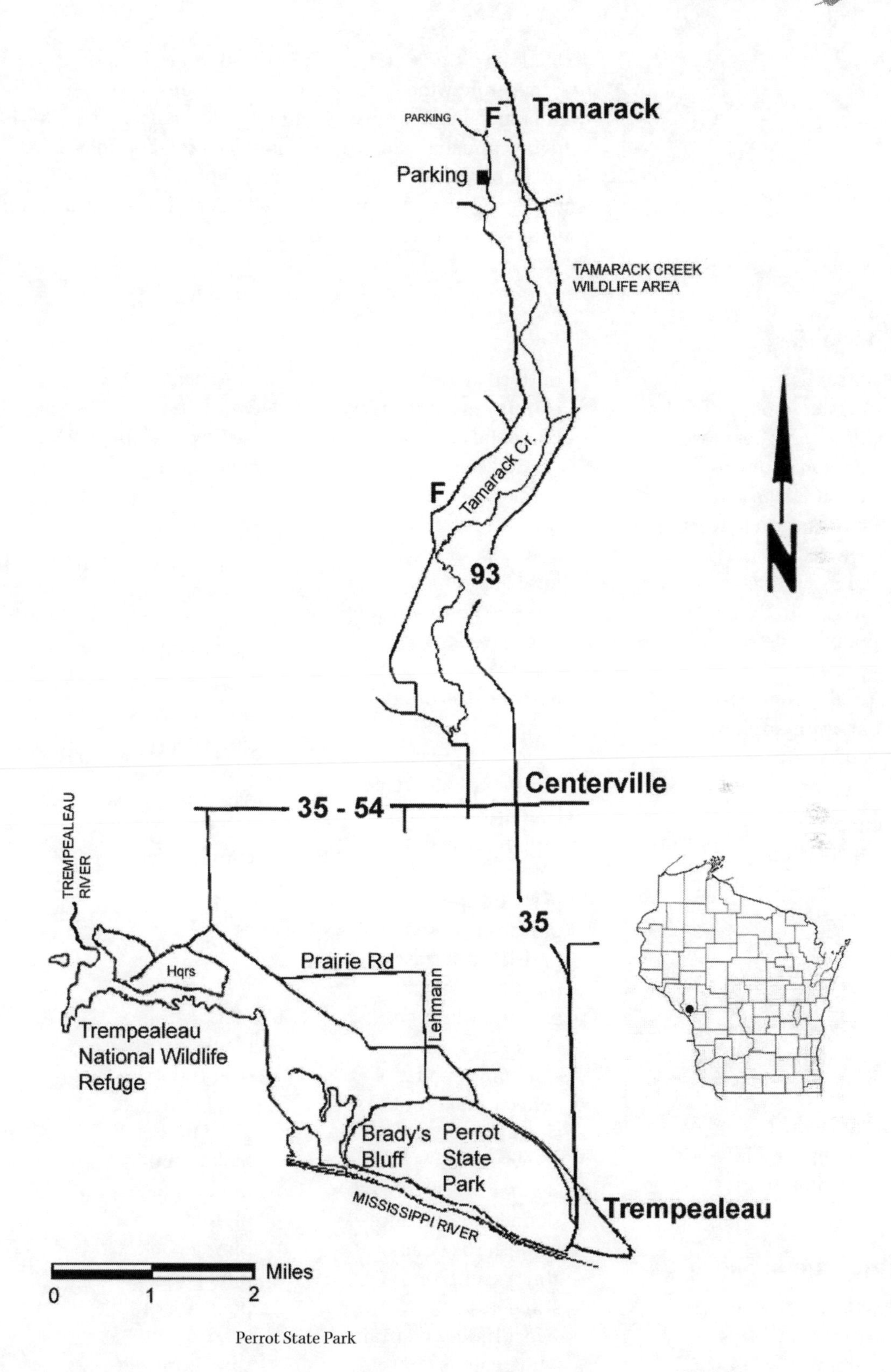

Perrot State Park

strip along Tamarack Creek nearly six miles long and up to one-half mile wide. The bog contains many species usually found much farther north, including a few balsam firs. This is a highly unusual habitat for nonglaciated sections of the state. Tamarack Creek can be reached by taking either Highway 93 or County Highway F north from Trempealeau. You will know you are there when the tamaracks appear.

Specialties

Plants

clustered poppy mallow
downy prairie clover
dragon sagewort
Great Plains ladies' tresses
hairy umbrellawort
jeweled shooting star
marsh horsetail
mosquito fern
pomme-de-prairie
prairie larkspur
prairie sagewort
snowy campion
sullivantia

Animals

Bell's vireo
common moorhen
king rail
lark sparrow
least bittern
osprey
yellow-crowned night heron

Other Features

fossils (brachiopods, trilobites)

Uncommon Species

Plants

American lotus
aromatic aster
butterfly weed
clammy ground cherry
cream false indigo
crested sedge
crooked aster
Davis sedge
downy paintbrush
false dragonhead
fire-on-the-mountain
flax-leaved aster
goat's rue
hairy beard-tongue
hairy grama
hairy lens-grass
Kentucky coffee tree
leafcup
oval milkweed
pale-spike lobelia
partridge pea
prairie crowfoot
prairie satingrass
prairie wedgegrass
purple cliff-brake
rattlebox
sand milkweed
Seneca snakeroot
short green milkweed
showy orchis
slender beard-tongue
slender false foxglove
smooth rose mallow
starry campion
stiff gentian
sweet Indian plantain
white camas
wild bean
wolfberry
yellow false foxglove
yellow stargrass

Insects

bicolored grasshopper
capsule moth
coral hairstreak
dainty sulphur
dusted skipper
gorgone checkerspot
great copper
olive hairstreak
sand-bar grasshopper
sedge skipper
short-winged green grasshopper
Sprague's pygarctica
tawny emperor

Mussels

black sandshell
deer toe
fawn foot
giant floater
hickory nut
monkey face
pigtoe
pimpleback
pink papershell
pocketbook
spike
threehorn

Reptiles and Amphibians
- false map turtle
- Ouachita map turtle
- timber rattlesnake

Birds (Breeding)
- orchard oriole
- osprey
- prothonotary warbler

Mammals
- badger
- big brown bat
- plains pocket mouse
- prairie mole
- southern flying squirrel

Wildcat Mountain State Park

Communities: Southern Red Oak–Mixed Forest, Floodplain Forest, Northern Hardwood–Hemlock Forest, River, Cliffs

As seen from the specialties list, cliff communities and their associated plants form the bulk of the unique features that can be seen at Wildcat Mountain State Park and elsewhere along the Kickapoo River. The cliffs harbor species endemic to this part of the state, such as musk-root. Most of the Kickapoo area's flora and fauna can be sampled in Wildcat Mountain State Park.

This 3,400-acre park includes highly irregular topography that is typical of the driftless area, such as an abundance of cliffs containing relict populations of plant species with northern affinities. Stop at the park headquarters to pick up a map of the trails and features. Two

Lapland rosebay (illustration by Jim McEvoy; courtesy of Wisconsin Department of Natural Resources)

Relicts

Relict communities of the glacial and post-glacial periods are still found along the Kickapoo River. Special conditions are needed for these natural communities and species to persist. Steep cliffs with cool dripping water and a northerly aspect provide most of the conditions needed by the plants. Some of these plants have their closest population either in boreal Canada or all the way up in the tundra. A great example is the Lapland rosebay. This arctic species is a member of the rhododendrons. Its closest populations are 900 miles away in the alpine areas of New England. Astonishingly, two populations occur in Wisconsin, one on a single north-facing cliff along the Kickapoo.

Some of these northern species have been separated from their closest populations long enough to allow for genetic variation. This may have occurred in the cliff cudweed, whose Wisconsin populations are isolated from the main habitat farther south. Future studies will be needed to verify that possibility.

Another possible relict population in the driftless area is pure speculation on my part, but it's one that I think needs scientific investigation. Every winter at four locations within the driftless area, several golden eagles overwinter. This species is most commonly associated with the western mountains, but there's a healthy population summering in the Canadian arctic. Part of that population consistently overwinters in Wisconsin, with four to six golden eagles found annually near the upper Kickapoo River between La Farge and Ontario. I speculate that ancestors of these birds lived throughout the year when the driftless area was tundra. As the glacier retreated north along with the tundra, this population of golden eagles moved north to nest but still maintained its connections to its homeland by wintering in the driftless area.

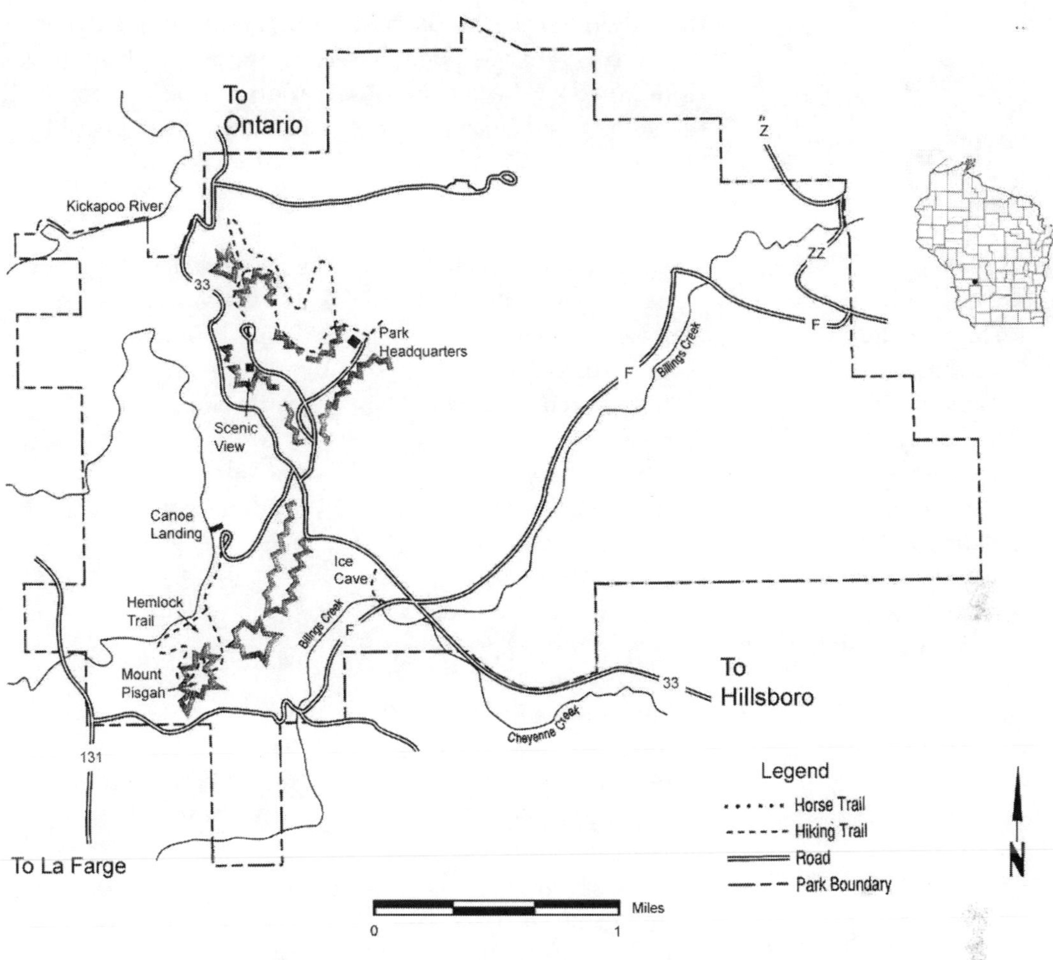

Wildcat Mountain State Park

trails in particular capture the essence of the area. First, the short trail to the Ice Cave brings you into a small relict forest of yellow birch and hemlock. The other is the Mt. Pisgah Trail, which contains hemlock relicts. Many larger stands of relict forest along the trail contain the specialty cliff plants, plus an abundance of fern species. The plant species found at the summit of this trail are far different from those found at the base. Exposed cliff areas near the peak have many dry-prairie species.

Although you can easily study the varied natural communities found in Wildcat Mountain State Park, you will need to go onto the Kickapoo River to study the specialized cliff communities. The area along the Kickapoo River between Ontario and La Farge contains nearly 30 different cliffs, each with different species due to differences in exposure, orientation, steepness, and wetness. These cliff communities are very hard to study, requiring special procedures and equipment. Fortunately, this is exactly what has saved these cliffs from intrusion. Much of the species observation has to be done with binoculars, and close-up photography is nearly impossible.

Specialties

Plants

birds-eye primrose
broad beech fern
cliff cudweed
cliff goldenrod
cliff saxifrage
cream gentian
Hooker's orchid
Lapland rosebay
limestone oak fern
muskroot
northern monkshood
rock clubmoss
sullivantia
woodland boneset

Animals

golden eagle
Kentucky warbler
worm-eating warbler

Uncommon Species

Plants

big-leaved aster
bird's-foot violet
bladdernut
blue phlox
bristly sarsaparilla
broad-leaved goldenrod
bunchberry
Carey's sedge
crested sedge
Davis sedge
Dutchman's breeches
early horse gentian
false rue anemone
few-fruited sedge
five-parted toothwort
ginseng
green dragon
hairy goldenrod
hairy sedge
hop sedge
large-leaved shinleaf
large straw sedge
late coralroot
late horse gentian
long-awned wood grass
nannyberry
narrow beech fern
partridgeberry
pipsissewa
poke milkweed
purple cliff-brake
purple giant hyssop
purple spring cress
putty root
red elderberry
Robin's fleabane
Rocky Mountain sedge
rose twisted stalk
sessile bellwort
showy orchid
silky wild rye
silvery glade fern
skunk cabbage
slender cliff-brake
slender lip-fern
stellate sedge
tall bellflower
Virginia bluebells
walking fern
white camas
whorled loosestrife
wild ginger
wintergreen
yellow jewelweed

Mushrooms

white-egg bird's nest

Insects

autumn yellow-winged grasshopper

Reptiles and Amphibians

four-toed salamander

Birds (Breeding)

Acadian flycatcher
Bell's vireo
cerulean warbler
field sparrow
red-shouldered hawk
tufted titmouse
whip-poor-will

Mammals

eastern pipistrelle
gray fox
southern flying squirrel

Wyalusing State Park

Communities: Southern Oak Forest, Southern Red Oak–Mixed Forest, Sugar Maple–Basswood Forest, Floodplain Forest, River, Cliffs

Located at the junction of Wisconsin's two great rivers, Wyalusing has been an important landmark for thousands of years. The park covers 2,575 acres of varied habitats ranging from the river bottoms to the bluff tops more than 500 feet above the valley floor. By far, the best way to observe the park's natural wonders is by hiking the 10 miles of trails. Each trail has its own degree of difficulty. Each also has its own biological specialties. There are bluff top trails, river bottom trails, trails along the bluff base, trails going nearly straight up (or down), and trails leading to ice caves. Each trail traverses different communities, so by hiking all the trails, it's possible to see every community and every species.

All the southern Wisconsin forest communities can be found within the park's boundaries. The oldest stands are on the north-facing slopes looking toward the Wisconsin River. By hiking the Immigrant Trail, you can easily see areas of huge black walnut canopy. At the base of the hills, both upland and floodplain species can be found. Further up the slope is the Flint Ledge Trail, which follows the bluff base and is a very good place for finding jewelled shooting star, sullivantia, and other cliff species.

Sullivantia (photograph by Thomas A. Meyer; used by permission)

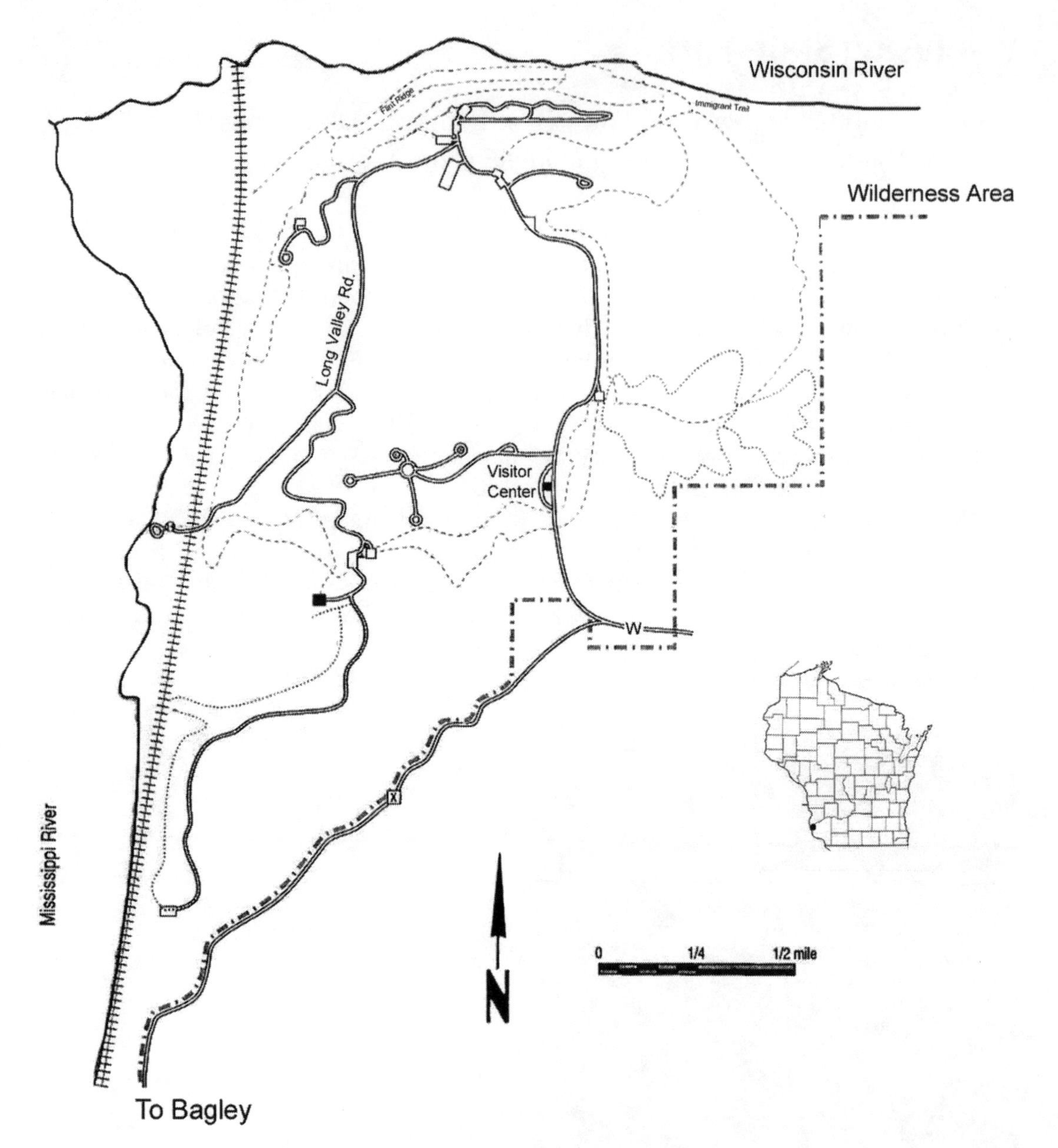

Wyalusing State Park

Though not a trail, Long Valley Road is a good place to see plants and, especially, unusual birds. As you walk down Long Valley, notice the exotic white pines. These trees were planted here in 1918 from the first state nursery. These pines and the rest of the valley annually produce southern birds whose range barely makes it to Wisconsin. Most consistent among these rarities has been the yellow-throated warbler and the Carolina wren. This road is probably the easiest place in the state to see a Kentucky warbler.

Black rat snake (Endangered Resource file photograh; courtesy of Wisconsin Department of Natural Resources)

Black Rat Snake

Springtime hikes on Wyalusing's trails can be the best way to observe one of the most intriguing reptiles in the area. The black rat snake can usually be found by walking the trails in mid-May. This species is a large, heavy-bodied snake measuring up to six feet long. A constrictor, the black rat snake feeds on mammals and birds. Although found more readily on the ground, this species is a real tree climber. It spends more time in the trees than any other Wisconsin snake, and when disturbed, it will often crawl into a tree hole. When hiking the park's trails, keep one eye on the ground and the other in the trees to find this fascinating species.

Specialties

Plants

broad beech fern
cliff saxifrage
cream gentian
ginseng
glade fern
golden seal
great Indian plantain
jewelled shooting star
mullein foxglove
nodding pogonia
pale purple coneflower
sullivantia
yellow giant hyssop
yerba de Tajo

Animals

Bell's vireo
black rat snake
Kentucky warbler
obovate-winged grasshopper
prairie ring-necked snake
short-winged grasshopper
winter wren
worm-eating warbler

Uncommon Species

Plants

American lotus
bladdernut
bur sedge
butterfly weed
cardinal flower
chinquapin oak
cliff goldenrod
common hemicarpha
common hop
Culver's root
downy paintbrush
downy rattlesnake plantain
dragonhead
dragon sagewort
Dutchman's breeches
early horse gentian
false aster
false dragonhead

false loosestrife
false pimpernel
fog fruit
Goldie's fern
great St. John's wort
great waterleaf
green dragon
groundnut
hairy lip-fern
indigo bush
Kentucky coffee tree
large straw sedge
large yellow lady's slipper
late coralroot
late horse gentian
leafcup
long-beaked sedge
Muskingum sedge
nodding trillium
pale-spike lobelia
partridge pea
poke milkweed
prairie alum-root
purple giant hyssop
purple joe-pye weed
purple prairie clover
putty root
rue anemone
Seneca snakeroot
Short's aster
showy orchis
silvery glade fern
small sundrops
smooth false foxglove
sneezeweed
squirrel corn
walking fern
water plantain
water purslane
water stargrass
white bear sedge
white camas
white water lily
wild bean
wild yam
wood betony
woodland sedge
woodland thistle

Insects

afflicted dagger
bombardier beetle
columbine dusky wing
coral hairstreak
dainty sulphur
giant cicada killer
green-legged grasshopper
honey locust moth
pleasant dagger
royal walnut moth
six-spotted leafhopper
snout butterfly
speared dagger
tawny emperor
ulalume underwing

Reptiles and Amphibians

false map turtle
map turtle
timber rattlesnake

Birds (Breeding)

black-crowned night heron
Carolina wren
dickcissel
eastern screech owl
field sparrow
great egret
Henslow's sparrow
northern bobwhite
prothonotary warbler
red-shouldered hawk
tufted titmouse
whip-poor-will

Birds (Migrants)

golden eagle
peregrine falcon
Swainson's hawk

Mammals

gray fox
prairie deer mouse
southern flying squirrel

Northwest Wisconsin

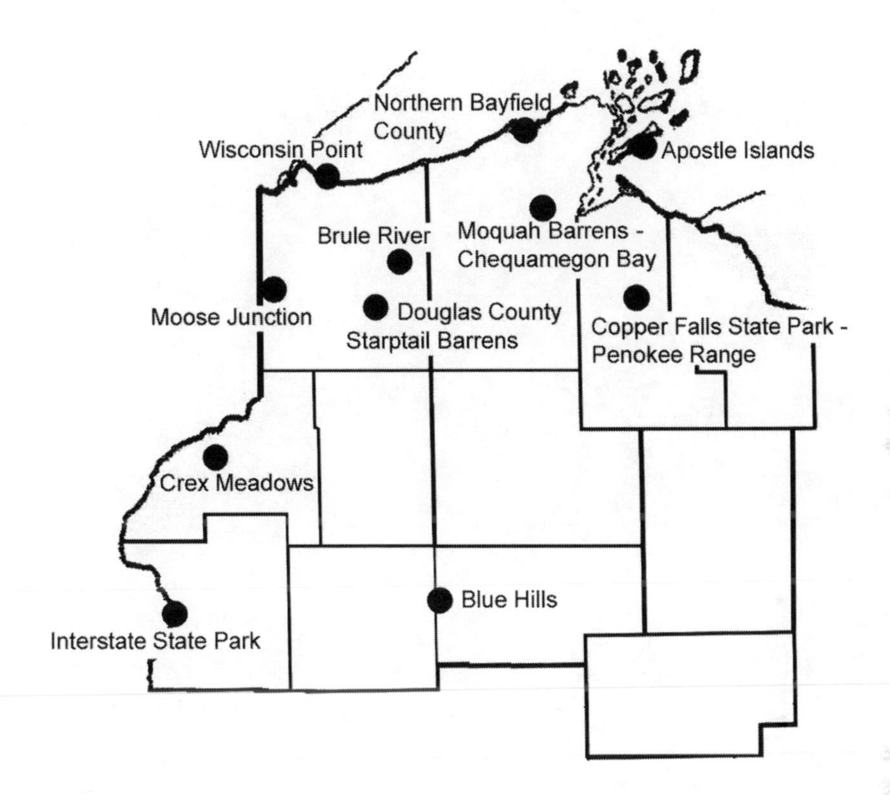

Apostle Islands

Communities: Dry Pine Forest, White Pine–Hardwood Forest, Northern Hardwood–Hemlock Forest, Forested Swamps, Great Lakes Shoreland Forest, Shallow Marsh, Lake Dune, Lake Beach, Cliffs

The easiest island to explore is Madeline Island, the largest of the 22 Apostle Islands and the only one permanently inhabited. Big Bay State Park on Madeline Island has more than 2,300 acres of varied terrain to explore, including five miles of hiking trails, a mile of sand beach, and sandstone cliffs and caves at Big Bay Point. Behind the sand beach (technically a baymouth bar) is an exemplary bog. About 120 acres in size, the bog formed when the barrier baymouth bar blocked water currents, and it consists of a lagoon, an open bog mat, and a swamp conifer forest. The bog has an extensive area of quaking sphagnum and supports one of the richest floras in the Lake Superior region.

The Apostle Islands National Lakeshore, which protects 21 of the islands, is a premier destination for many explorers. Visitors can view the islands at their most picturesque when they appear as emeralds emerging from the tranquil blue waters. At other times, storms pound with such force that you can feel nature's power penetrate to your bones. A multitude of reasons may bring you to the islands, but regardless of your reason, your first stop should be the visitors center in Bayfield. Here you can obtain information from park personnel regarding the island's features and how to access them.

The islands of the National Lakeshore can be reached by boat from Bayfield. Several alternatives are available: you can take the Park Service shuttle boat to Stockton Island for camping or exploring, you can take one of several excursion boats, you can rent a boat, or you can launch your own boat.

Taking the shuttle boat to Stockton, Oak, or Basswood Islands is an easy way for you to camp, explore, hike the trails, and to get a feel for these islands. Stockton has many rewards, with seven miles of trails and a 1½-mile sand beach at Julian Bay. Mixed forest covers Stockton Island, and, as on Madeline Island, it has an outstanding bog that is rich in rare species. This bog is part of the sand-ridge and swale topography of the Presque Isle Point tombolo.

A tombolo is a geological feature that forms when sand builds between a small island and a larger piece of land or between two islands. In Wisconsin, tombolos occur in the Apostle Islands and nowhere else. By searching the swales of this tombolo, you might find the rare linear-leaved sundew and unusual sedges. From Presque Isle Point or the Julian Bay area, you can see large numbers of herring gulls that breed a few miles to the east on three-acre Gull Island. Upwards of 75 percent of the nesting herring gulls in the area use Gull Island. Any human use of Gull Island during the breeding season is strictly prohibited.

You will need a private boat or a rental boat through a local outfitter to visit the other islands. Some excursion boats stop at certain islands, but the timing and location of your stay is under the control of the tour provider. Every island in the chain has a certain appeal and history. Before visiting any island, check with park rangers for current conditions.

Each island has a different flora and fauna. Oak and Basswood Islands, which are close to the mainland, share many characteristics. Oak Island is the tallest island. In the deep ravines on the island's north-facing slopes, a few broad-leaved twayblades might be found, but they are

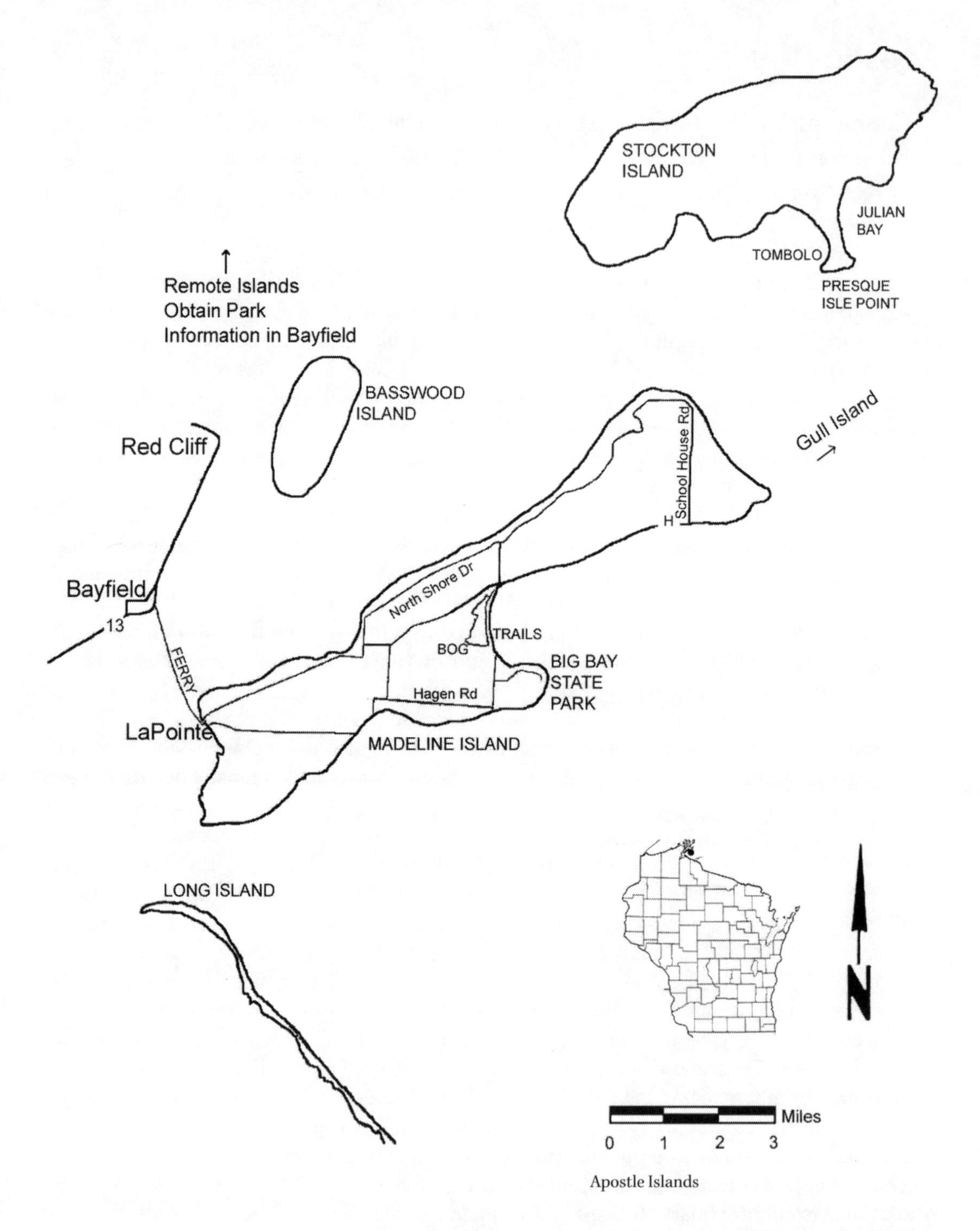

Apostle Islands

not easy to locate. Oak and Stockton Islands are the only islands that have bears. Deer live on few of the islands, which allows American yew (a favorite winter deer food) to become the dominant shrub on many islands. American yew gives the forest an appearance very different from the mainland. The

Devils Island (Endangered Resource file photograph; courtesy of Wisconsin Department of Natural Resources)

Lichen-covered rock in splash zone (Endangered Resource file photograph; courtesy of Wisconsin Department of Natural Resources)

Wave Splash Community

The species living on the north-facing cliffs of the outermost islands endure rigorous environmental conditions. The open great lake can impose brutal forces of wind and wave on the cliffs. This pounding surf can dislodge the most stubborn root system. The water temperature of the crashing waves also keeps these cliffs very cold. In addition, these cliffs receive very little direct sunlight. All of these factors create an environment that permits relict species from the ice age to persist.

Several boreal plant species, such as the sedges, rushes, and mosses of the far north, are found on or near the cliffs of the outermost islands. One species, the butterwort, is found only in the direct splash zone (see color insert). This parasitic plant receives additional nutrition from insects that land on its sticky leaves and are dissolved and absorbed into the plant.

plants are often so dominant that an impenetrable thicket forms.

Devils, North Twin, and Outer Islands, which are far into Lake Superior, are probably the most significant from the naturalist's point of view. Devils Island has a wave-battered coast with cliffs occupying most of the shoreline. Its uplands are flat and poorly drained, forming an unusual upland conifer bog. The plants are very similar to boreal Canada and so are the birds. This is the only known nesting area for pine grosbeak and Wilson's warbler in Wisconsin. Both species normally nest hundreds of miles farther north. The heath rush, another species found here and nowhere else in Wisconsin, is native to Greenland. Road-building equipment previously stationed in Greenland most likely brought the plant here.

Water from the uplands eventually seeps through the cracks in the sandstone. This seepage along with wave battering has provided the north coast of Devils Island with the cold, wet conditions perfect for many species of sedge that are normally found in the arctic.

Outer Island has a sand spit on the south side of the island that allows easy access by boat. The spit itself has a rich assortment of beach plants and probably the largest concentration of migratory birds in Wisconsin. Literally tens of thousands of birds stack up here in spring and fall. The island has aspen–birch forest dominating the southwest quarter. The rest of the island is northern hardwood interspersed with boreal forest and hemlock stands. An area of old-growth hemlocks occurs on the island's northwest edge, with hemlocks up to four feet in diameter. The recently discovered woodland cudweed grows near the island's center. Also, summer populations of some boreal warbler species have been found.

Bear Island is another gem, but it is a difficult island to visit. The island is high, and it escaped inundation in the flooding that occurred when the glaciers melted. Because of its height, a small part of the island was jutting out of glacial Lake Duluth during the high water. When Bear Island was tiny, a small bog formed along its shore, and as the water receded, the bog remained. The summit of Bear Island has a five-acre perched bog surrounded by a 6,000-year-old beach ridge. On the slopes north and east of the summit bog is a 30-acre virgin hemlock–hardwood stand with trees that originated in the 1700s.

The most southern part of the Apostle Islands is the 2½-mile long, ¼-mile wide Long Island. Since the late 1970s, the

Piping plover (Endangered Resource file photograph; courtesy of Wisconsin Department of Natural Resources)

former island is now the tip of a long sand spit at the mouth of Chequamegon Bay because sand deposits connected the island with the peninsula. The long peninsula is the last stronghold in Wisconsin for the endangered piping plover. Because this bird is very rare, island visitation is prohibited during the nesting season. If you visit the island during May or in August and September, you might still see the piping plover plus many other shorebirds, including all of the "large type" shore birds that are regularly found in Wisconsin, such as the whimbrel, the marbled godwit, and the willet.

Specialties

Plants

beautiful sedge
black hawthorn
bog reed-grass
broad-leaved twayblade
butterwort
Chilean sweet cicely
coast sedge
dragon's mouth
drooping sedge
English sundew
fir clubmoss
flat-leaved willow
hairgrass
hair-like sedge
heath rush
lake cress
least moonwort
lenticular sedge
limestone oak fern
linear-leaved sundew
Michaux's sedge
mingan moonwort
moonwort
mountain cranberry
pale sedge
rayless ragwort
Robbin's spike-rush
round-leaved orchid
satiny willow
slender bog arrow-grass
small purple bladderwort
sooty beak-rush
sparse-flowered sedge
spike trisetum
spreading wood-fern
swamp grass-of-parnassus
tufted hairgrass
woodland cudweed

Animals

blackpoll warbler
palm warbler
pine grosbeak
piping plover
Tremblay's salamander
Wilson's warbler

Uncommon Species

Plants

adder's tongue fern
alpine rush
arrowhead coltsfoot
beaked sedge
beaked willow
birds-eye primrose
blanketflower
blueleaf willow
blunt-leaved orchid
bog St. John's wort
bracted orchid
Canada yew
common bog arrow-grass
daisy-leaved grape fern
dwarf scouring-rush
false melic grass
fir clubmoss
golden corydalis
grass pink
great-spurred violet
green adder's mouth
green-flowered shinleaf
hooded ladies' tresses
hooked crowfoot
hook-spurred violet
marsh St. John's wort
mountain holly
northern heart-leaved aster
necklace sedge
northern cranesbill
red-stemmed gentian
round-leaved orchid
slender lip-fern
stitchwort
striped coralroot
swollen sedge
tall lungwort
tesselated rattlesnake plantain
three-toothed cinquefoil
twig rush
western rattlesnake plantain
white mandrain

Insects
- bunch-grass grasshopper
- laurentian skipper
- spot-winged grasshopper

Reptiles and Amphibians
- four-toed salamander
- spotted salamander

Birds (Summer)
- black-backed woodpecker
- black-throated blue warbler
- gray jay
- merlin
- northern goshawk
- red crossbill
- white-winged crossbill

Birds (Significant Migrants)
- Harris' sparrow
- LeConte's sparrow
- long-billed dowitcher
- long-tailed duck
- peregrine falcon
- piping plover
- red-necked grebe
- red-throated loon
- whimbrel
- willet

Birds (Winter)
- common redpoll
- gray jay
- northern shrike
- pine grosbeak
- red crossbill
- white-winged crossbill

Mammals
- bobcat
- fisher
- lemming mouse
- meadow jumping mouse
- red-backed vole

Blue Hills

Communities: White Pine–Hardwood Forest, Northern Hardwood–Hemlock Forest, River

The Blue Hills is the name given to a large monadnock in eastern Barron, northwest Rusk, and extending into southwestern Sawyer Counties. A monadnock is a ridge of hard rock that resisted the pressure and grinding of the glacier. Today, these Blue Hills extend above the surrounding plains. At an elevation of 1,650 feet, the hills are the predominant landform for miles. Second-growth forest of aspen, birch, red maple, and red oak now tops the former mountain.

A convenient starting point to begin exploring is the city of Rice Lake. Before starting on the tour of the hills, you could take a short side trip to Bandli County Park. This park is 6 1/2 miles southeast of Rice Lake on County M. Look for a small sign pointing the way to the park. The park has a 20-acre remnant of old-growth white pine with some trees that reach 30 inches in diameter. The understory is diverse but has many disturbance species.

To reach the Blue Hills from Bandli Park, zig-zag through the numbered roads to 19 1/2 Avenue. Barron County's town roads are numbered by mile and half-mile, starting with one in the south and west corner of the County. To go directly to the hills, proceed east out of Rice Lake on County C for about seven miles to 19 1/2 Avenue, then south to a parking area. From here you can begin exploring the back country. You will be exploring the Blue Hills on logging roads, poorly marked trails, or by obtaining the appropriate topographic maps and setting out cross-country.

These hills are very important to the Ojibwe. Here they still quarry catlinite, a soft mineral from which they carve pipes for ceremonial use. Pipes made from Blue Hills catlinite turned up occasionally across North America before European settlement, indicating a robust ancient commerce and trade. Scattered quarries can be found throughout the area, but examples of the catlinite can be seen along logging roads. The Ojibwe craft the stone pipes from freshly

Squashberry (Endangered Resource file photograph; courtesy of Wisconsin Department of Natural Resources)

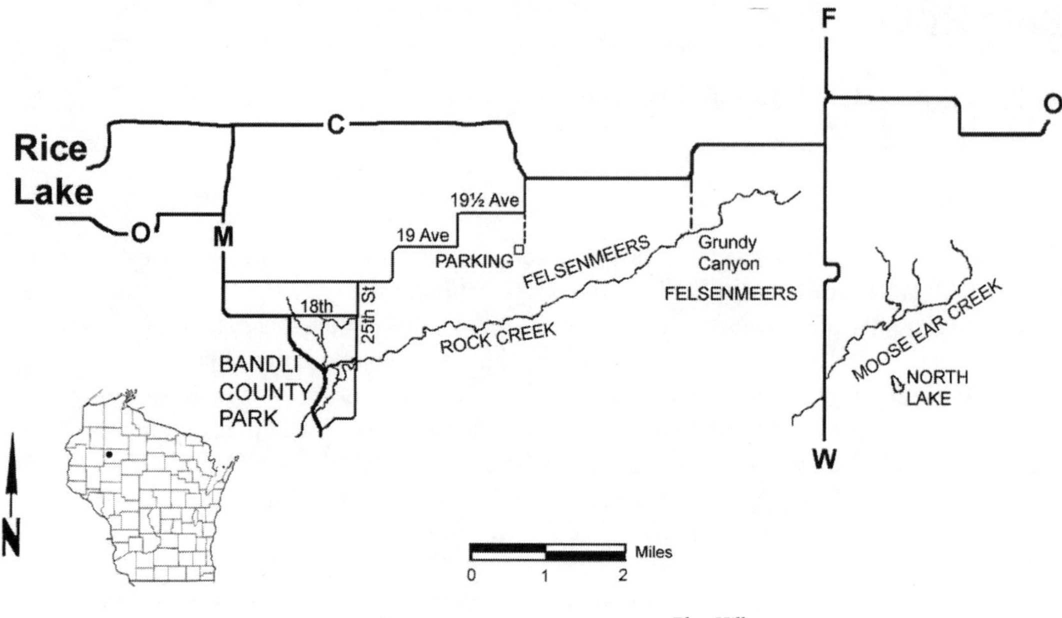

Blue Hills

quarried rock, which is still pliable. Once the exposed rock oxidizes, it becomes too hard to be carved. Removal or quarrying of catlinite is restricted.

Your exploration resumes by returning to County Highway C going east; at the Rusk County line it becomes County Highway O. One mile into Rusk County is Mansfield Road, which can be used to access Grundy Canyon and the extraordinary Rock Creek watershed. Grundy Canyon is a 30-foot deep cut through quartzite that supports a diverse cliff community. By working your way downstream on foot, you will encounter Signe Falls, which consists of a three-step cascade through a 125-foot deep ravine. Rock Creek continues winding southwestward into Barron County. The stream has exceptional clarity and is home to the uncommon red pondweed. The cliffs and talus slopes along the stream give this scenic area a mountainous feel. The canyon concentrates cool air, which makes these slopes ideal for the cold-loving hawthorn-leaved gooseberry and squashberry.

One mile to the south of Grundy Canyon is the Blue Hills felsenmeer area. A felsenmeer is a German word meaning "sea of rocks." There are several felsenmeers in the Blue Hills. They are similar in appearance to the talus slopes found in mountainous areas, but there are no mountains from which rocks can fall to form talus.

The only way that you can access the felsenmeers is by overland travel, which is precisely why the area has special ap-

Felsenmeer (photograph by author)

Felsenmeers

Felsenmeers developed under the extremes of freezing and thawing. For a long period, the Wisconsin glacier perched right on the crest of the Blue Hills, with tundra and spruce forest stretching to the southern horizon. Severe fluctuations in temperature occurred due to the glacier's position, the south-facing slopes, and cold air drainage. The quartzite cracked along cleavage lines, and over time, a felsenmeer developed.

Today these felsenmeers are the Blue Hills' most obvious glacial feature, although ardent geologists may be able to identify other mountainous glacial features, such as hanging valleys, cirques, and tarns, which are less obvious because they have weathered (and are only mountains in miniature). The Rock Creek and Moose Ear Creek drainages contain excellent examples of these geological features.

peal to many. This is not an area for the wary or timid naturalist. You will get a real taste of exploring and solitude. Caution is the key because felsenmeers have sharp edges, and the rocks are shifting and slippery.

Another exceptional area to explore is the headwaters of Moose Ear Creek. The watershed is entirely within the Rusk County Forest. You can access the headwaters by a long but rewarding trek across country. The most obvious spot to start is where the creek crosses County Highway W. Just walk upstream to an entirely forested watershed. The stream contains the rare red fountain moss and has numerous singing Louisiana waterthrushes near their breeding-range limit in the state. The young forest contains enough mature trees to harbor good populations of the state-threatened cerulean warbler. The creek begins as seeps from the hillside, and the upper reaches provide habitat for the animals described in the River Communities chapter in part 1.

If you want to explore further in the hills, there are less strenuous but interesting places to visit. To reach these areas continue east on County O for an additional eight miles, until County O intersects with the Old Blue Hills Trail, where you can enjoy a more leisurely exploration by continuing north on the trail. Turn at Perch Lake Road and continue one mile to a campground. Perch Lake has several interesting aquatic plants, including several pondweeds and arrowheads.

Returning to the Old Blue Hills Trail, continue north to the next road, which leads to Three Lakes bog. There is easy access to the bog, with hiking trails leading to the tamaracks and spruce. Another exciting area is the Big Weigor Creek headwaters that can be reached going north again on Old Blue Hills Trail about one-quarter mile until it crosses the creek, or hiking the Four Hills Trail directly opposite the road to Three Lakes bog.

Naturalists usually bypass the Blue Hills, yet some of our next big discoveries might come from such little-used areas.

Specialties

Plants

- assiniboine sedge
- Canadian gooseberry
- red pondweed
- showy lady's slipper
- squashberry
- Wolf's bluegrass

Animals

- black-backed woodpecker
- blackburnian warbler
- cerulean warbler
- Louisiana waterthrush
- red-shouldered hawk

Other Features

- felsenmeers
- monadnock
- mountain glacial features
- perched bogs
- pipestone

Uncommon Species

Plants

- arrow-leaved violet
- beaked sedge
- bracted orchid
- Carolina cranesbill
- drooping wood sedge
- dwarf ginseng
- dwarf serviceberry
- fringed sedge
- graceful sedge
- Jacob's ladder
- large yellow lady's slipper
- mosses–lichens on felsenmeers
- narrow beech fern
- necklace sedge
- rock elm
- Rocky Mountain sedge
- short-fruited rock cress
- showy orchis
- silvery glade fern
- tall agrimony
- white-edged sedge

Insects

- Acadian hairstreak
- several caddisfly species
- striped hairstreak

Reptiles and Amphibians

- pickerel frog

Birds (Breeding)

- golden-winged warbler
- hooded merganser
- mourning warbler
- olive-sided flycatcher

Mammals

- black bear
- fisher
- gray fox
- long-tailed weasel
- meadow jumping mouse
- northern flying squirrel

Brule River

Communities: Northern Hardwood–Hemlock Forest, Forested Swamps, Great Lakes Shoreland Forest, Alder Thicket, Shallow Marsh, River

Throughout its history, the Brule River has been an outlet of the glaciers, a highway for Native Americans, a thoroughfare for the French, a trout stream of presidents, and the best canoeing stream in the state. What more can be said of this river of history, river of memories? Perhaps the best thing that can be said is that this is a naturalist's paradise. The essence of everything that is northern Wisconsin can be found in some form in its waters or along its banks.

The Brule, officially the Bois Brule River, starts sluggishly as a stream near Solon Springs. In the upper reaches, the stream is a slow-moving, muck-bottomed body meandering among alder thickets and conifer swamps with magnificent bird communities. Along County P, at the end of Stone Chimney Road and at Stones Bridges, a bird watcher can see the best northern Wisconsin has to offer. The banks are slowly tapering and vegetated with northern forest bogs, alder thickets, and cedar swamp communities.

As the river continues, the banks become steeper and the flow quickens. The lower reaches of the river are a series of rapids, falls, and pools, finally slowing near the mouth as it enters Lake Superior. There is a small sand spit at the mouth with a small estuary behind the bar. The flora also changes as the river does, and many microhabitats occur on the steep banks. As you approach the mouth, note the regenerating boreal forest of white spruce and white pine.

The best way to experience the Brule River is to choose a section and explore. This exploring should not be limited to summer. Nearly every year one of three northern owl species shows up somewhere along the river forest during winter. No matter what the season, no matter what the section, you will not be disappointed.

A totally different landscape is found to the south of Stone's Bridge. This land was once pine barrens but today's landscape is much different due to fire suppression and pulp production. Stands of jack pine or areas recently harvested and subsequently replanted red pine stretch for miles. The canopy of trees looks much different than the pine barrens of old. The groundlayer

Bois Brule

The glacial spillway of the Bois Brule River represents the rich diversity that can be found all along the river. Melting water from the receding glacier spilled over a land dam and flowed south. Over 10 miles in length, the glacial spillway offers habitat for most of northern Wisconsin's forest birds. Not surprisingly, the plant community is equally diverse, with numerous rare species, including calypso orchid and Lapland buttercup, a boreal species previously unknown to Wisconsin until its discovery in the early 1990s. Glacier-deposited sand surrounds the spillway. These areas absorb water that percolates down to the point of resistance and then flows sideways. This flow emanates throughout the spillway as seeps and springs.

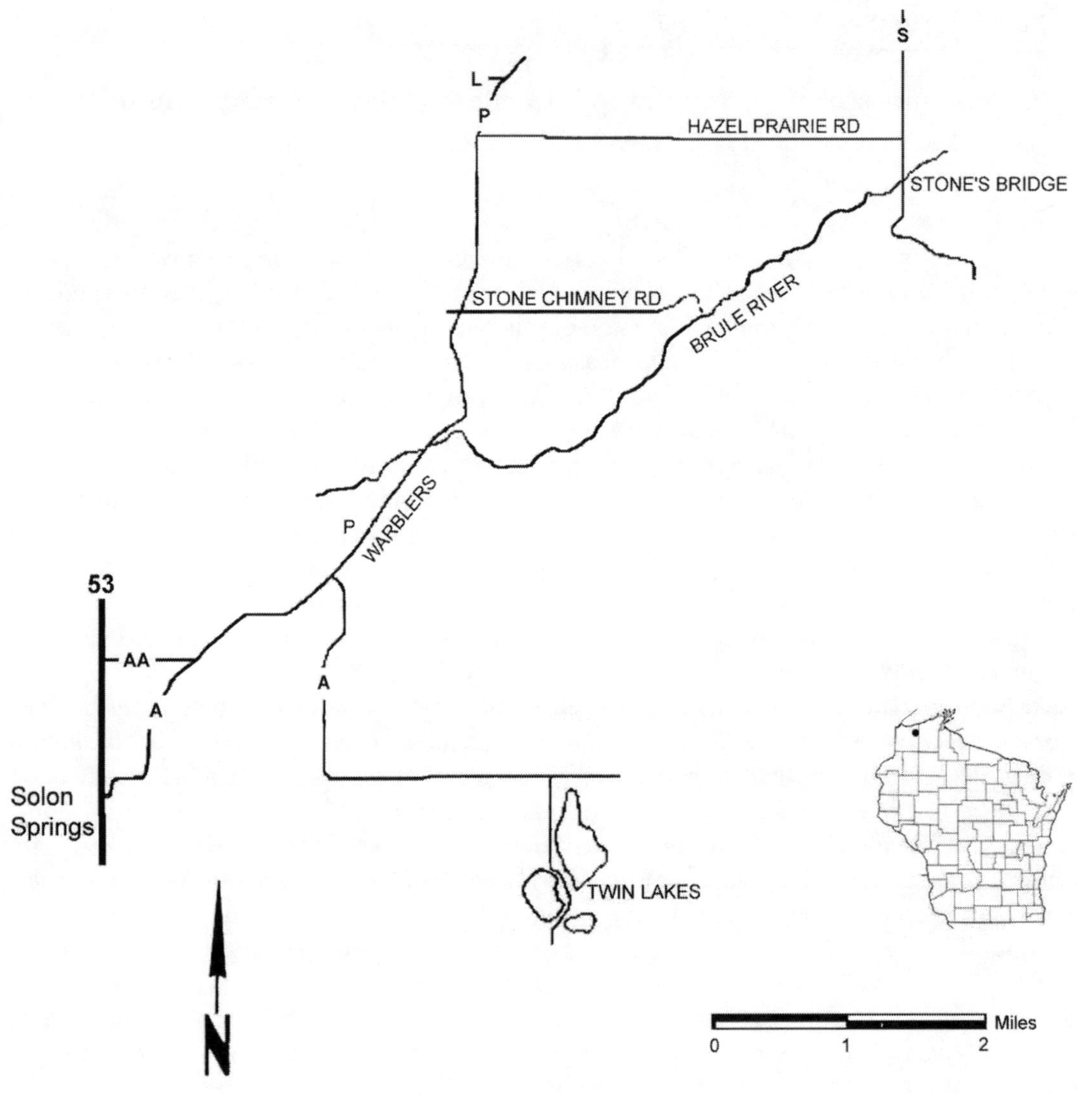

Brule River

under the influence of dense red pines has few species. There are, however, areas still managed for jack pine that contain a diverse component of groundlayer species.

While the flora is exciting to view, the real star of the area is the Connecticut warbler. As many as 20 pairs use the area around Highways S and A. Not one field guide lists shrubby jack pine forest as Connecticut warbler habitat, but such is the case in Wisconsin—a vivid reminder of why we need better field observers.

Another unusual habitat is found a little farther east on Highway A and south on Twin Lakes Road. The Twin Lakes have excellent examples of the interior beach community.

The fluctuating water levels and ice pushing up the bottom materials create a sparsely vegetated beach area. These beaches have plant and animal species much different from those found on the Great Lakes. Species must be able to tolerate years of inundation until the water levels drop to expose the interior beach. Two unimproved boat or lake accesses on the south ends of the lakes are great places to study these interior beach communities.

Specialties

Plants

calypso orchid
Canadian black currant
fir clubmoss
Lapland buttercup
large water starwort
least moonwort
purple virgin's bower
sheathed sedge
showy lady's slipper
white bog orchid

Animals

black-backed woodpecker
Connecticut warbler
fisher
great gray owl
hawk owl
merlin
pine marten
wood turtle

Uncommon Species

Plants

Canada yew
fen willow-herb
golden saxifrage
great St. John's wort
heartleaf twayblade
marsh cinquefoil
marsh St. John's wort
northern cranesbill
one-flowered pyrola
purple avens
purple fringed orchid
purple-fringed Riccia
small green fringed orchid
striped coralroot
tall white violet
water moss
white bog orchid
white water crowfoot
wolfberry

Insects

arctic skipper
laurentian skipper

Reptiles and Amphibians

spotted salamander

Birds (Summer)

Cape May warbler
evening grosbeak
golden-winged warbler
Lincoln's sparrow
long-eared owl
olive-sided flycatcher
red crossbill
white-winged crossbill

Birds (Winter)

hoary redpoll
northern goshawk
pine grosbeak
white-winged crossbill

Mammals

black bear
bobcat
northern flying squirrel
otter
southern saddle-back shrew
star-nose mole
water shrew

Copper Falls State Park–Penokee Range

Communities: White Pine–Hardwood Forest, Northern Hardwood–Hemlock Forest, River, Cliffs

The Penokee or Gogebie range in Wisconsin is 53 miles long and up to a mile wide, running from southwest of Mellen, northeastward to just south of Hurley. The range rises from 100 to 600 feet above the surrounding countryside. The Penokee range is a monadnock made of metamorphic rocks, which the glacier did not grind into soil, with some areas that contain iron deposits. Nine water gaps puncture the range. During the past 10,000 years, erosion created these gaps. Streams cutting through the rocks have formed narrow gorges, many with waterfalls and rapids. Running parallel to the range a few miles to the north is an area of ridges known as the Keweenaw Trap. Between are lowlands forming a broad valley of broken slate.

The range has an interesting history of mineral exploration, including what was once the deepest active iron mine in the world. The geology is also very interesting. If you explore tailings areas, you can find many small specimens of the area's distinctive minerals. Most of these areas are in the eastern portion of the range.

The better-known areas of the range are near Mellen, and the best-known spot is Copper Falls State Park. A second-growth sugar maple forest dominates the park, but there are areas of large pine along the Bad River. The herbaceous layer is only moderately rich. The real highlights are the waterfall gorges and the associated natural communities. There is a 100-foot-deep gorge just downstream from the falls with an exciting fern and moss flora. Unfortunately, the area can only be explored by expert use of mountaineering equipment. To see a small sample of these cliff communities, inspect the Devil's Gate. Conglomerate with a narrow band

Iron County Mountains

Iron County Forest is managed mostly for timber products, recreation, and game species, but the enlightened residents of the county established a thousand-acre block as a forest preserve. This block lies just north and west of Iron Belt, between Hoyt Fire Lane and Weber Lake. Information and maps of the Iron County Forest are available at the Iron County Visitors Center, county offices in Hurley, or at information displays found at most motels in the vicinity.

The land is very rugged, with features that are almost mountain-like. The land is also completely forested, which provides habitat for those species that require larger blocks of forest. Old logging roads and trails still traverse the block, offering entry to many users. The forest has a rich assortment of plants, including a multitude of sedge species. Even more botanically exciting are the balds. These areas of bedrock at the surface have species normally found much farther south on the south-facing slopes or much farther north on the north-facing slopes, such as white mandarin and pearlwort.

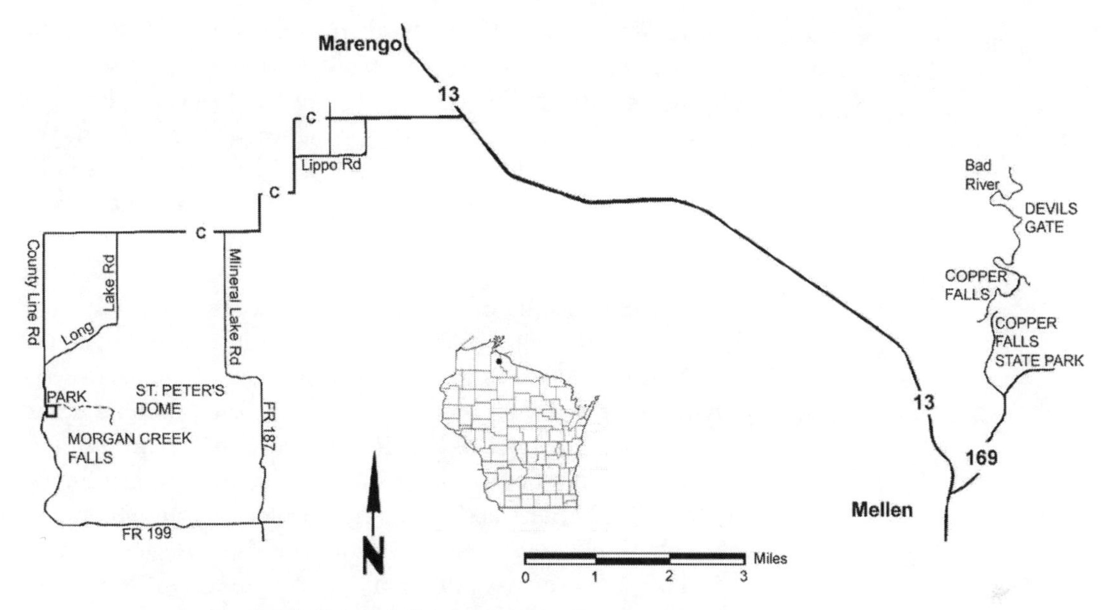

Copper Falls State Park–Penokee Range

Canyon area at Copper Falls (illustration by Jim McEvoy; courtesy of Wisconsin Department of Natural Resources)

of shale is exposed here. There are many miles of trails in the park, some leading to remote backpacking campsites.

Morgan Creek Falls is 12 miles west of Mellen and 2 miles north. Here there is a 30-foot drop in Morgan Creek. Along the small cliffs near the falls and gorge area the rare Braun's holly fern and many other fern species can be found. Morgan Creek Falls and surrounding lands are part of the Chequamegon National Forest.

If you continue on the trail after you reach the falls, you will eventually wind your way to the granite summit of St. Peter's Dome. This is a landmark easily seen 15 miles away in Ashland. It is always interesting to study the lichen flora and the associated rock outcrop flora growing on the summit. The spreading woodfern grows just to the southwest of Morgan Creek Falls. You'll also see many warblers if you explore the big block of woods to the south of the falls.

For something very different, visit Augustine Lake between August and October. Here you'll see a creature that most amateur naturalists have never seen. The stranger is a ½-inch-diameter freshwater jellyfish. Apparently, there is a robust population here; Augustine Lake is one of only a handful of lakes in the state that consistently produce the medusa stage. Augustine Lake is eight miles east of Glidden on County N (refer to a state highway map or gazetteer to locate the boat access).

Specialties

Plants

assiniboine sedge
Braun's holly fern
fragrant fern
great toothwort
large-leaved sandwort
maidenhair spleenwort
male fern
northern oak fern
pale sedge
Smith's melic grass
spreading woodfern
western cliff-fern

Animals

freshwater jellyfish
lynx
Tremblay's salamander

Other Features

goethite
jaspilite
kidney ore
needle ore
specular hematite
water gaps

Uncommon Species

Plants

broad-leaved toothwort
bronze sedge
Canada yew
crested wood-fern
dwarf ginseng
early coralroot
five-parted toothwort
fringed polygala
fringed sedge
goblin fern
golden saxifrage
kidney-leaved violet
least moonwort
leather-leaved grape fern
long-beaked sedge
narrow beech fern
New England violet
northern starwort
northern St. John's wort
northern sweet coltsfoot
pearlwort
purple virgin's bower
round-leaved orchid
rusty cliff-fern
stitchwort

striped coralroot
swollen sedge
white mandarin
witch-hazel

Plants among Granite Outcrops

bearberry
cliff saxifrage
cow-wheat
early goldenrod
fringed polygala
gray goldenrod
hairgrass
Hooker's orchid
huckleberry
maidenhair spleenwort
orange grass
pink corydalis
prairie violet
rock moss
showy goldenrod
slender ladies' tresses
three-toothed cinquefoil

Insects

arctic skipper
laurentian skipper
northern wingless grasshopper

Reptiles and Amphibians

spotted salamander
wood turtle

Birds (Breeding)

evening grosbeak
golden-winged warbler
gray jay

Mammals

black bear
bobcat
fisher
hoary bat
meadow jumping mouse
northern flying squirrel
timber wolf

Crex Meadows

Communities: Northern Sedge Meadow, Oak Barrens, Pine Barrens, Shallow Marsh

Crex Meadows, a state-owned site, contains just over 30,000 acres of brush prairie and marsh. Most of the area is open to the public with the exception of a 2,300-acre refuge in the center. This center refuge can still be easily viewed from the surrounding roads, so in essence, all 30,000 acres are available to the public. The refuge contains food patches for geese, cranes, and other waterfowl.

The underlying sands were part of glacial Lake Grantsburg. Through earlier drainage schemes, attempts to farm the area failed because the sands were acidic and unproductive. Large areas were converted to wiregrass sedge that became the material for making indoor carpets. The grass carpet business boomed until synthetic materials became available. Later, there were more attempts to farm the land, but these also failed.

Today, Crex Meadows is a protected site. Its 11,000 acres of marsh are supported by 18 miles of dikes. A management experiment in the 1950s converted the oak forests into brush prairie by cutting and prescribed burning. Even after many years of burning, thousands of oak grubs

Nelson's sharp-tailed sparrow (illustration by Jim McEvoy; courtesy of Wisconsin Department of Natural Resources)

Reed Lake Meadow

The wiregrass sedge that was so desired by carpet makers has developed into sedge meadows that harbor several rare bird species, including the yellow rail and Nelson's sharp-tailed sparrow. Any serious birder in the state has visited the pump-house location in hopes of hearing or seeing these rare birds. The pump house is on Main Dike Road at the south end of the Reed Lake meadow that stretches more than three miles from north to south.

This is the most reliable place in the state to observe the aforementioned birds and the LeConte's sparrow. Here they can all be heard from the same location. Other natural wonders, like the wing flaps of a flushed short-eared owl, the courtship behavior of sharp-tailed grouse, or the subtleties of Karner blue butterflies, can be enjoyed near the pump house.

The bottom line, for those so inclined, is that bird-watching activity brings in large amounts of money to the Grantsburg area. The specialized sparrow and yellow rail birding is just the tip of the iceberg. Other birds, such as ducks, geese, eagles, and trumpeter swans, bring in the bulk of observers and money in excess of $600,000 per year. It is important to remember that birding is good business. Protection of wonderful habitats like Crex Meadows can provide substantial financial benefit to area residents.

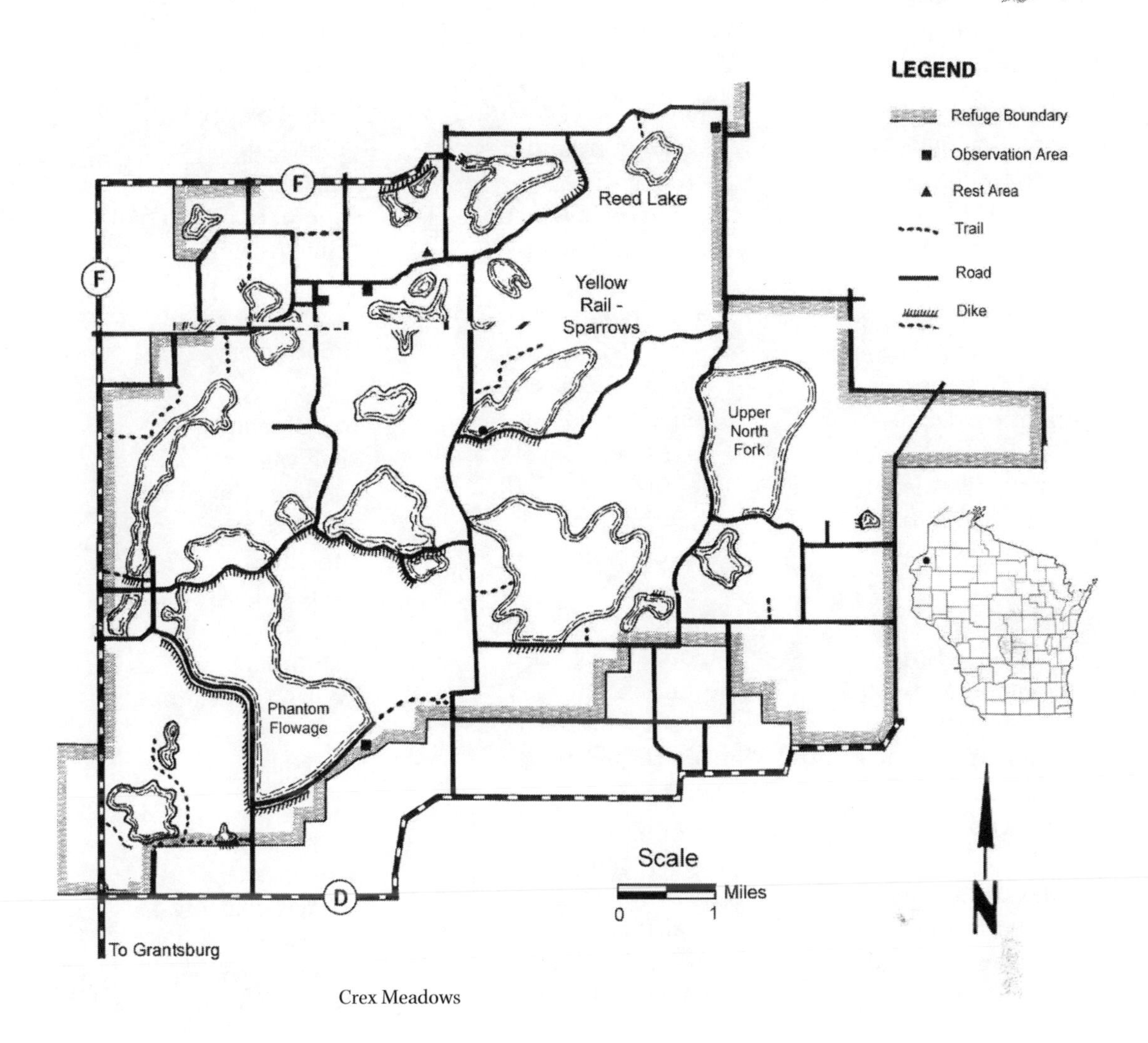

Crex Meadows

persist. When the original clearing took place in the 1950s, there were enough prairie plants persisting in the area to stock the openings created. There are well over 200 true prairie plants at Crex. The opening of the forest and its perpetuation as brush prairie has provided excellent habitat for sharp-tailed grouse and other bird species. The combined balance of large acres of marsh and prairie is a magnet for birds, with as many as 245 species of birds recorded.

If there is only one wetland area that you can visit, I would recommend Crex. There is no other wetland where you can see so many species of plants and animals at close range. The dike system allows access to the entire area. Though use of the area is sometimes heavy, there is no apparent impact on wildlife viewing. Nature study groups, individuals, hunting parties, or locals just out for a drive view the wildlife together. No matter what the season, anyone visiting the area is in for a treat.

Specialties

Plants

adder's tongue fern
cross milkwort
dotted blazing star
downy prairie clover
Farwell's water milfoil
hairy beard-tongue
hairy umbrellawort
oval milkweed
prairie larkspur
sand violet
small yellow lady's slipper
Torrey's threesquare

Animals

bald eagle
Blanding's turtle
common sooty wing (butterfly)
Franklin's ground squirrel
golden eagle
Karner blue butterfly
LeConte's sparrow
Mexican sulphur (butterfly)
mottled dusky wing (butterfly)
Nelson's sharp-tailed sparrow
northern prairie skink
osprey
prairie ringlet (butterfly)
sharp-tailed grouse
trumpeter swan
yellow rail

Uncommon Species

Plants

arrow-leaved aster
bird's-foot violet
blue giant hyssop
blunt-leaved orchid
bog goldenrod
bracted orchid
browned sedge
butterfly weed
clammy ground cherry
crooked aster
downy willow-herb
drooping woodreed
false melic grass
gray goldenrod
great St. John's wort
Houghton's sedge
lance-leaved violet
long-leaved bluets
lupine
New Jersey tea
northern heart-leaved aster
nodding ladies' tresses
porcupine grass
prairie alum-root
prairie blue-eyed grass
prairie brome
prairie larkspur
prairie onion
prairie phlox
prairie smoke
prairie sunflower
prairie violet
purple milkwort
ragged fringed orchid
red-stemmed gentian
rice cutgrass
short green milkweed
showy goldenrod
silvery glade fern
sky blue aster
stiff goldenrod
upland wild timothy
venus looking glass
western sunflower
white prairie clover

Mushrooms

pixie cup

Insects

bog copper
brown elfin
cobweb skipper
dion skipper
dusted skipper
gorgone checkerspot
Henry's elfin
purplish copper
seaside grasshopper
spot-winged grasshopper

Birds (Breeding)

gadwall
golden-winged warbler
lesser scaup
northern pintail
redhead
ring-necked duck
whip-poor-will
Wilson's phalarope

Birds (Migrants)

American golden plover
eared grebe
Forster's tern
Harris' sparrow
Hudsonian godwit
marbled godwit
merlin
red-necked grebe
western grebe
white-fronted goose
willet

Birds (Winter)

common redpoll
golden eagle
hoary redpoll
northern shrike
short-eared owl
snow bunting

Mammals

badger
bobcat
otter
water shrew

Douglas County Sharptail Barrens

Communities: Dry Pine Forest, Pine Barrens

The Douglas County Sharptail Barrens, consisting of 3,800 acres, is located a few miles to the north of Gordon, just west of Highway 51 on County M. A jack pine savanna managed for sharp-tailed grouse is the main feature of the barrens. The topography is slightly rolling with acid sands that are remnants of the Wisconsin glacier. You can get an idea of what the original pine barrens landscape looked like here. Descriptions from the original land surveyors indicate that the distance between large trees back then was similar to today. These pine barrens originally covered more than 2,000,000 acres in Wisconsin and were shaped by regular fires. Without fire, these barrens slowly transform to jack pine–red pine forest. Management of the barrens now includes regular burns, and the barrens composition is retained.

You will see the full suite of barrens species expected for this latitude on this publicly owned site. Open grassland birds abound, and even some far northern species such as the Tennessee warbler have been recorded in the summer. You should check the depressions because they often have species not found on the rest of the barrens, such as the adder's tongue fern, which is occasionally found in the deepest of the depressions.

An example of virgin red pine forest can be viewed at Lucius Woods County Park. Located within the town of Solon Springs, this 15-acre park contains virgin red and white pines. The towering pines are awe-inspiring, giving the observer the feel of the northern Wisconsin pinery. Several northern warbler species nest in the park.

To the southwest of Solon Springs between Gordon and Wascott is a vast area that burned in 1977. The huge wildfire was devastating, with many acres of forest products consumed. The apparent destruction actually led to regeneration of jack pines and an unexpected event: the appearance of the federally endangered Kirtland's warbler.

West of Wascott on County Highway T (refer to a road map), the effects of a large fire can

Sharp-Tailed Grouse

Sharp-tailed grouse are birds of the northern barrens and plains. They seem to do best in areas that are open but have some brush and patches of woods. This bird formerly lived in most of the prairie, savanna, and barrens areas of the state. Huge populations provided winter food for early settlers.

More recently this grouse species has declined precipitously in the state. Only 10 populations are now considered healthy, and the smaller scattered groups have little chance to persist. The decline is directly attributable to changes in land use. Prior to European settlement, these birds lived on open prairie and savannas. After settlement, the prairies and savannas were plowed, and the northern forest was mostly cut. This cut-over land provided excellent habitat for the sharp-tailed grouse for decades. Now the northern cut-over lands are becoming forest again, and the grouse are found only in large managed open areas.

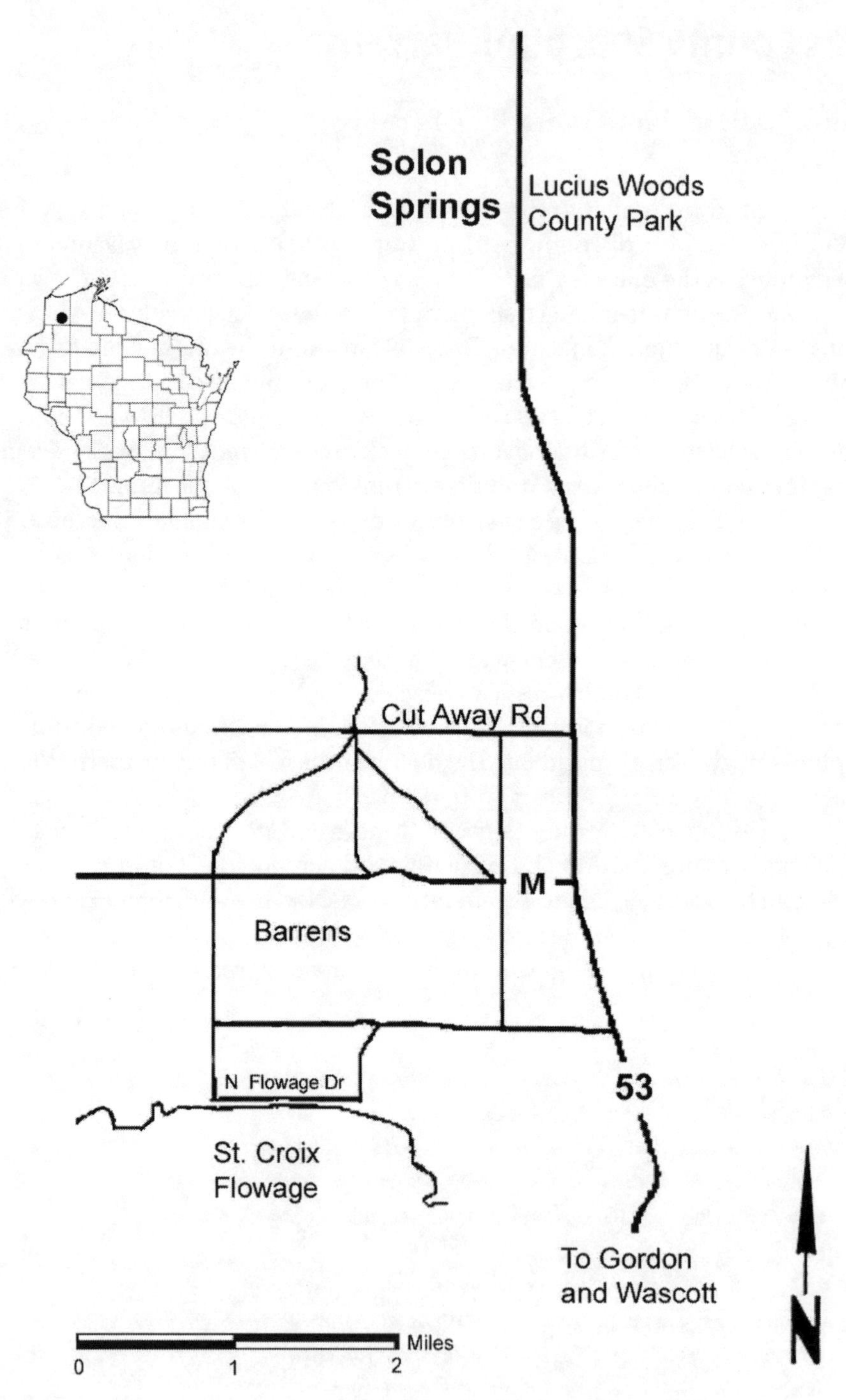

Douglas County Sharptail Barrens

be seen. A few areas show patches of older trees that escaped burning because of the nuances of wildfire. Many areas have been planted in red pine, but areas of dense jack pine seeded in naturally after the fire. The young jack pines (8 to 15 years after the fire) provided habitat for the Kirtland's warbler, but now they have grown beyond the preferred height for that species.

Specialties

Plants

adder's tongue fern
arrowhead coltsfoot
Lapland buttercup
Missouri rock cress
mountain cranberry
Torrey's threesquare

Animals

buck moth
Connecticut warbler
northern prairie skink
sharp-tailed grouse
upland sandpiper
western tailed blue (butterfly)

Uncommon Species

Plants

bearberry
blue giant hyssop
broom sedge
cow-wheat
Crawford's sedge
dragonhead
green adder's mouth
Hooker's orchid
hook-spurred violet
Houghton's sedge
June grass
large-leaved avens
scarlet paintbrush
spotted coralroot
swamp dewberry
upland wild timothy
wild comfrey

Mushrooms

burn site shield cup

Insects

gorgone checkerspot
prairie ringlet

Birds (Breeding)

evening grosbeak
field sparrow
golden-winged warbler
long-eared owl
mourning warbler
red crossbill
sandhill crane

Birds (Winter)

hoary redpoll
pine grosbeak
white-winged crossbill

Mammals

badger
black bear
bog lemming
ermine
fisher
long-tailed weasel
pocket gopher
porcupine
woodland deer mouse

Interstate State Park

Communities: Southern Red Oak–Mixed Forest, Sugar Maple–Basswood Forest, Floodplain Forest, River, Cliffs

For anyone with an interest in Wisconsin's glacial history, a visit to Interstate State Park is essential. The slow process of geology is brought to life in this park. Bedrock geology along with transport of glacial material determined what the park is today. The bedrock is basalt that forms when lava hardens. Basalt is a very hard rock, but there are cracks in the apparently solid rock, permitting more erosion than expected. The extraordinary removal of the basalt came by way of silt-laden glacial waters of Lake Duluth, which flowed to the sea south through the park. Because there was a heavy silt load, the abrasive nature of these waters helped erode the rock at an accelerated pace. These waters carried not only silt but also boulders. Powerful whirlpools captured these boulders in a permanent spin cycle. There was so much force generated that they bored holes into the lava, forming the famous potholes, which are large smooth-walled cylinders ground into the basalt.

While justly known for its geology and spectacular 200-foot cliffs, Interstate State Park's 1,325 acres are a transition area, a crossroads for many southern species that reach their northern limit here and for many northern species that reach their southern limit here.

Silk moths are especially prominent in the area. With the right equipment, black lights, and capture buckets, you can see all five of the large colorful moths listed in the uncommon species list at night in June and early July.

The conditions for sustaining life on these rocks are very poor. There's virtually no soil, and where there is soil, it's very acidic. Over time an unusual plant community called an acid bedrock glade developed. This community contains some prairie plants, cliff plants, and woodland species, as well as several distinctive plants. Most of these are pioneering lichen species, although some, like pink corydalis, are more abundant here than anywhere else in the state. A good bedrock glade community can be observed along the Potholes Trail.

Pothole (photograph by Julie Fox Martin; courtesy of Wisconsin Department of Natural Resources)

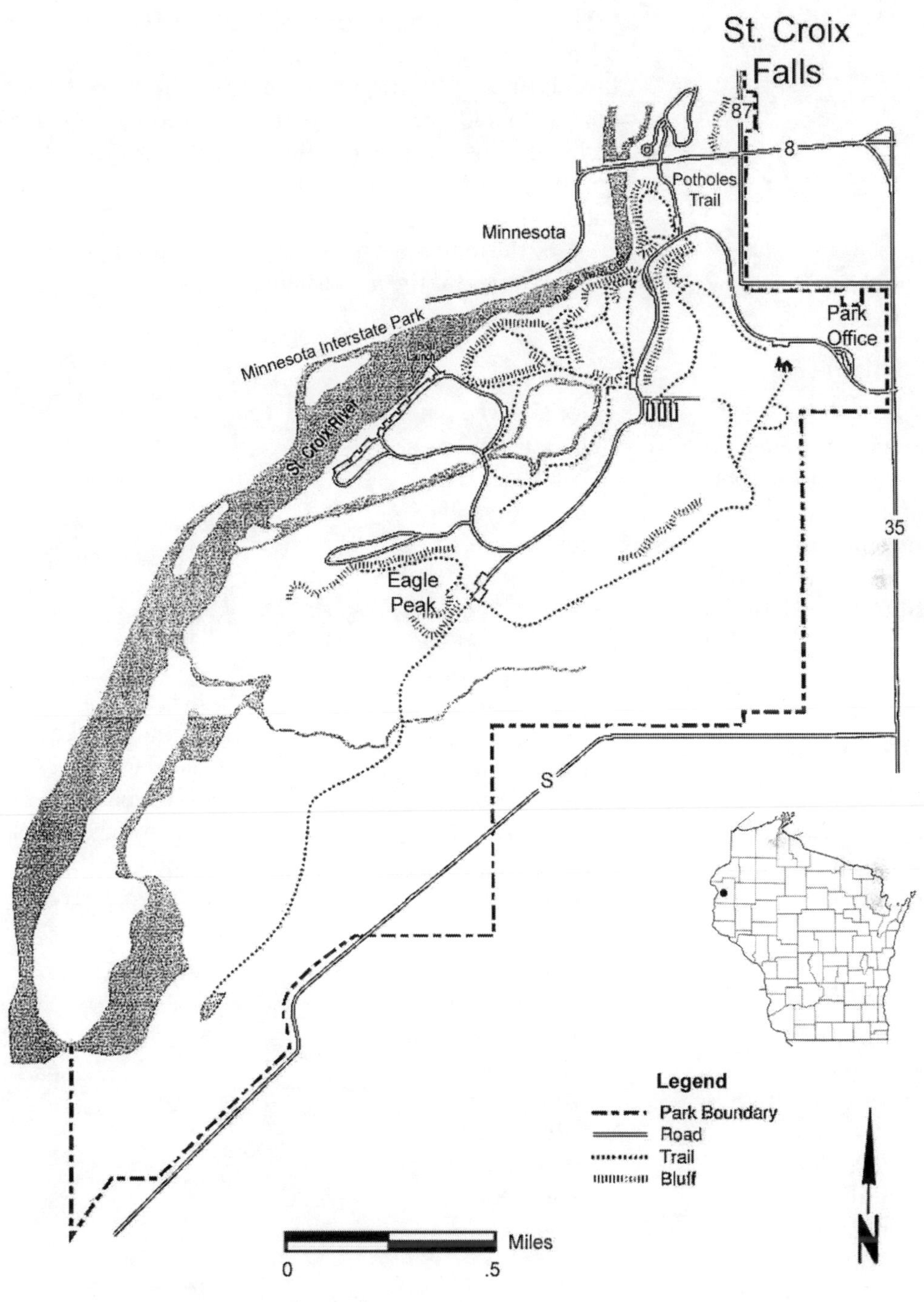

Interstate State Park

The deep waters coursing through The Dalles harbor some of the rarest species on earth. This section of the St. Croix River is the only location in the state for the winged mapleleaf mussel. In addition to this species many other rare mussels, including the federally endangered Higgins' eye mussel, are found here. You need special knowledge of freshwater mussel habitat and diving gear to study these animals. You can, however, see the incredible freshwater mussel diversity by searching the shores for the multishaped shells.

Specialties

Plants
assiniboine sedge
downy prairie clover
fame flower
fragrant fern
veined meadow-rue
western cliff-fern

Animals
five-lined skink
freshwater mussels
northern prairie skink

Other Features
potholes

Uncommon Species

Plants
bearberry
birds-eye primrose
bristly sarsaparilla
broom sedge
butternut
Carolina spring beauty
dwarf ginseng
dwarf serviceberry
false dragonhead
golden alexanders
hairy grama
hairy lip-fern
indigo bush
June grass
long-leaved bluets
New Jersey tea
oak fern
pink corydalis
pipsissewa
prairie alum-root
prairie blue-eyed grass
prairie dropseed
purple prairie clover
purple fringed orchid
rice cutgrass
round-leaved pyrola
rusty cliff-fern
short green milkweed
showy tick-trefoil
silky wild rye
slender cliff-brake
slender false foxglove
stiff sandwort
swamp saxifrage
upland wild timothy
white aster
witch-hazel
yellow pimpernel
yellow trout lily

Insects
cecropia moth
Columbian moth
luna moth
polyphemus moth
promethia moth
silvery blue

Fish
gilt darter

Reptiles and Amphibians
four-toed salamander
spotted salamander
wood turtle

Birds (Breeding)
blue-gray gnatcatcher
cerulean warbler
eastern bluebird
golden-winged warbler
mourning warbler
red-shouldered hawk

Birds (Migrants)
Harris' sparrow
merlin
white-crowned sparrow

Mammals
badger
bog lemming
gray fox
long-tailed weasel
silver-haired bat
southern flying squirrel
star-nose mole

Moose Junction

Communities: White Pine–Hardwood Forest, Northern Hardwood–Hemlock Forest, Northern Forested Bogs and Fens, Rivers, Cliffs

This area in west central Douglas County is nearly devoid of people, and that's the way the residents like it. Some animals like it that way, too, because they do best where human population is sparse. Featured in the site are large blocks of uninhabited land needed by big animals. Between Chaffey and Moose Junction is a huge area of mostly second-growth forest and immense bogs. This area covers more than 100 square miles and is nearly roadless, with only peripheral roads with dead-end spurs and Highway 35, which bisects the area. Second-growth areas don't usually lend themselves well to a complete flora because the full range of forest age classes is not present, especially the oldest ages of forest. The fauna, however, seems to depend more on habitat size than age. Highly mobile large mammals, such as wolves and moose, and birds requiring large tracts of land thrive in this sparsely populated landscape that has few disturbances and ample food.

Finding these large mobile animals is another story. The size of the area is the best protector of these animals. Only extremely experienced wilderness explorers should attempt any deep penetration into the interior. For those with less backwoods knowledge, you can still get close to the action by traveling the roads on the periphery, and it's possible to glimpse a moose or timber wolf from the road.

One possible route around the edge would be to take County Highway M east from Moose Junction 2.1 miles to Empire Swamp Road. Turn north and continue on this road to the intersection with County Highway A, then go west on A to Summit Tower Road (this intersection is where A turns 90 degrees toward the north). Stay on Summit Tower Road to Highway 35. This route penetrates deep into wolf territory. The land near the roads has an abundance of black ash and the species associated with black ash swamps, such as northern waterthrush.

At this intersection, you can turn north toward Chaffey, or if you are truly adventurous, walk straight ahead into the immense Belden Swamp. This swamp, a hybrid community of

Moose Junction (illustration by Jim McEvoy; courtesy of Wisconsin Department of Natural Resources)

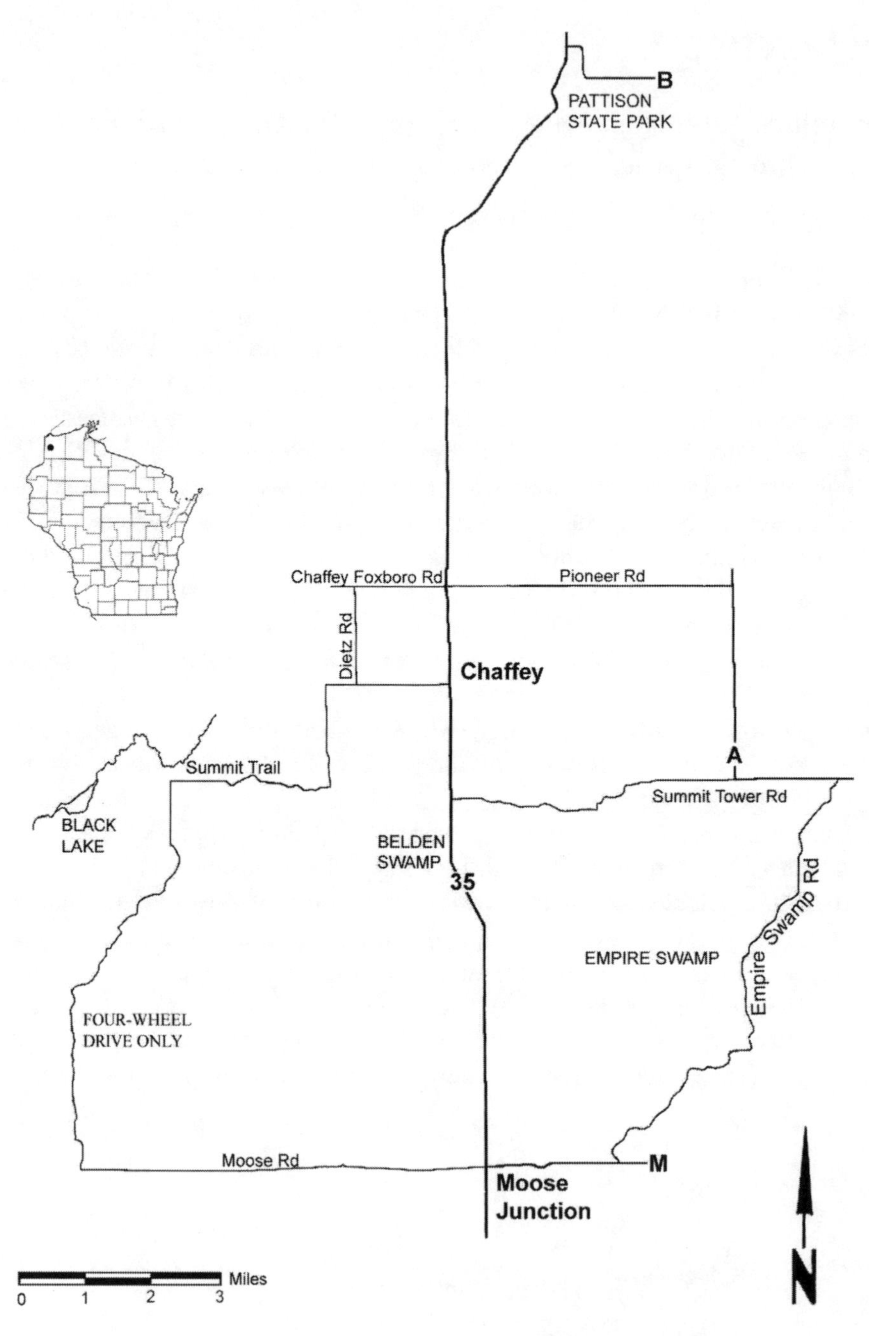

Interstate State Park

northern forested bog, open bog, and poor fen, is a designated state natural area. All the northern forested bog and fen birds can be found, including the great gray owl. In the poor fen, great numbers of LeConte's sparrow nest.

At Chaffey, go west on Summit Trail. This road meanders near the Minnesota border. After approximately six miles, you

Timber wolf (Endangered Resource file photograph; courtesy of Wisconsin Department of Natural Resources)

will be near Black Lake. This is a vast undisturbed bog area on the Minnesota border. Also, this is probably the best area in Wisconsin for hearing the timber wolf; sighting one is not easy. To continue the tour, stay on Summit Trail to the junction with Moose Road, then go east to Moose Junction. A warning is in order: All roads in this area, except county and state highways, can become impassable in spring and wet weather. Even in dry conditions, potholes and ruts can make travel hazardous. The Summit Trail is passable with four-wheel-drive vehicles only, and even they might get stuck.

To reach Pattison State Park, continue north on Highway 35. Before going to Pattison, the adventurous botanist may consider a side trip to Foxboro to visit the only Wisconsin home of the floating marsh marigold. Northeast of Foxboro along the railroad tracks the floating marsh marigold grows at its only known location in Wisconsin (the town and tracks are easily found on topographic maps). Differing from its cousin, the common marsh marigold, this plant blooms in midsummer with white to pale purple flowers.

Pattison is a beautiful park with many trails. However, there are no unusual plants or animals living here, except for a good variety of ferns on the cliff faces of the gorges near both Little and Big Manitou Falls.

Specialties

Plants

boreal cotton-grass
floating marsh marigold
fragrant fern
hybrid cliff-fern
mingan moonwort
small shinleaf
tall lungwort
Vasey's rush
western cliff-fern

Animals

black-backed woodpecker
bog copper (butterfly)
chryus arctic (butterfly)
Connecticut warbler
freija fritillary (butterfly)
frigga fritillary (butterfly)
great gray owl
jutta arctic (butterfly)
lynx
moose
red-disked alpine (butterfly)
red-shouldered hawk
three-toed woodpecker
timber wolf
titania fritillary (butterfly)

Uncommon Species

Plants

bog sedge
bronze sedge
brownish sedge
Carolina spring beauty
dragon's mouth
Dutchman's breeches
dwarf ginseng
fibrous-rooted sedge
fringed sedge
golden saxifrage
green adder's mouth
mountain cranberry
northern St. John's wort
purple virgin's bower
small green fringed orchid
wild calla
wiregrass sedge
yellow trout lily

Birds (Breeding)

evening grosbeak
golden-winged warbler
Lincoln's sparrow
long-eared owl
merlin
northern goshawk
palm warbler
saw-whet owl
Swainson's thrush

Birds (Winter)

Bohemian waxwing
hoary redpoll
northern hawk-owl
pine grosbeak
red crossbill
white-winged crossbill

Mammals

black bear
bobcat
fisher
woodland deer mouse

Moquah Barrens–Chequamegon Bay

Communities: Dry Pine Forest, Pine Barrens, Alder Thicket

Moquah Barrens and Chequamegon Bay are vastly different but are included together here because of their proximity to each other. Moquah Barrens, as with other barrens areas, is associated with glacial sands. It was established by the Chequamegon National Forest as a research area in the early 1930s to study the effects of fire suppression on a barrens. Researchers use the barrens as a natural control to gather valuable information to compare to that on managed lands. Stations have been set up to photograph long-term vegetation changes. Much of the barrens area converted to forest, but some openings are very long-lived. The Connecticut warbler is a species that has benefited from the succession to forest. This species is most often found in bogs, but it can also be seen under mature jack pines.

Outside the research area, the Chequamegon National Forest manages about 8,000 acres of pine barrens. The management objective is to reestablish the natural process of fire. At Moquah Barrens there is a conscious attempt to have the tree composition and spacing similar to the original barrens. Large open areas, scattered trees, small groves, and open woodland were all characteristics of pine barrens. This area should get better with time.

Dramatically different natural communities thrive on the shores of Chequamegon Bay. Areas of marsh develop where small streams enter the bay. The rising and falling waters of the bay influence the vegetation. These seiches, which are similar to ocean tides, can fluctuate several inches. Winds, the earth's rotation, and the moon's gravitational pull all influence the magnitude of these tidal features. Exceedingly small on most lakes in Wisconsin, these seiches

Auricled Twaybalde

A very rare orchid species is found sparingly along the shores of Chequamegon Bay. The auricled twayblade is a small (3 to 10 inches high) orchid that is not very showy. This small green plant with yellowish green flowers is very hard to find. Its habitat is the area where alder grows at the spot where water meets the land. In the Chequamegon Bay, the shore is protected from large waves that would scour the shore and prevent alder establishment. When the coast is somewhat protected, the scouring action is lessened. Ice and wind remove some vegetation every year, and this creates small areas of raw vegetation-free sands. It's on these sands under alder where the auricled twayblade grows. A small green plant growing under dense alder lapped by cold water—it's no wonder the plant has not been seen by many.

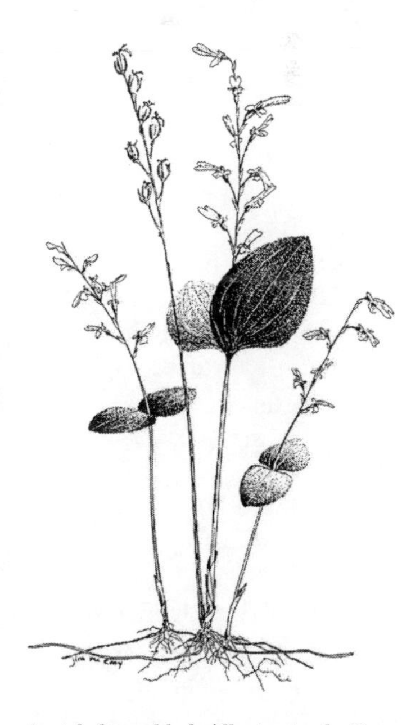

Auricled twayblade (illustration by Jim McEvoy; courtesy of Wisconsin Department of Natural Resources)

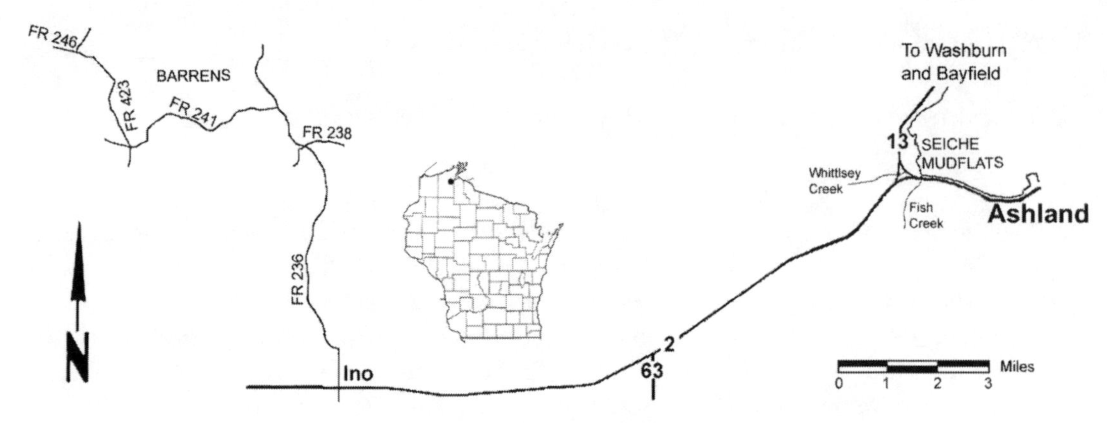

Moquah Barrens–Chequamegon Bay

can be significant on huge bodies such as Lake Superior. These seiches have an influence on the stream mouths, forming miniature estuarine systems.

Fish and Whittlsey Creek enter the bay just west of Ashland and offer the best spot for viewing these tidal ebbs and flows. The influence is such that at low seiche (tide), hundreds of shorebirds can be seen probing the mud flats, whereas at high seiche, not a bird can be found because these flats are flooded.

Farther north between Washburn and Bayfield is another much smaller area of seiche influence that lies between the Sioux and Onion Rivers. Estuaries form at the river mouths with parallel beach bogs between. These estuarine systems along western Chequamegon have very unusual flora, and shore birds use them as refueling stops.

Specialties

Plants

auricled twayblade
dragon's mouth
hairgrass
tall lungwort
ternate grape fern

Animals

Connecticut warbler
migrating shore birds
sharp-tailed grouse

Other Features

estuaries
tides (seiches)

Uncommon Species

Plants

bearberry
blue giant hyssop
clammy cudweed
cow-wheat
early coralroot
forget-me-not
fringed puccoon
green adder's mouth
northern heart-leaved aster
oval milkweed
pipsissewa
rough blazing star
showy goldenrod
sky blue aster
slender ladies' tresses
spotted coralroot
squawroot
tesselated rattlesnake plantain

three-toothed cinquefoil
wild comfrey

Tidal Vegetation

arum-leaved arrowhead
beach grass
bog bedstraw
cinnamon willow-herb
drooping woodreed
fringed sedge
great St. John's wort
heartleaf twayblade
leather-leaved grape fern
mare's tail
marsh cinquefoil
narrow-leaved sundew
northern bugleweed
ragged fringed orchid
Robbin's spike-rush
round-leaved sundew
rusty cotton-grass
small green fringed orchid
turtlehead
yellow water crowfoot

Insects

prairie ringlet

Birds (Breeding)

field sparrow
green-backed heron
Lincoln's sparrow
mute swan†
upland sandpiper

Birds (Migrants)

bald eagle
Caspian tern
common goldeneye
common loon
dunlin
horned grebe
Hudsonian godwit
least sandpiper
long-billed dowitcher
long-tailed duck
marbled godwit
red knot
red-necked grebe
red-throated loon
short-billed dowitcher
solitary sandpiper
surf scoter
western grebe
white-rumped sandpiper
white-winged scoter
willet

Mammals

badger
black bear
ermine
pocket gopher
porcupine
star-nose mole

Northern Bayfield County

Communities: Northern Hardwood–Hemlock Forest, Forested Swamps, Northern Forested Bogs and Fens, Great Lakes Shoreland Forest, Alder Thicket, Shallow Marsh, Lake Dune, Lake Beach, River

Northern Bayfield County is a long but very narrow band extending from Port Wing to Little Sand Bay. This area is greatly influenced by the cooling effects of Lake Superior. Foremost in this area are the bays, with their associated features, such as baymouth bars, sand spits, and estuary-like lakes, and headlands with their cliffs and ledges that are more resistant to the erosion forces of water and wind; all lay within a few miles of the lakeshore. The general cover type here is second-growth boreal forest on the upland sites. On the wetter sites, especially in bays, are bogs, fens, and wet forest. The bars protecting the bays are mostly beach and dune communities, but some have low areas that form interdunal swale bogs with highly unusual flora.

A good way to explore this area is to drive Highway 13, making stops at various bays and headlands. A logical place to start is Port Wing. The Flag River enters Lake Superior here and with it, of course, are the associated sandbars. Behind this barrier is an area of marshy sphagnum bog and sedge bog. Of greater importance is an area of old-growth boreal forest on some higher ground within the bay. Big Pete Road bisects this state-owned area, affording an excellent opportunity for study.

To continue, follow Highway 13 northeast to Herbster, then make a departure from the main highway by taking Bark Point Road to the tip of Bark Point. There are sandstone cliffs

Michaux's sedge (illustration by Jim McEvoy; courtesy of Wisconsin Department of Natural Resources)

Lake Superior Bays

The estuaries and enclosed bays of the Lake Superior shoreline have specialized conditions that allow a distinctive set of species to come together to form a rare natural community. In most northern areas the cool climate does not permit the complete decomposition of plant material. Furthermore, without mineral-water influence, sphagnum mosses thrive and bogs develop.

The situation in these Lake Superior bays is unique in Wisconsin. The cold waters of the huge lake would seemingly be ideal for moss development. However, these waters are also mineral rich, which gives a competitive advantage to a different set of species, such as coast and Michaux's sedges. These species must be able to tolerate cold water and short growing seasons, so many northern (boreal) species thrive in these bays. Many are found nowhere else in the state. Most of the species of special interest are sedges, which is why the natural community is referred to as the Lake Superior sedge fen.

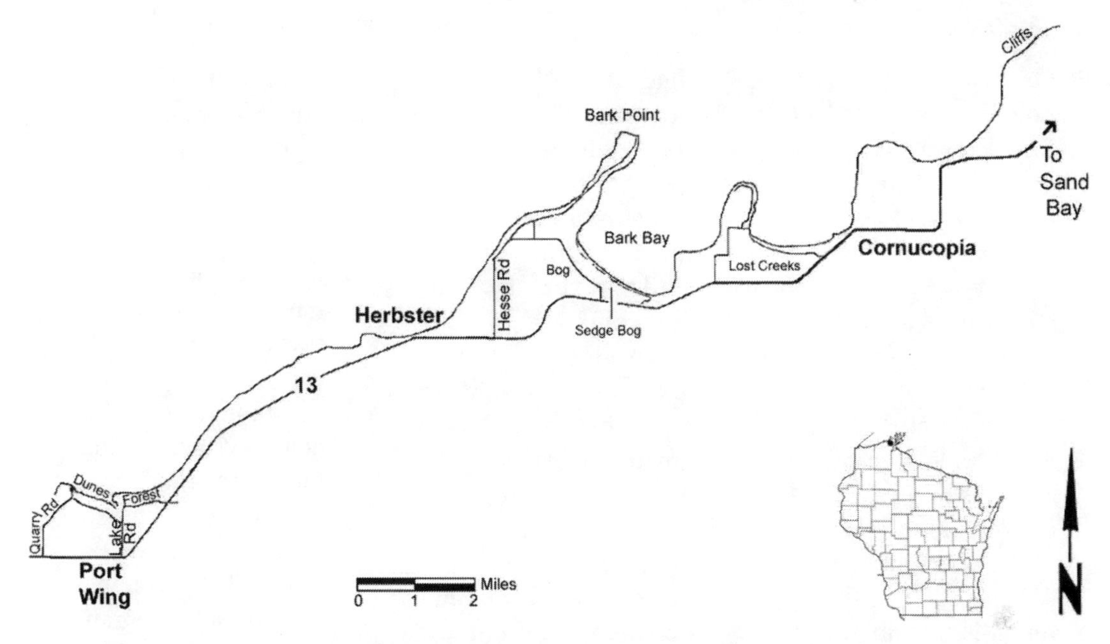

Northern Bayfield County

here along with excellent views from which you can observe water birds. Also along the point are several small ravines that formerly held mingan moonwort, and some might still do so today. It should be noted that along these promontories large groups of raptors sometimes congregate, including at times several peregrine falcons.

Upon returning from the point, take Bark Bay Road to the east. The road is on an old beach ridge that is mostly public land on both sides. On the lake side is another beach farther out that forms nearly a mile of untrammeled beach. Behind this beach is a baymouth lake, sedge fen, and floating bog. This area is exceptional for its size and wildness. The beach can only be reached by boat or canoe, which gives its inhabitants protection from casual disturbance.

Back onto Highway 13, continue on to Cornucopia, where similar habitats exist. West of Cornucopia lies the baymouth bars holding back the Lost Creeks. The wetlands behind this bar are unusual in that they contain large numbers of lake cress, a species until recently thought extirpated from the state. Most of the Lost Creeks area is publicly owned.

Follow the signs north of County Highway K to Sand Bay, where the Apostle Islands National Lakeshore headquarters is located. The baymouth bog at Sand Bay is very similar to other baymouth bogs in the area and harbors many orchids. You should also check with National Park Service personnel about trails to the cliffs west of the headquarters.

Specialties

Plants
broad-leaved twayblade
Chilean sweet cicely
dragon's mouth
fly honeysuckle
hairgrass
lake cress
large-leaved avens
limestone oak fern
livid sedge
marsh horsetail
marsh willow-herb
Michaux's sedge
mingan moonwort
mountain cranberry
Robbin's spike-rush
round-leaved orchid
sparse-flowered sedge
spike trisetum

Animals
speckled rangeland grasshopper

Other Features
baymouth bars
lake cliffs

Uncommon Species

Plants
arrow-leaved violet
bog bedstraw
bristly sedge
cinnamon willow-herb
common bog arrow-grass
daisy-leaved grape fern
drooping woodreed
dwarf mistletoe
dwarf rattlesnake plantain
dwarf scouring-rush
golden saxifrage
grass pink
green adder's mouth
green-flowered shinleaf
green-spurred violet
green twayblade
leather-leaved grape fern
mare's tail
marsh cinquefoil
narrow-leaved sundew
northern cranesbill
one-flowered pyrola
slender ladies' tresses
sooty beak-rush
striped coralroot
tall lungwort
tesselated rattlesnake plantain
western rattlesnake plantain
wolfberry
yellow water crowfoot

Lichens
goat's horn lichen

Insects
arctic skipper
confused hapola
cracker grasshopper
great tiger moth
Huron short-winged grasshopper
laurentian skipper

Birds (Breeding)
bay-breasted warbler
brown creeper
clay-colored sparrow
common tern
golden-crowned kinglet
long-eared owl
mourning warbler
northern goshawk
olive-sided flycatcher
red-breasted merganser
red crossbill
saw-whet owl
Swainson's thrush
white-winged crossbill
yellow-bellied flycatcher

Birds (Migrants)
long-tailed duck
red-throated loon
white-winged scoter

Mammals
bobcat
gray fox
meadow jumping mouse
northern flying squirrel
star-nose mole
woodland jumping mouse

Wisconsin Point

Communities: White Pine–Mixed Forest, Alder Thicket, Shallow Marsh, Lake Dune, Lake Beach

A visit to Wisconsin Point will give you a feel for natural communities that grow on sand spits. You may also experience a truly phenomenal annual event of migrant bird concentrations if your timing is right. To reach Wisconsin Point, take Moccasin Mike Road at the south edge of the city of Superior. The road is well marked and only goes east. The first few miles you will see scrubby aspen growth, but then, as the road turns north, you will drive through an extensive wetland at the head of Allouez Bay. Beyond the wetland, which is quite diverse in plant species such as arrowhead, sedges, and leatherleaf, the whole of Wisconsin Point will appear before you.

Nearing the sand spit, there is a field of grass on the right where the city dump used to be. Now planted with bluegrass and smooth brome, which act as a vegetation cover to prevent erosion, this field is a very good spot to observe wayward transient birds with grassland affinities, such as LeConte's sparrow and Nelson's sharp-tailed sparrow. To the south, the sand spit adjoins the clay banks of Lake Superior, which are typical of the south shore of the lake. The clay banks also have an interesting moss flora that seems to thrive on disturbance.

For the next 2½ miles, Wisconsin Point juts into Lake Superior. Very narrow near its base, it widens to 700 yards near the tip. Parallel communities of lake beach and lake dune are on the lake side of the point. These water-deposited sands have the second largest beach–dune communities on the Wisconsin coast of Lake Superior.

The dunes lack many species typical of Lake Michigan beaches. Toward the bay side are stands of deciduous forest, dominated by trembling aspen and paper birch. These poorly developed communities have sparse herbaceous layers. Near the tip of the sand spit is a stand of mixed forest, probably more typical of the original vegetation. White pine, red pine, and paper birch are the dominant tree species. At the apex of the peninsula is a very unusual dune heath

Aerial view of Wisconsin Point (photograph by Eric Epstein; used by permission)

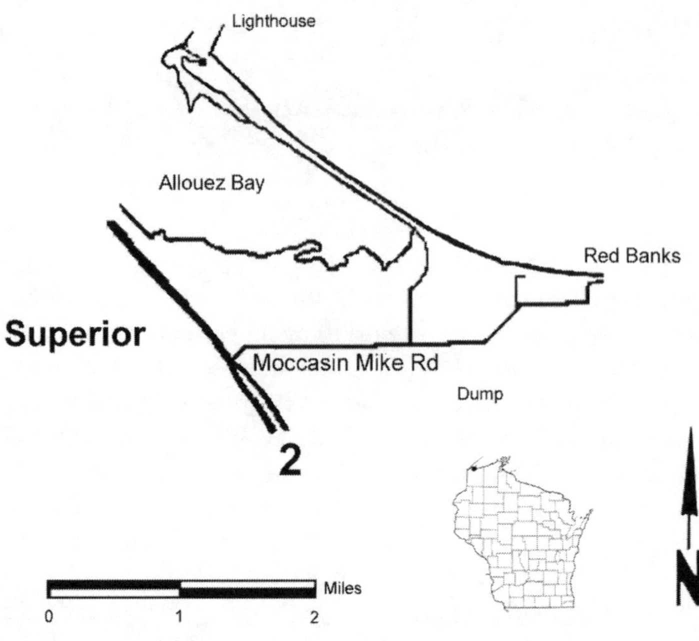

Wisconsin Point

community with a mat-forming groundlayer composed mostly of false heather, bearberry, creeping juniper, and three-toothed cinquefoil. Although the vegetation is sparse, there are several rare plants here. Also because of the proximity with the international harbor area, this region has produced many exotic species, such as gumweed and scarlet pimpernel.

For birders, rare, uncommon, and exotic birds are almost commonplace along this sand spit. Most bird species refuse to fly over large bodies of water such as Lake Superior and concentrate at the western end of the lake. Foul weather, thunderstorms, and dense fog seem to enhance this natural concentration. With right combinations of meteorological events, such as an intense low-pressure system, Wisconsin Point can be transformed from a good spot to view birds into a place where they seem to be dripping off the trees.

Jaegers

Birders call a piece of land where birds concentrate year after year a migration trap. These places will often harbor birds rarely seen in the state because these rare birds tend to follow or at least show up with the common birds. Most of the time these rare birds will appear once and then never again. Occasionally, the focused activity of the birders at these migration traps will shed new light on the phenomena of bird migration.

Such is the case with our three jaeger species, especially the parasitic jaeger. Prior to the 1970s, these species were thought restricted to the oceans during migration. Birders visiting the migrant traps in the Duluth–Superior area began finding jaegers in the fall (primarily September). The excitement of a possible rare species observation brought more birders, who added to our knowledge base and confirmed the annual nature of these jaeger migrations.

Specialties

Plants

- arrowhead coltsfoot
- Canadian gooseberry
- clustered bur-reed
- hairgrass
- neat spike-rush
- seaside crowfoot
- small yellow water crowfoot
- swamp grass-of-parnassus
- Vasey's rush
- veined meadow-rue
- white-flowered ground cherry

Animals

- gyrfalcon
- jaegers
- piping plover

Uncommon Species

Plants

- adder's tongue fern
- American willow-herb
- beaked willow
- blue giant hyssop
- blunt-leaved orchid
- Canada St. John's wort
- caraway
- Dewey's sedge
- Drummond's rock cress
- dwarf ginseng
- early coralroot
- false dragonhead
- green adder's mouth
- green-flowered shinleaf
- gumweed
- hairy goldenrod
- hook-spurred violet
- kidney-leaved violet
- marsh cinquefoil
- marsh pea
- Maxmillian's sunflower
- moccasin flower
- narrow-leaved collinsia
- nodding trillium
- pink shinleaf
- prairie ragwort
- red-stemmed gentian
- sand dropseed
- scarlet pimpernel
- showy mountain ash
- slender false foxglove
- slender spurge
- small bur-reed
- smooth false foxglove
- striped coralroot
- swamp aster
- tall lungwort
- three-toothed cinquefoil
- variegated scouring-rush
- veined meadow-rue
- Virginia wild rye
- water sedge
- wild four o'clock
- wiregrass sedge
- yellow wood violet

Insects

- cracker grasshopper
- graceful sedge grasshopper
- Huron short-winged grasshopper
- night-wandering dagger

Birds (Migrants)

- American avocet
- American golden plover
- Arctic tern
- Baird's sandpiper
- bald eagle
- bay-breasted warbler
- black-bellied plover
- blackburnian warbler
- blackpoll warbler
- black scoter
- black-throated blue warbler
- buff-breasted sandpiper
- Cape May warbler
- Caspian tern
- Connecticut warbler
- Forster's tern
- Franklin's gull
- gadwall
- golden-winged warbler
- greater scaup
- harlequin duck
- Harris' sparrow
- horned grebe
- long-billed dowitcher
- long-tailed duck
- marbled godwit
- merlin
- mourning warbler
- northern goshawk
- northern shoveler
- olive-sided flycatcher
- orange-crowned warbler
- osprey
- Pacific loon
- peregrine falcon
- Philadelphia vireo
- redhead
- red knot
- red-necked phalarope
- red-throated loon
- ruddy turnstone
- rusty blackbird
- sanderling
- semipalmated plover
- short-billed dowitcher
- stilt sandpiper
- surf scoter
- water pipit
- whimbrel
- white-crowned sparrow
- white pelican
- white-winged scoter
- willet
- yellow-bellied flycatcher

Birds (Winter)

- Bohemian waxwing
- glaucous gull
- gyrfalcon
- snowy owl
- Thayer's gull

Northeast Wisconsin

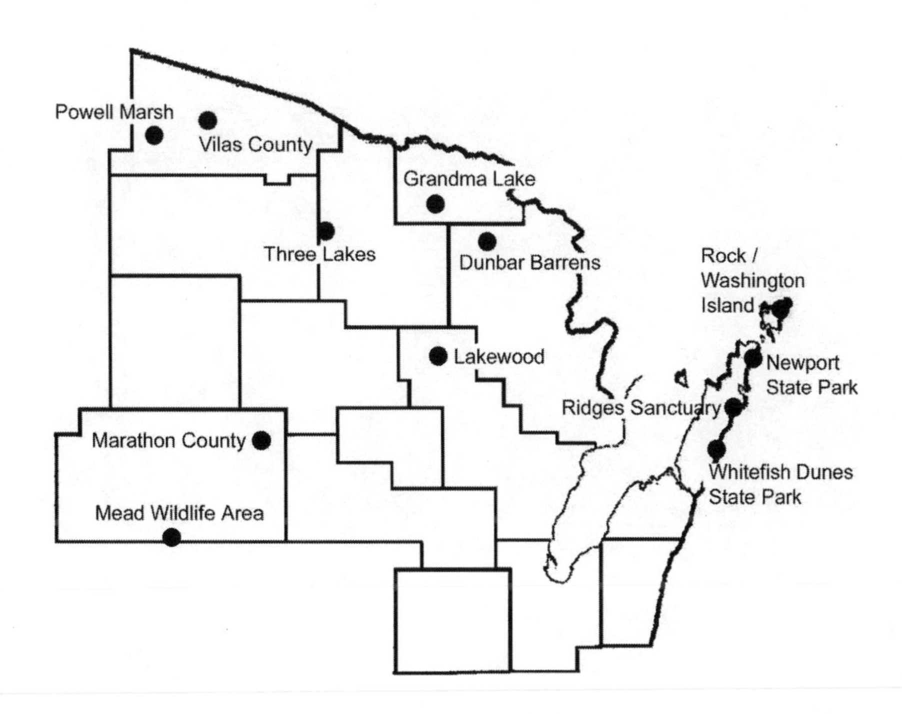

Dunbar Barrens–Grandma Lake

Communities: Pine Barrens, Northern Forested Bogs and Fens

Dunbar Barrens lies in extreme northern Marinette County. This 1,500-acre pine barrens is a few miles west of Dunbar and north of Highway 8. As with other barrens areas, these barrens formed on glacially deposited sands, and they require management to keep them as pine barrens. Aspen, oak, and jack pine forest surround the barrens openings. There are some low areas that remain open because of the action of frost in these pockets.

To the northwest of Dunbar are areas of dwarf bilberry and the associated northern blue butterfly, whose sole food happens to be this rare shrub. Most of the bilberries occur in forest openings in northwest Marinette County and south central Florence County. I am not giving exact locations for this species because of the sensitivity of the bilberry and especially the butterfly. The truly dedicated conservation-minded explorer can find the species using deductive reasoning and information found in the pine barrens community chapter in part 1.

Upland Sandpiper

The 1,500 acres of Dunbar Barrens do not come close to the 10,000 acres needed to sustain a viable population of sharp-tailed grouse. However, it is large enough to sustain several pairs of upland sandpiper year after year. These so-called shore birds are always found in grassy areas, but if the grass is too sparse and short, too dense or tall, or too brushy, these world-class travelers will abandon the area as a nesting site. They usually arrive in southern Wisconsin between April 25 and May 5 and in northern Wisconsin between May 5 and May 15. Upon arrival the males begin their courtship whistling. The sound, often likened to a "wolf whistle" given by World War II soldiers, can be heard for several miles. They can be seen in May and June whistling and flying around with their stiff-wing flight.

Upland sandpiper (illustration by Jim McEvoy; courtesy of Wisconsin Department of Natural Resources)

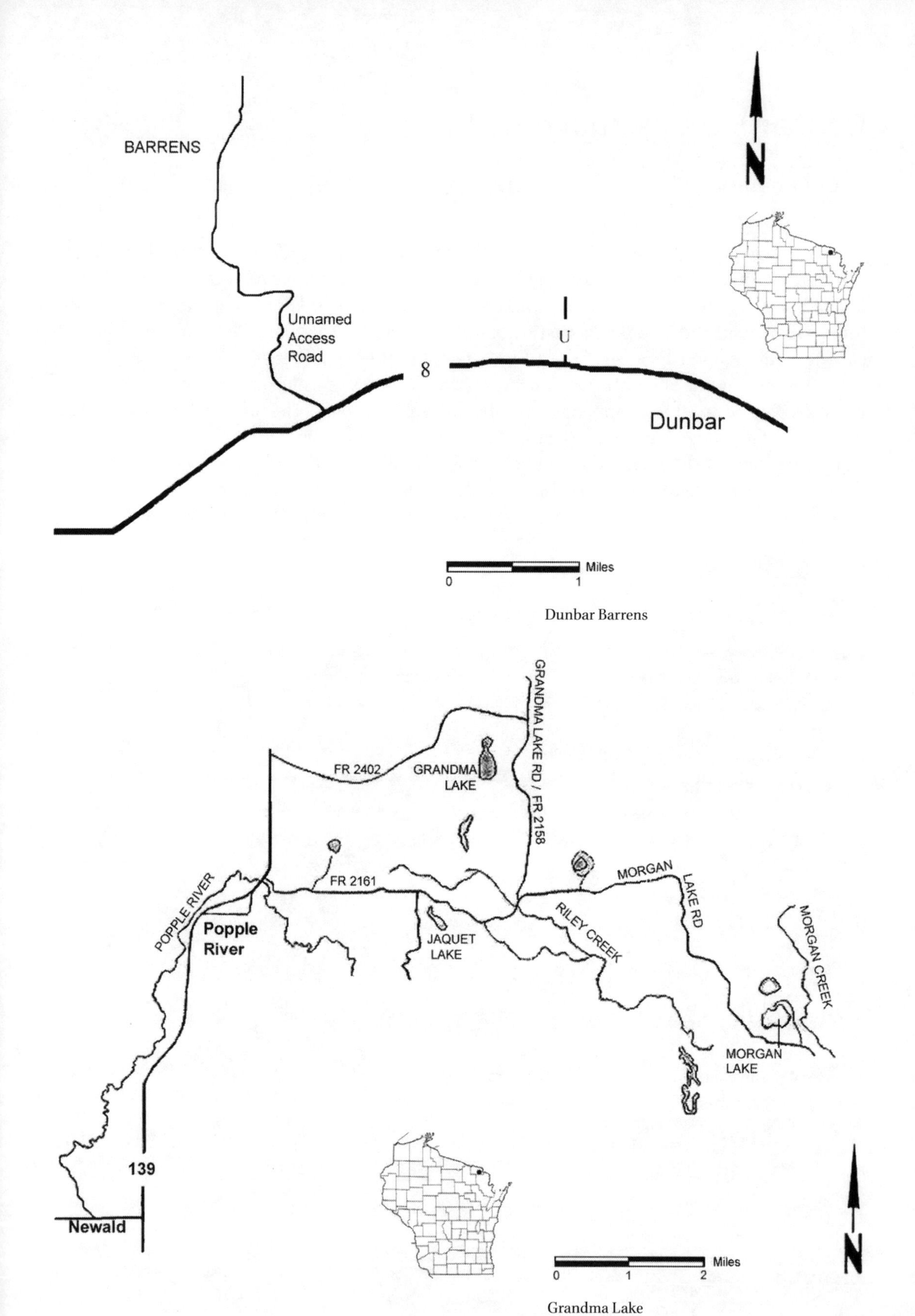

Dunbar Barrens

Grandma Lake

The very interesting Grandma Lake area has just begun to be explored by naturalists. To reach the area, take FR 2161 from Popple River. The first major intersection on 2161 is FR2158 or Grandma Lake Road. Take the road north 1½ miles to Grandma Lake. This is a soft seepage lake with neutral water surrounded by a wet bog and black spruce swamp. Several uncommon sedges find a home here, including the bog rush; this is one of its only two known Wisconsin locations. This bog has thousands of bog rush plants growing, probably because of the higher mineral content of the water than in most other bogs.

Just north of Morgan Lake there is another very nice bog lake with a floating sphagnum mat, which is unusual in a drainage lake.

Continuing on 2158 north for six more miles, you will find another plant species, the western Jacob's ladder at its only Wisconsin location. It is quite common in a cedar area that was cut in parallel rows during the early 1970s. It grows with the rare marsh valerian in some nearby bogs. These recently documented rare-plant locations prove that we do not know everything about Wisconsin's biota and discoveries can still be made. Again, I am not giving exact locations because of the sensitivity of the species, but an experienced naturalist can find them.

Specialties

Plants

- bog rush
- common bog arrow-grass
- dragon's mouth
- dwarf bilberry
- livid sedge
- marsh valerian
- showy lady's slipper
- slender bog arrow-grass
- sooty beak-rush
- sparse-flowered sedge
- western Jacob's ladder

Animals

- chryxus arctic (butterfly)
- gorgone checkerspot (butterfly)
- northern blue butterfly
- sharp-tailed grouse
- upland sandpiper

Uncommon Species

Plants

- bearberry
- bentgrass
- blue giant hyssop
- bog clubmoss
- bog St. John's wort
- bracted orchid
- browned sedge
- conifer cotton-grass
- early goldenrod
- false dandelion
- golden ragwort
- grass pink
- gray goldenrod
- hair beak-rush
- hairy beard-tongue
- hook-spurred violet
- Houghton's sedge
- Jacob's ladder
- low cudweed
- northern heart-leaved aster
- northern sweet coltsfoot
- one-flowered pyrola
- pinesap
- pipewort
- pond sedge
- prairie brome
- purple milkwort
- purple virgin's bower
- rock sandwort
- rough blazing star

rough-leaved ricegrass
round-leaved shinleaf
running clubmoss
rusty cotton-grass
sand cherry
showy goldenrod
silvery sedge
sky blue aster
slender knotweed
slender ladies' tresses
smooth blue aster
spotted coralroot
starved panic grass
thimbleweed
thin-scale cotton-grass
trailing arbutus
triangle grape fern
upland wild timothy
veined meadow-rue
white adder's mouth
white bog orchid
whorled loosestrife
wild calla
wild comfrey

Mushrooms

cannon fungus

Insects

cobweb skipper
dion skipper
Henry's elfin
hoary elfin
northwest red-winged grasshopper
pepper and salt skipper
pine elfin
roadside skipper

Mammals

badger
black bear
bobcat

Lakewood Area

Communities: White Pine–Hardwood Forest, Northern Hardwood–Hemlock Forest, Forested Swamps, Northern Forested Bogs and Fens, Shallow Marsh, Lakes, Rivers

Located in northwestern Oconto County, Lakewood is centrally located for visits to a number of sites in the area. All the sites described here are within the Nicolet National Forest, where past land use varies from extreme degradation to small areas of superior quality. The Forest Service roads provide easy access and usually have a name and a forest road number.

To the southwest of Lakewood is an area of many small lakes, good quality forests, and exceptional wetlands. To get a sample of this region, take Archibald Lake Road (FR2121) south to Cathedral Drive, then turn north. After about one-half mile, you enter the cathedral of pines. Here lies an area of virgin red and white pine super canopy forest. It is a

Little Goblin Fern

The strangest of the area's plants has to be the little goblin fern. This member of the grape ferns rises one to three inches above the ground, and many times it does not even emerge from leaf litter. A species of the Midwest, it is only found in the Great Lakes forests of Minnesota, Wisconsin, and Michigan. Even in this limited range, it is only found on certain soils. Another factor in locating this uncommon species is the age of the forest—if the forest is too young or too old the plant will usually not be found. It prefers hardwood forest between 60 and 100 years old and medium-textured, nutrient-rich soils. Formerly thought very rare, it is now considered uncommon because of its miniature size and inability to penetrate the uppermost leaf litter.

Little goblin fern (courtesy of USDA Forest Service)

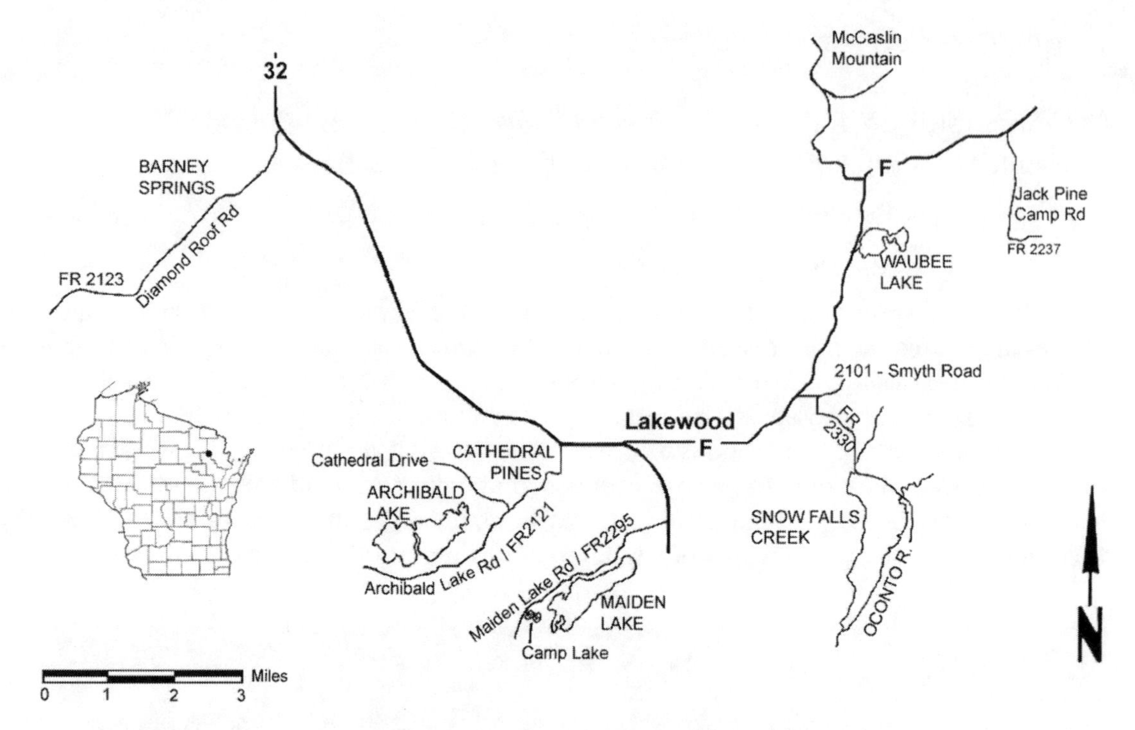

Lakewood Area

pleasing place, but the groundlayer is sparse. If you return to Archibald Lake Road and continue southwest, you will see an excellent area of northern hardwood forest. The area about one mile south of Cathedral Drive has the best ground flora.

There are several lakes and associated shorelands of exceptional quality found just to the south, which are easily reached by taking north Maiden Lake Road (FR2295) just south of Lakewood. Drive past Maiden Lake to Camp Lake. Here is a wonderful little lake with many unusual plants growing on the bog mat surrounding the lake. By continuing southwest, you will encounter several small lakes with bogs and conifer swamps surrounding them, all with interesting flora. Pine, Bass, Cave, and Upper Island Lakes are the best of these small lakes (consult a topographic map or gazetteer for details).

Different habitats can be found by traveling east of Lakewood on County F. After about three miles, turn east on Smyth Road (FR2101), then shortly south on Sullivan Springs Road (FR2330). After 1½ miles, the road crosses Snow Falls Creek. To locate the older northern hardwood–hemlock forest, follow the creek downstream. This area has an abundant rich herbaceous layer and is especially exciting in spring with numerous blooming flowers. By returning to County F, you will encounter a vastly different area because of the drier soils and rare species. Continue northeast for 5½ miles to Jack Pine Camp Road (FR2237). By turning south for about one-half mile, you will see a heath-like area of dwarf bilberry, which occurs at its southern range limit.

Still different areas can be found by traveling north on Highway 32 to a point just south of the forest county line. This is the western edge of McCaslin Mountain, a large quartzite monadnock stretching east for 10 miles. This portion of the mountain is important because of the large high-quality quartz crystals that can be found here.

If you turn south on Diamond Roof Road (FR2123) and continue into Langlade County, you will drive through many habitats, all worthy of exploration. It is just north of this road that the very rare foamflower still exists. Fortunately, the population is a ways off the road, helping to ensure that curiosity seekers will not be a problem for the flower's survival.

Dwarf bilberry and northern blue butterfly (illustration by Jim McEvoy; used by permission)

Specialties

Plants

- calypso orchid
- Crawe's sedge
- dragon's mouth
- dwarf bilberry
- foamflower
- Hooker's orchid
- Indian cucumber-root
- least moonwort
- little goblin fern
- northern bog sedge
- purple bladderwort
- showy lady's slipper
- slender bog arrow-grass
- small purple bladderwort
- small round-leaved orchis
- sparse-flowered sedge
- white bog orchid
- withe-rod

Animals

- northern blue butterfly
- northern ribbon snake
- two-spotted skipper (butterfly)

Other Features

- monadnock
- quartz crystals
- specular hematite

Uncommon Species

Plants

- Allegheny vine
- arrow-leaved aster
- bearberry
- beech drops
- blue phlox
- blunt-leaved orchid
- bog lobelia
- bristly crowfoot
- broad beech fern
- broad-leaved toothwort
- Canada yew
- Carolina spring beauty
- creeping rattlesnake plantain
- crooked aster
- cup plant
- dissected grape fern
- dwarf ginseng
- dwarf rattlesnake plantain
- false heather
- false melic grass
- fringed polygala
- Goldie's fern
- grass pink
- great-spurred violet
- green-flowered shinleaf
- hairy agrimony
- heartleaf twayblade
- Kalm's St. John's wort
- large-leaved shinleaf
- large yellow lady's slipper
- marsh bluebell
- marsh St. John's wort
- moonseed
- New Jersey tea

- nodding ladies' tresses
- nodding trillium
- pipewort
- purple fringed orchid
- red honeysuckle
- round-leaved orchid
- round-leaved shinleaf
- silvery glade fern
- slender false foxglove
- small green fringed orchid
- spurred gentian
- squirrel corn
- striped coralroot
- triangle grape fern

Insects

- arctic skipper
- chestnut schizura
- green marvel
- olive and black carpet
- pepper and salt skipper
- silvery blue

Birds (Breeding)

- black-throated blue warbler
- golden-winged warbler
- saw-whet owl

Birds (Winter)

- evening grosbeak
- red crossbill
- white-winged crossbill

Mammals

- American pygmy shrew
- black bear
- bobcat
- bog lemming
- hoary bat
- meadow jumping mouse
- northern flying squirrel
- otter
- star-nose mole
- woodland jumping mouse

Marathon County

Communities: Northern Hardwood–Hemlock Forest, Rivers, Cliffs

From the list of rare specialty species, you could assume that Marathon County has little to offer biologically, an assumption that is totally untrue. It has much to offer, enough so that a small part of the county, Mead Wildlife Area and Big Eau Pleine Park, has an entire chapter devoted to describing its exceptional natural values. The geology, however, is magnificent and may detract attention from the living world.

Places like the Dells of the Eau Claire County Park, with its fine trail system, wonderful hardwood forest, and marvelous spring wildflower display, are still secondary to the geology. Precambrian rhyolite, tilted nearly vertical, interacts with the Eau Claire River, forming terraces with many small potholes.

The county's exceptional geology can also be observed at Rib Mountain State Park. The mountain is a quartzite monadnock that has dominated the landscape for millennia. Monadnocks are very resistant rocks that the glaciers could not wear down. Today, Rib Mountain dominates the scenery of Wausau. Plant species, especially forested ground flora, are exceptionally rich in Marathon County, for which the bedrock geology is responsible.

The more botanically inclined should consider a visit to the Plover River Fishery Area. Located in the far northeastern corner of the county, this public land, which was established for fishing, contains some of the best ground flora in all of northern Wisconsin. The parking areas are well marked on maps and by signs. Just park your vehicle and wander the woods. A full range of spring ephemeral plants can be seen, including Carolina spring beauty, trilliums, Virginia and appendaged waterleaf, and wild ginger. The luna moth, a large and distinctive-shaped species, can be found in June. This moth feeds on a variety of tree species, and it is especially numerous in the Plover River Fishery Area.

Quarries are everywhere; many are still active, but most are abandoned. A large percentage of the area is available for prospecting by the amateur. The larger quarries don't allow access, except for scheduled tours. Each quarry and rock cut along the roadways is different geologically. The variety can keep a budding geologist occupied for months.

Luna moth (courtesy of Ohio Department of Natural Resources)

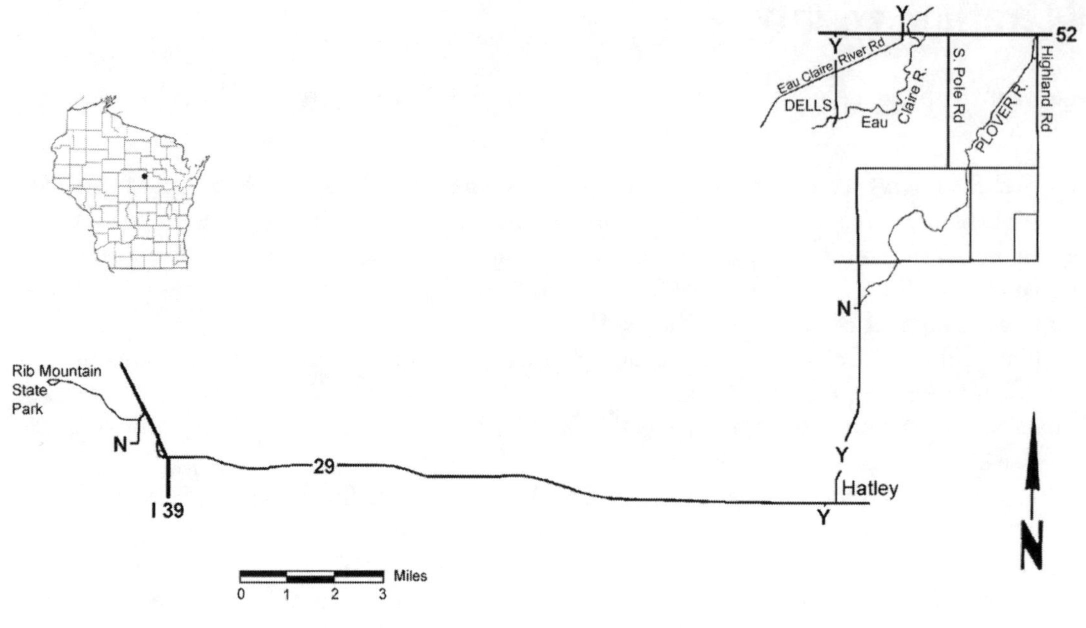

Marathon County

Specialties

Plants

- blunt-lobed grape fern
- Missouri rock cress

Other Features

- acmite
- cordierite
- epidote
- garnet
- jade
- labradorite (moonstone)
- monadnocks
- purple flourite
- pyrolusite
- quartz
- red granite
- smoky quartz
- staurolite
- syenite
- unakite
- zircon

Uncommon Species

Plants

- American fly honeysuckle
- arrow-leaved violet
- Bishop's cap
- blue beech
- bristly sarsaparilla
- broad-leaved toothwort
- bronze sedge
- butternut
- Canada yew
- Carolina spring beauty
- daisy-leaved grape fern
- Dutchman's breeches
- dwarf ginseng
- dwarf serviceberry
- fescue sedge
- fireweed
- fringed sedge
- grass-leaved goldenrod
- great waterleaf
- ground-pine
- Indian pipe
- kidney-leaved violet
- lance-leaved violet
- large-flowered trillium
- leatherwood
- long-awned wood grass
- marginal wood-fern
- Missouri rock cress
- moonseed
- necklace sedge
- New England violet
- nodding fescue
- northern starwort
- oak fern
- pointed tick-trefoil
- prairie wedgegrass
- purple sneezeweed
- rock elm
- round-leaved shinleaf
- sessile bellwort
- shining clubmoss
- small enchanter's nightshade

- spotted coralroot
- squirrel corn
- stalked sedge
- swollen sedge
- tall northern bog orchid
- triangle grape fern
- Virginia waterleaf
- white baneberry
- wild leek
- winterberry
- witch-hazel
- yellow trout lily

Reptiles and Amphibians

- spotted salamander
- wood turtle

Birds (Breeding)

- brown creeper
- evening grosbeak
- golden-winged warbler
- mourning warbler

Mammals

- ermine
- gray fox
- least weasel
- long-tailed weasel
- meadow jumping mouse
- northern flying squirrel
- otter
- snowshoe hare

Mead Wildlife Area

Communities: Northern Hardwood–Hemlock Forest, Northern Sedge Meadow, Alder Thicket, Shallow Marsh, Cattail Marsh

Mead Wildlife Area encompasses more than 26,000 acres on the Little Eau Pleine River. Mead only vaguely resembles the original landscape, although significant areas of sedge meadow and forested bogs still persist. You would think that after ditching, draining, burning, clear cutting, and blasting with explosives, the area would be a wasteland with few natural resources of interest left, but just the opposite is true. These are all techniques employed by wildlife management specialists to enhance the habitat. Admittedly, these techniques are used to benefit certain targeted species, but those species, including the greater prairie chicken, the double-crested cormorant, the eastern bluebird, and ducks, have reaped the benefits.

If there is one place in this state to learn what traditional wildlife-management theory and practice is all about, this is the place. Some specific management practices include building artificial poles for nesting cormorants and herons and installing nearly 400 wood duck houses, more than 500 squirrel boxes, and more than 50 bluebird houses. Besides this management, more than 38 miles of constructed ditches and dikes maintain shallow water areas, 700 blasted potholes enhance waterfowl production, clear cut patches of timber allow rapid aspen production for ruffed grouse, and the burning and mowing of 7,000 acres benefit prairie chickens.

There have been controversies over why the area should be managed at all. One argument is just to let the species develop as they will since they seem to be doing all right on their own. The critical point is that these species are doing well now because they have been managed for years. If nothing had been done a long time ago, many species would not be with us today.

A new and more urgent task is the rescue of thousands of nongame species, such as grasshopper sparrow and short-eared owl, from a similar fate. This situation appears to be as

Franklin's ground squirrel (courtesy of Missouri Department of Conservation)

Franklin's Ground Squirrel

Franklin's ground squirrel is a species of the tallgrass prairie. It thrived in the former prairie, where it preferred the edges between prairie and wooded areas. This is a secretive species and not easily observed, unlike its cousin the thirteen-lined ground squirrel, which is abundantly obvious along our roadways. The populations seem to fluctuate, with good numbers in some years and hardly any the next year. The largest grassy areas have the best populations.

The little ground squirrel digs numerous burrows into the soil, especially banks. These burrows might have many chambers and be up to eight feet deep. As with other ground squirrels, the Franklin's ground squirrel feeds on both plant and animal foods. This species is scarce in Wisconsin, due most likely to loss of grassland.

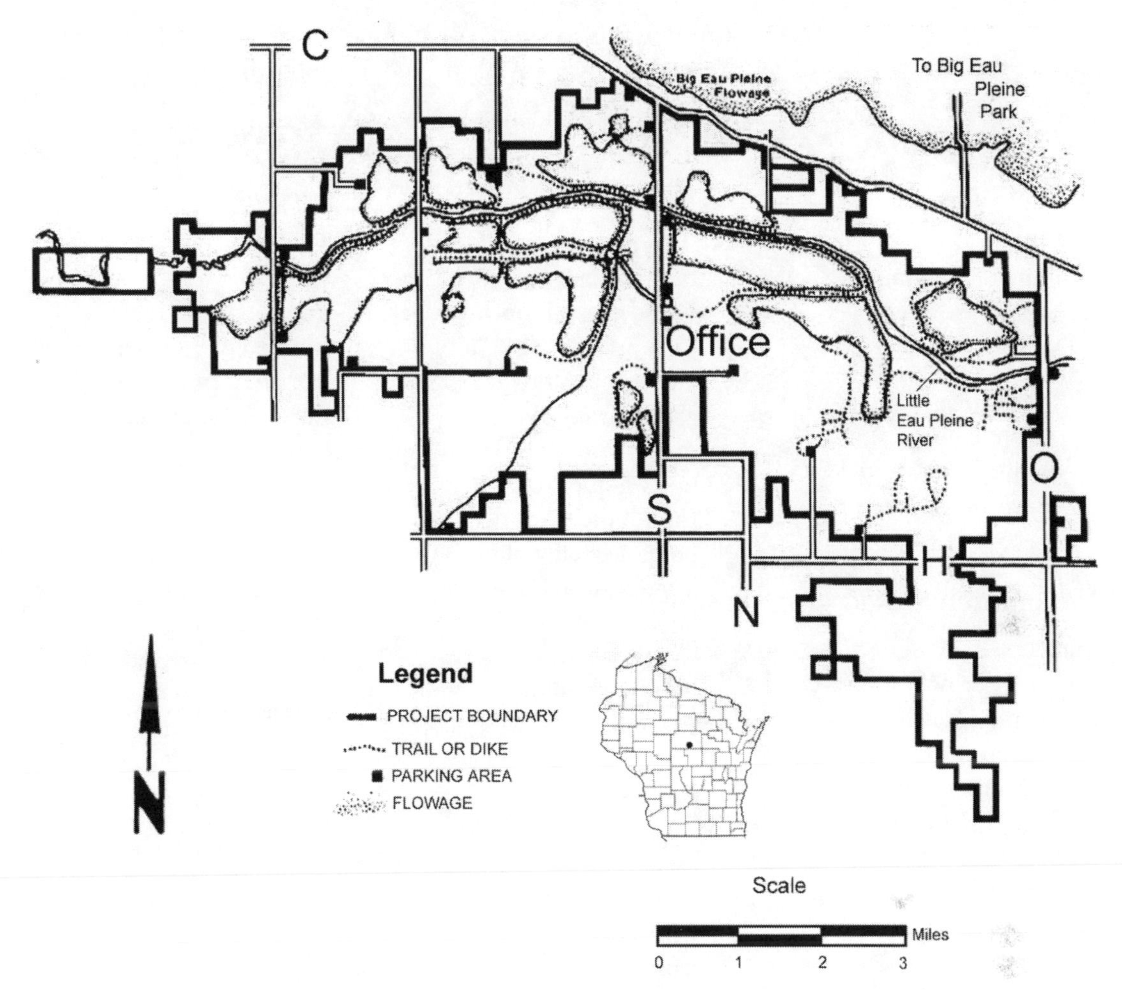

Mead Wildlife Area

hopeless as the game situation was in years past, but with areas like Mead experimentally leading the way, new techniques can be applied to the species that need help now. Significant debate over the ethical nature of these individual species prescriptions as opposed to managing an entire ecosystem should take place. It can only make us better stewards of the land.

These management goals and practices are extensive, as is the opportunity to observe the workings of the managers. More than 70 miles of dikes and trails are available to observe the wildlife and management practices.

From a management perspective, Big Eau Pleine Park is completely different from Mead. This park covers 5,000 acres on a peninsula jutting into the Big Eau Pleine reservoir. It has an excellent old-growth stand of hemlock hardwoods near the tip of the peninsula. A well-designed trail leads the visitor through this magnificent stand of trees. This park is a recreational facility easily located by following the signs after crossing the reservoir.

Specialties

Plants
- hairy beard-tongue

Animals
- Franklin's ground squirrel
- greater prairie chicken
- sharp-tailed grouse

Uncommon Species

Plants
- bristly crowfoot
- common satingrass
- crested wood-fern
- dark-green bulrush
- dwarf mistletoe
- fox sedge
- fringed brome
- golden alexanders
- marsh bluegrass
- marsh cinquefoil
- moccasin flower
- mountain holly
- pickerelweed
- pitcher plant
- swamp aster
- water purslane
- wild calla
- yellow water crowfoot

Big Eau Pleine Park Plants
- blue cohosh
- broad-leaved toothwort
- Carolina spring beauty
- Dutchman's breeches
- early coralroot
- great waterleaf
- hairy goldenrod
- Indian pipe
- lance-leaved violet
- large-flowered trillium
- large yellow lady's slipper
- New England violet
- nodding trillium
- shining clubmoss
- small Bishop's cap
- spotted coralroot
- squirrel corn
- twin flower
- wild ginger
- wild leek
- witch-hazel

Insects
- unarmed wainscot

Birds (Breeding)
- American widgeon
- black-crowned night heron
- field sparrow
- gadwall
- green-winged teal
- hooded merganser
- northern pintail
- northern shoveler
- osprey
- redhead
- ring-necked duck
- short-eared owl
- upland sandpiper
- Virginia rail

Birds (Migrants)
- American golden plover
- bald eagle
- horned grebe
- short-billed dowitcher
- solitary sandpiper

Birds (Winter)
- northern shrike
- short-eared owl
- snow bunting

Mammals
- badger
- gray fox
- otter

Newport State Park

Communities: Northern Hardwood–Hemlock Forest, Forested Swamps, Great Lakes Shoreland Forest, Lake Beach

Newport State Park, located at the tip of the Door Peninsula, is managed much differently than most parks: it operates as semi-wilderness, with walk-in access to campsites. The 2,200 acres of the park are open to hiking, backpack camping, nature study, and cross-country skiing. Within the park are 28 miles of trails and 11 miles of shoreline. This shoreline alternates between rocky ledges and sandy beaches. Off shore are islands that barely reach above the crashing waves. These islands have large nesting colonies of gulls and cormorants. These waters are also important as migratory routes for waterfowl.

Newport has many northern-type habitats scattered throughout the park. By hiking the trails, you can pass from natural community to natural community in short distances. The most interesting places are the boreal forest natural area, the dunes area, and the old-growth forests east of Europe Lake. You will also find an abundance of solitude. The few people that you encounter here are usually looking for a similar experience. To find the specialties here, you should explore each habitat.

Bedrock Beach

An unusual natural community appears near the tip of Door County. Flat areas of dolomite bedrock extend a great distance into Lake Michigan. Near the shore, these areas are either exposed or covered with water, depending on the lake's level. This fluctuation prevents trees from becoming established and eliminates competition from many other plants that cannot tolerate these extreme conditions.

These wet bedrock beaches are vegetated with specialized species that can tolerate years of inundation followed by years of exposure to the air, with disturbance from occasional storm surges and waves. Several rare plants are found on these bedrock beaches. One of the most ubiquitous and colorful species is the birds-eye primrose, a small plant with several yellow-centered purplish flowers that bloom in May and June.

All of the peninsulas at Newport have Niagara dolomite bedrock with these extreme conditions for plant growth. Twenty feet above the present lake level there is another ridge of exposed dolomite with ledges that were the shoreline of ancient Lake Nipissing. Uncommon plant species, such as Allegheny vine, are found on this ledge more often than anywhere else in the park.

Birds-eye primrose (photograph by Thomas A. Meyer, used by permission)

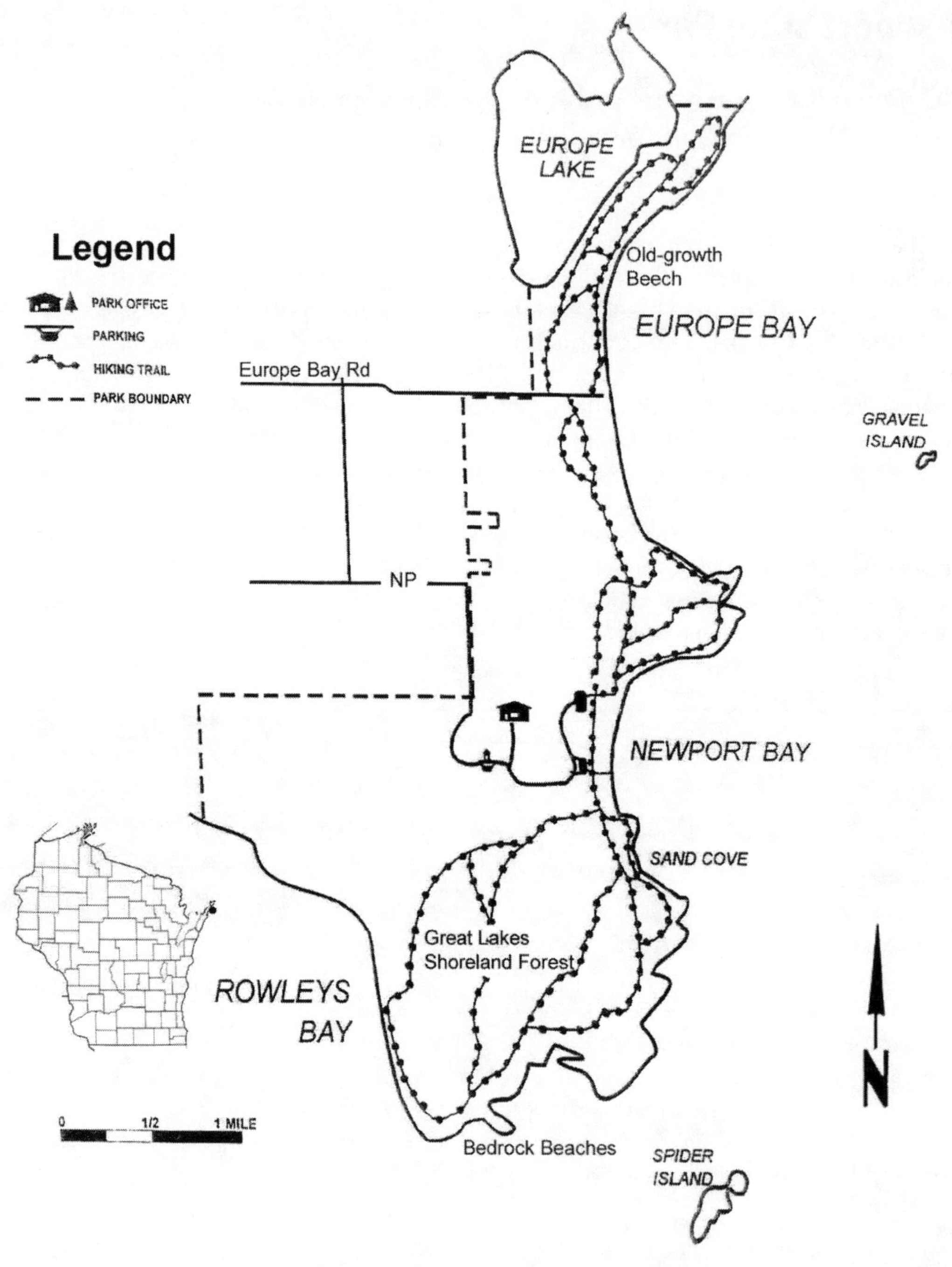

Newport State Park

Specialties

Plants

- Allegheny vine
- Crawe's sedge
- dune goldenrod
- dwarf lake iris
- Garber's sedge
- hairgrass
- Lake Huron tansy
- long-spurred violet
- low calamint
- maidenhair spleenwort
- pinedrops
- sand tickseed
- thickspike wheatgrass
- variegated scouring-rush
- western fescue

Animals

- white-M hairstreak (butterfly)

Uncommon Species

Plants

- alder buckthorn
- alpine rush
- American fly honeysuckle
- bog lobelia
- bracted orchid
- bristly sarsaparilla
- broad-leaved goldenrod
- Chilean sweet cicely
- cow-wheat
- cursed crowfoot
- Dewey's sedge
- drooping wood sedge
- dwarf rattlesnake plantain
- dwarf scouring-rush
- few-flowered spike-rush
- fox sedge
- fringed polygala
- golden corydalis
- golden sedge
- great-spurred violet
- green-flowered shinleaf
- herb robert
- hooked crowfoot
- hop sedge
- large yellow lady's slipper
- marsh cinquefoil
- meadow spikemoss
- needle spike-rush
- northern sweet coltsfoot
- one-flowered broomrape
- one-flowered pyrola
- plantain-leaved sedge
- purple false oats
- purple fringed orchid
- rabbit-berry
- round-leaved orchid
- scarlet paintbrush
- showy goldenrod
- smooth false foxglove
- snowberry
- spotted coralroot
- spurred gentian
- striped coralroot
- swamp buttercup
- tall white violet
- walking fern
- white camas
- wild comfrey

Lichens

- black and blue lichen
- elegant lichen

Birds (Breeding)

- black scoter
- Caspian tern
- common merganser
- common tern
- double-crested cormorant
- mourning warbler
- saw-whet owl

Mammals

- ermine
- long-tailed weasel
- porcupine
- snowshoe hare

Powell Marsh

Communities: Alder Thicket, Northern Sedge Meadow, Shallow Marsh

Covering 13,000 acres, Powell Marsh (in name only) is a large wetland area in a predominantly forested area. It is a broad shallow bog with water levels controlled by dikes. The surrounding uplands are forest lands growing on sand deposits. Poor drainage and fire combined to determine the species composition. These fires occasionally burned through the peat layers right down to the sands. When vegetation reclaimed the land, the area had bog plants on the peat and sedges on the sands, forming an unusual hybrid community.

The marsh now has seven dikes that control water levels. By varying water levels, managers try to optimize use of certain areas by certain species. These variations in water levels have allowed some areas to be ideal for LeConte's sparrows. At least 25 territorial males sing each June. Yellow rails and Nelson's sharp-tailed sparrows use the area, but both species are not present every year. The best means of observing or hearing these species is by walking the control structures anywhere from pre-dawn to one hour after sunrise. A real treat in our natural world is spending an early morning on a marsh. At Powell Marsh, this treat becomes an obsession, with people who make yearly pilgrimages to drink the vocal nectar. You can really appreciate how species respond to their habitat by visiting when they're most active.

Yellow rail (courtesy of Wisconsin Department of Natural Resources)

Yellow Rail

Yellow rails are probably the most secretive and least-known bird species in the state. These six-inch-long birds live in large open fen areas. Wiregrass sedge and thin grasses provide the ideal habitat. They seldom fly or even flush, unlike sora and Virginia rails. These birds stay hidden in the sedge where they forage for snails and insects. They arrive on the breeding grounds in mid- to late May and establish territories. Their repetitive tic,tic-tic,tic,tic calls can be imitated by tapping two rocks or two quarters together. Occasionally, this method will draw the birds close enough for you to see them.

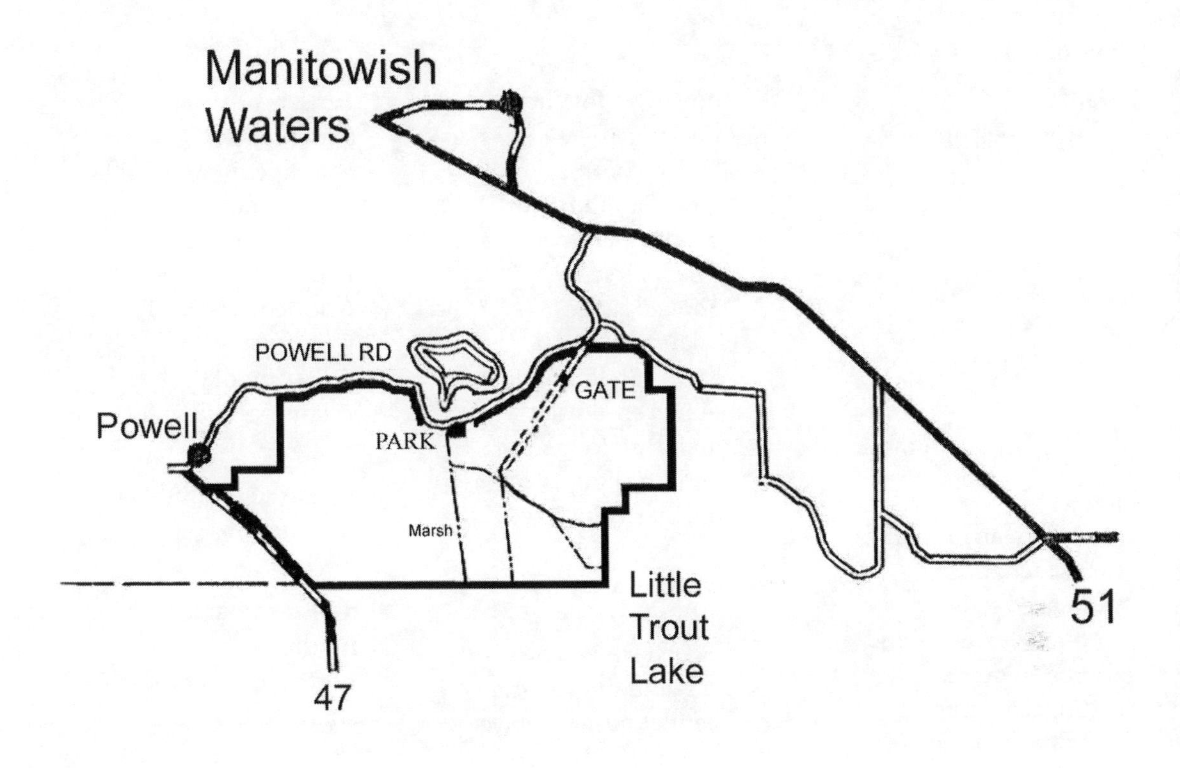

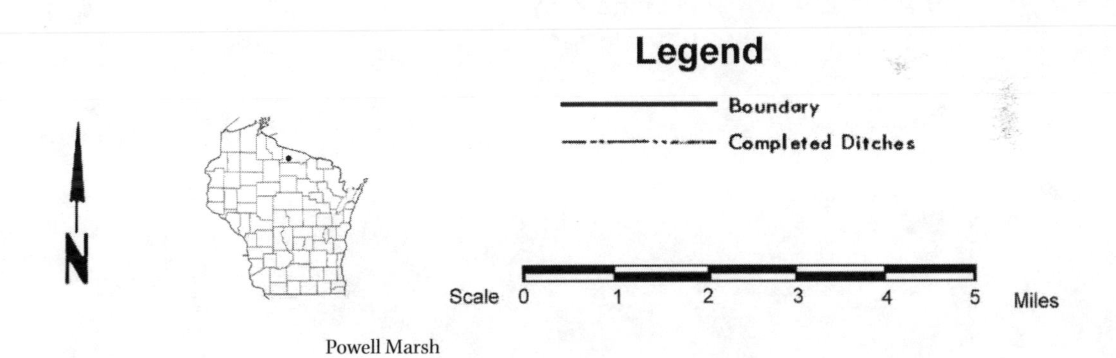

Powell Marsh

Specialties

Plants

white bog orchid

Animals

LeConte's sparrow
Nelson's sharp-tailed sparrow
yellow rail

Uncommon Species

Plants

American brooklime
awl sedge
beaked sedge
blunt-leaved orchid
bog goldenrod
bog reed-grass
bottle gentian
bristle-stalked sedge
broom sedge
bulblet water hemlock
bull-head pond-lily
downy willow-herb
fox sedge
fringed brome
great St. John's wort
great water dock
Hayden's sedge
hooked sedge
inland sedge
Kalm's St. John's wort
marsh speedwell
marsh St. John's wort
nodding ladies' tresses
northern clustered sedge
pond sedge
prairie brome
purple avens
ragged fringed orchid
red-stemmed gentian
slender false foxglove
swamp aster
tall cotton-grass
tall northern bog orchid
tall sunflower
tufted loosestrife
turtlehead
water lobelia
water pennywort
water sedge
water smartweed
white water crowfoot
whorled loosestrife
wool grass
yellow monkey flower
yellow pond-lily

Reptiles and Amphibians

mink frog

Birds (Breeding)

merlin
northern goshawk
sharp-tailed grouse

Mammals

bobcat
otter
star-nose mole

Ridges Sanctuary

Communities: Forested Swamps, Northern Forested Bogs and Fens, Great Lakes Shoreland Forest, Shallow Marsh, Lake Dune, Lake Beach

Ridges Sanctuary, a nonprofit conservation organization that opens part of its land to visitation by the public, the University of Wisconsin–Green Bay's Toft Point, and adjacent Department of Natural Resources Mud Lake Wildlife Area are internationally known as botanical treasure houses. They occupy more than 2,500 acres of richly varied communities just north of Bailey's Harbor. Representation of nearly every northern Wisconsin plant community occurs in the low ridge–swale topography. The preserved lands cover nearly all the interior of Ridges peninsula from the beaches at Bailey's Harbor to a baymouth bar lake on Moonlight Bay, and from County Q nearly to the tip.

The ridge–swale topography formed as receding lake levels with the last glacier exposed dune ridges. The farther from Bailey's Harbor you go, the older the ridges and swales become. Near the beach are raw wet sandy swales with very unusual wet beach flora, especially sedges. Next are ridges of pine and juniper with some balsam fir. A little bit older, these ridges have more spruce and fir with tamaracks in the swales. Still older ridges have a thick dark white

Plant Poaching

Even with all Ridges Sanctuary's notoriety, public use, and caring stewards with watchful eyes, plants end up missing every year. Visitors come from all over the world to view the flora. Many people fall in love with the beauty and aesthetic qualities of the plants and the area. Unfortunately, some people dig the plants up to take home. More often than not, the plant's new environment is not conducive to its survival and it dies.

A few years ago, I was entering the parking area of the sanctuary, and I saw only one other car in the lot. Its hatchback was open, and the back contained seven 12-by-18-inch boxes filled with plants. I confronted the owners of the car and asked them where they had dug the plants. Their response was "out there, just off the trails." I told them this was a sanctuary and the plants were protected, to which they responded by saying that there were lots of plants out there and that they could take as many as they wanted. By the time I had contacted the manager, the car was gone.

A conundrum exists when promoting the natural world. We do inform people like the above couple who consider their personal rights paramount and can justify their actions regardless of the damage done to the natural world. These kinds of people will always be around. My hope, however, is that a much larger number of people will emerge who care about the natural condition of our world and will by their number intimidate the selfish into not digging.

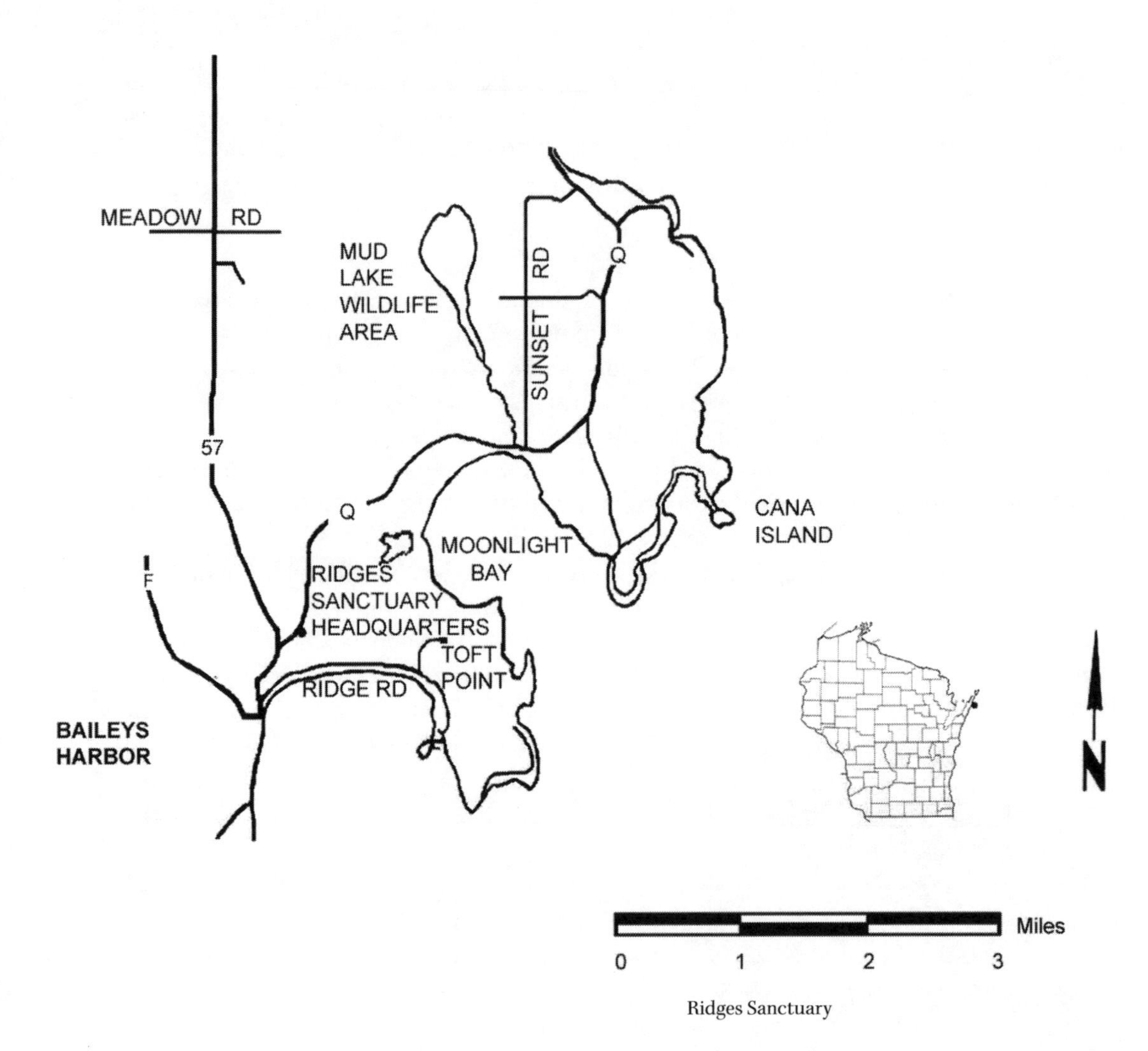

Ridges Sanctuary

cedar forest. On the Moonlight Bay side of the peninsula are large areas of sedge meadow, northern hardwood–hemlock forest, and some dense white cedar–spruce bogs.

Because this is a wildflower sanctuary, most of the area is off-limits to the public, as it should be to protect the outstanding assemblage of rare plants. However, an excellent area of trails is maintained near the Ridges headquarters just off County Q north of Bailey's Harbor. Your first stop should be the nature center to pick up a trail guide and ask the staff questions. Your next stops should be in the hundreds at various places along the trails. These trails cover most of the habitats, and nearly all of the rare plants can be observed somewhere in this area. This sanctuary has strict rules of conduct, and the trails are continually monitored by sanctuary members. These rules have allowed for maximizing preservation while promoting some usage. The Ridges Sanctuary is a na-

Hine's emerald dragonfly (photograph by William Smith; used by permission)

tional natural landmark and one of our truly exemplary natural areas.

Mud Lake Wildlife Area contains a large white cedar swamp. The lake itself connects to Moonlight Bay by a narrow stream. The stream and parking areas provide access points to the interior, where at the center there is an almost impenetrable thicket of cedar and alder that provides needed protection for some rare species, such as the federally endangered Hine's emerald dragonfly.

Specialties

Plants

adder's tongue fern
beautiful sedge
calypso orchid
Crawe's sedge
cuckoo flower
dune goldenrod
dwarf lake iris
Garber's sedge
hair-like sedge
Hooker's orchid
moonwort
northern bog sedge
northern spikemoss
purple false oats
ram's-head lady's slipper
sheathed sedge
showy lady's slipper
slender bog arrow-grass
small flowered grass-of-parnassus
spoon-leaf moonwort
tufted bulrush
tufted hairgrass
variegated scouring-rush
white bog orchid

Animals

Hine's emerald dragonfly

Uncommon Species

Plants

beaked sedge
bearberry
Bebb's sedge
birds-eye primrose
bog lobelia
bristle-leaved sedge
brown sedge
bulblet fern
common bog arrow-grass
cow-wheat
daisy-leaved grape fern
drooping wood sedge

dwarf rattlesnake plantain
fringed gentian
fringed polygala
golden corydalis
golden sedge
grass pink
green adder's mouth
green-flowered pyrola
heartleaf twayblade
herb-robert
large yellow lady's slipper
marsh cinquefoil
nodding ladies' tresses
Ohio goldenrod
one-flowered broomrape
rabbit-berry
slender ladies' tresses
smooth false foxglove
snake-mouth orchid
spurred gentian
striped coralroot

Insects

bog copper
narrow-winged grasshopper
orange-barred sulphur

Reptiles and Amphibians

four-toed salamander
spotted salamander

Birds (Breeding)

green heron
olive-sided flycatcher
ruby-crowned kinglet

Birds (Migrants)

bay-breasted warbler
black-bellied plover
blackpoll warbler
Bonaparte's gull
Cape May warbler
common tern
dunlin
fox sparrow
gray-cheeked thrush
least sandpiper
orange-crowned warbler
red knot
red-necked phalarope
ruddy turnstone
sanderling
semipalmated plover
semipalmated sandpiper
short-billed dowitcher
Swainson's thrush
whimbrel
white-crowned sparrow
white-rumped sandpiper
willet
Wilson's warbler

Rock–Washington Islands

Communities: Northern Hardwood–Hemlock Forest, Lake Dune, Lake Beach

Regular ferry service connects Washington Island with the mainland, and seasonal ferry service connects Washington Island with the uninhabited Rock Island. These commercial ferries operate when the water is ice-free but check first about times and seasonal schedules. Washington Island has many miles of roads with only a few areas of public land, while uninhabited Rock Island has many miles of trails and is virtually all public land.

Rock Island is a state park that can be reached regularly by the Karfi ferry out of Jackson Harbor. No cars are allowed and all supplies must be carried in, which adds to the unique natural values of this park. Rock Island has an interesting flora and fauna but harbors only a few of the specialties. Most of Rock Island is northern hardwood–hemlock forest, with scattered areas of conifer. The island comprises 905 acres, with 40 primitive campsites and more than 10 miles of trails. The trails circle the island and traverse the youngest portion of the forest. Off trail in the island's interior lies an old second-growth forest that is close to attaining old-growth structure. All told, the island is in an idyllic setting, a remote island far into an inland sea.

Washington Island, while permanently inhabited, is not without its natural attractions. Jackson Harbor Ridges, a state natural area owned by Washington Island Township, is fore-

Dwarf Lake Iris

The ridge and swale dunes provide the conditions needed by some of our rarest species. The dwarf lake iris, which is federally listed as a threatened species, thrives at Jackson Harbor. This species is endemic to the shores of Lake Michigan and Lake Huron. Its range is narrow, but where it grows, it grows profusely in dense colonies. The dunes have zones with different groups of species that prefer different locations. The dwarf lake iris does best where the trees are short and scattered. You won't find these plants in areas that are too open or too shaded.

Other species that thrive on these dunes and swales are also rare, such as dune goldenrod, low calamint, northern commandra, and slender bog arrow-grass, which grow abundantly on and between the dunes. These species appear to be stable at protected areas like Jackson Harbor, but these areas are few and far between. Destruction of a few dunes with development or sand removal could eliminate several species.

Northern commandra (illustration by Jim McEvoy; courtesy of Wisconsin Department of Natural Resources)

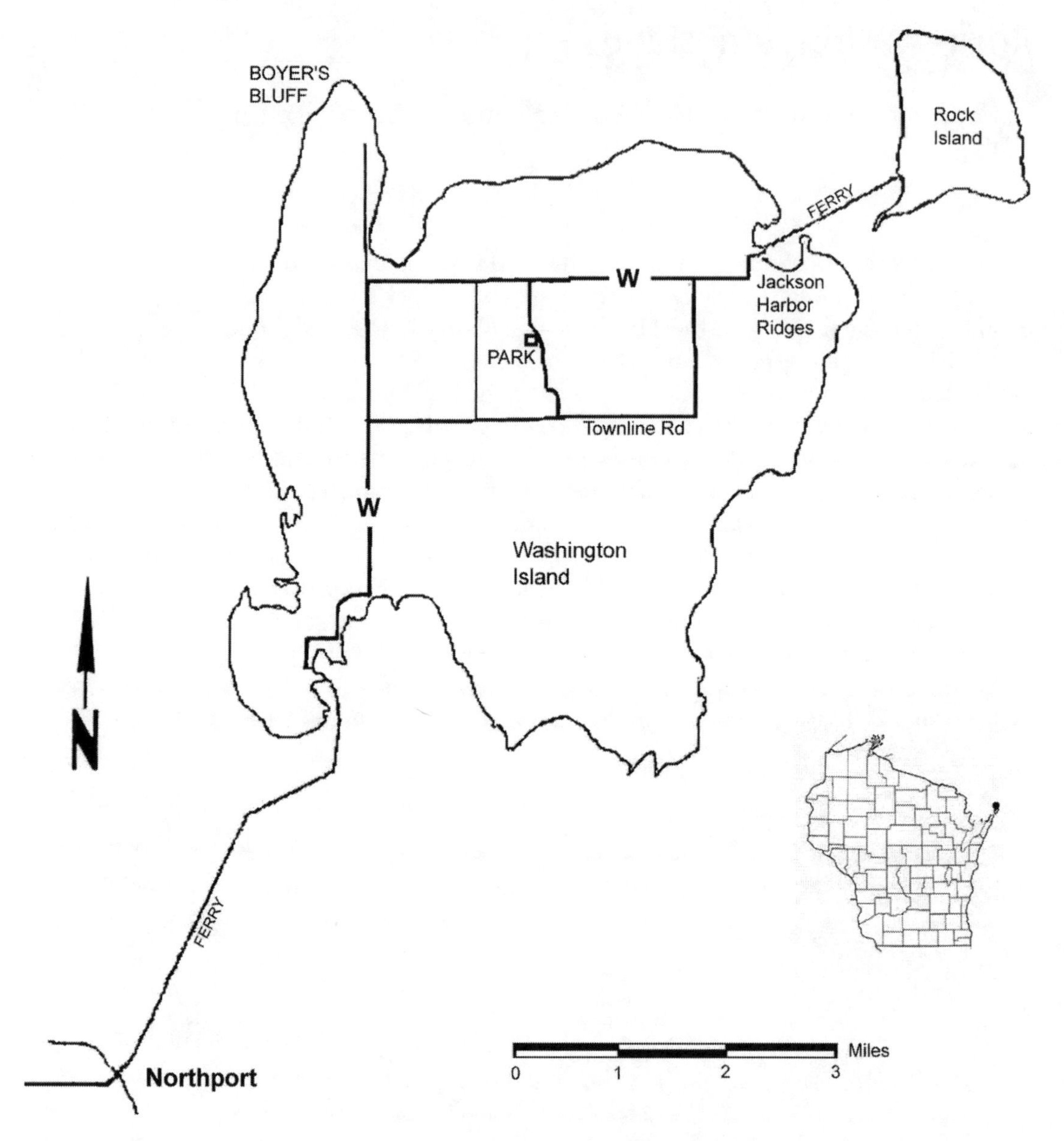

Rock–Washington Islands

most among the natural areas. These "ridges" form a northeast-facing complex of 12 ridges and swales. They occur in a narrow area where you can go from open beach to boreal forest within 400 feet. The combination of ridge–swale topography and advanced beach succession allows for a phenomenal species diversity. Developed trails provide the only access.

Another very critical habitat is the bluff area along the north shore of the island. These bluffs rise to 150 feet above the present lake level, although ancient lakes scoured various points on the bluffs. Boyer's Bluff has been above all glacial lake elevations and has provided refuge for some of Wisconsin's rarest plants. Near the center of the island is a small town park with steps leading to an observation tower. The cliffs along these steps harbor several rare plant species, such as the state-endangered green spleenwort. Elsewhere, the island contains forests, fields, conifer swamps, dune areas, and several parks. You can easily see all of these features by driving the roads on the island.

Specialties

Plants

- Chilean sweet cicely
- Crawe's sedge
- cuckoo flower
- dune goldenrod
- dune thistle
- dwarf lake iris
- few-flowered spike-rush
- Garber's sedge
- green spleenwort
- Lake Huron tansy
- low calamint
- northern comandra
- Ohio goldenrod
- one-flowered broomrape
- rock whitlow-grass
- slender bog arrow-grass
- small flowered grass-of-parnassus
- thickspike wheatgrass
- twin-stemmed bladderwort
- variegated scouring-rush
- western fescue

Other Features

- fossils

Uncommon Species

Plants

- Allegheny vine
- birds-eye primrose
- bog lobelia
- bristly black currant
- common whitlow-grass
- cow-wheat
- dwarf rattlesnake plantain
- early coralroot
- false asphodel
- fringed polygala
- green-flowered shinleaf
- herb-robert
- lanceolate whitlow-grass
- large yellow lady's slipper
- maidenhair spleenwort
- nodding ladies' tresses
- sand tickseed
- small green fringed orchid
- spurred gentian
- striped coralroot
- swamp loosestrife
- tall northern bog orchid
- walking fern
- white camas

Reptiles and Amphibians

- spotted salamander

Birds (Breeding)

- Caspian tern
- common merganser
- common tern
- eastern bluebird
- field sparrow
- mourning warbler
- Virginia rail

Mammals

- northern flying squirrel
- woodland deer mouse

Three Lakes Area

Communities: Northern Hardwood–Hemlock Forest, Forested Swamps, Northern Forested Bogs and Fens, Alder Thicket, Northern Sedge Meadow, Shallow Marsh, River

This area description centers on the town of Three Lakes because the location is readily accessible to some wonderful areas. A little conifer bog, well known to bird watchers, is just west of Three Lakes. This spruce–tamarack bog has a typical flora but also a certain attractiveness for northern birds. The real stand-out species is the boreal chickadee, with as many as 25 birds that congregate in winter-foraging flocks. This bog is one mile west of Three Lakes on County A, at the spot where the rounded corner replaced the still present but abandoned 90-degree corner. To the west is an extensive cranberry bog. Time will tell what effects this agricultural bog will have on the bird life.

The large Thunder Lake–Rice Lake wetland complex lies to the north. Take Highways 45 and 32 north from Three Lakes for about one mile to an access road for the wildlife area. Managed for waterfowl, these lakes have diverse faunas. Rice Lake's name is appropriate because a tremendous amount of wild rice grows there. Thunder Lake is much larger and seems to attract wayward birds, such as the anhingas, which has summered here in the past.

The Nicolet National Forest east of Three Lakes is ripe for exploration. There are many routes that can be followed; the forest roads are numbered and well maintained. If you are adventurous you can explore the expansive Headwaters Wilderness Area, which is so diverse that almost anything could show up.

If you have limited time, there is a short and easy but extremely productive route near the headwaters of the Pine River that highlights the best the forest has to offer. To reach it from Three Lakes, drive east on Highway 32 to FR2183. This road (sometimes called Lake Julia road) is about 6½ miles east of Three Lakes and just beyond FR2178. Proceed east on FR2183, or Scott Lake Road. Make sure you stay on this road because there are some spur roads in the

Spruce grouse (courtesy of Wisconsin Department of Natural Resources)

Spruce Grouse

Each lowland area possibly holds a spruce grouse or two, but the birds are not easy to find. The best way to locate a spruce grouse is to drive the roads at dawn, when one or two birds might be on the roads picking grit. Spruce grouse prefer extensive conifer areas. They tend to prefer conifer bog areas but will leave the bogs if the uplands are forested with nearly pure hardwoods. The spruce grouse remaining in Wisconsin are found in isolated pockets of conifer bogs, and the largest and best location for spruce grouse in the state is this headwaters area.

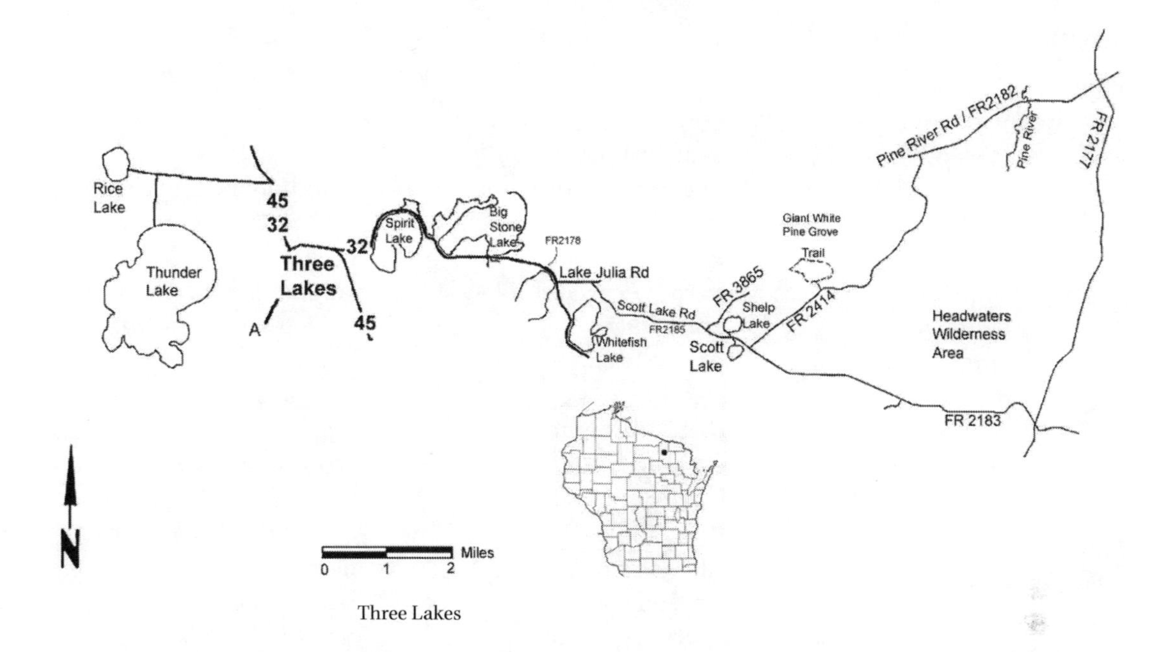

Three Lakes

first few miles. The well-marked Scott Lake–Shelp Lake Trail is the first stop. There is a boardwalk penetrating into the quaking bog of Shelp Lake. This trail allows close inspection of conifer and open bogs. To the south is a loop trail into the magnificent old-growth hemlocks that border Scott Lake. It is amazing to me how little these areas are used. I have been on these trails on weekend afternoons and not seen another person.

A short distance east of Scott and Shelp Lakes on FR2414, turn north to reach the Giant White Pine Grove. After traveling 1.65 miles north of 2183, you'll find a parking area at the trail head. Park here and walk west one-half mile on the woods road to the Giant White Pine Grove. You'll find a sugar maple–hemlock canopy with white pines up to 36 inches in diameter that form a super canopy. This spot gives the naturalist a picture of what northern Wisconsin hardwood–hemlock forests looked like before exploitation.

The route continues north on FR2414 to FR2182 then east to 2177 south, finally returning on 2183 to the start. The rest of this route goes through alternating uplands and wet swampy areas.

These low swampy areas are also home to a great variety of orchids; some of the rarest orchids in the state grow near this route. With time and patience, your forays into these swampy habitats will pay many rewards—perhaps a spruce grouse, a calypso orchid, or an "arctic" butterfly.

Specialties

Plants

calypso orchid
Clinton's wood-fern
dragon's mouth
little goblin fern
northern bog sedge
Robbin's spike-rush
showy lady's slipper
small round-leaved orchis
sparse-flowered sedge
white bog orchid

Animals

black-backed woodpecker
boreal chickadee
dorcas copper (butterfly)
freija fritillary (butterfly)
gray jay
jutta arctic (butterfly)
lynx
pine marten
red-disked alpine (butterfly)
spruce grouse
tawny crescent (butterfly)
Tremblay's salamander

Uncommon Species

Plants

blunt-leaved orchid
bog reed-grass
bog sedge
creeping sedge
delicate sedge
drooping woodreed
dwarf rattlesnake plantain
dwarf scouring-rush
fragrant white water lily
grass pink
green adder's mouth
heartleaf twayblade
marsh St. John's wort
moccasin flower
mountain holly
narrow-leaved sundew
pond sedge
prairie brome
purple avens
rusty cotton-grass
sheathed sedge
small green fringed orchid
snake-mouth orchid
spotted coralroot
spurred gentian
striped coralroot
water horsetail
water shield
wild calla
yellow sedge

Insects

purplish copper
tawny crescent

Fish

black nose dace
blacknose shiner
brook stickleback
finescale dace
mottled sculpin
northern red-belly dace
pearl dace

Reptiles and Amphibians

four-toed salamander

Birds (Breeding)

black-throated blue warbler
brown creeper
Cape May warbler
common loon
common merganser
Connecticut warbler
evening grosbeak
golden-crowned kinglet
golden-winged warbler
hooded merganser
Lincoln's sparrow
long-eared owl
mourning warbler
northern goshawk
olive-sided flycatcher
palm warbler
red crossbill
ruby-crowned kinglet
saw-whet owl
Swainson's thrush
Tennessee warbler
three-toed woodpecker
white-winged crossbill
yellow-bellied flycatcher

Birds (Winter)

Bohemian waxwing
common redpoll
pine grosbeak

Mammals

black bear
bobcat
fisher
hoary bat
long-tailed weasel
otter
star-nose mole
water shrew
woodland deer mouse

Vilas County

Communities: Dry Pine Forest, White Pine–Hardwood Forest, Northern Hardwood–Hemlock Forest, Forested Swamps, Northern Forested Bogs and Fens, Pine Barrens, Lakes

Vilas County is an expansive area with seemingly ideal natural habitats everywhere. The question is where to go to observe the best habitats? The Northern Highland American Legion State Forest offers more than 220,000 acres to explore, including all the stops in this chapter. If you prefer discovery, just pick any area and enjoy. But if you have only a limited amount of time or patience, the following route should help to show the range of diversity that Vilas has to offer.

A trip could begin north of Minoqua–Woodruff, east of Highway 51 and just south of County M at the well-known Black Tern Bog. This bog today appears misnamed because the large numbers of terns that used to nest here have not done so since the mid 1980s. Black Tern Bog has large numbers of pink orchids (dragon's mouth orchid, snake-mouth orchid, and grass pink). Though orchid locations are not usually published because of probable disturbances, Black Tern Bog is almost impossible to walk on, which affords these beautiful plants the best protection possible: inaccessibility. But you can observe orchids with binoculars from the road.

Continue the trip by traveling north on County M. Just north of the intersection with County N is Trout Lake conifer swamp, a fantastic white cedar swamp with a very diverse flora. The area is very swampy and full of mosquitoes, plus there is a good chance of coming face to face with a black bear. There is no way into the area except on foot. These facts are enough to keep the casual observers or the curiosity seekers away, but for the ardent naturalist, what a place to explore!

If you would like to get a taste of a swamp without the knee-deep mire, there is a beautiful trail at the north edge of Trout Lake. The well-marked trail is off County Highway M. This trail goes around two small bog ponds with their associated wetlands. Boardwalks traverse the wetter areas, although during the summer months visitors must contend with mosquitoes.

Continue north on M through Boulder Junction to County B. East of here are very different communities. From here travel east for 3½ miles to the intersection with High Lake Road. Take this road south about 1½ miles to Jute Lake Road, which extends eastward. By taking this road, which can be rough, you can find an extensive area of barrens and bracken grassland. About one mile east on this road, you will see Jute Lake to the north.

At the eastern end of this lake are logging roads going south across Garland Creek into the heart of the bracken grasslands near Johnson Lake. This is a large expanse extending eastward to several areas of headwater springs. Near these springs and along the associated streams are conifer bog formations with adjacent sandy uplands that have mostly aspen with scattered jack pine as a cover. The uplands are on a sandy plain with many wet pockets that allow great landscape diversity. With the exception of blueberry pickers, this area is virtually never visited by humans. The animal life is even less known than the plant life. On a trip into this area in June 1987, I discovered spruce grouse along Garland Creek and heard a yellow rail calling from

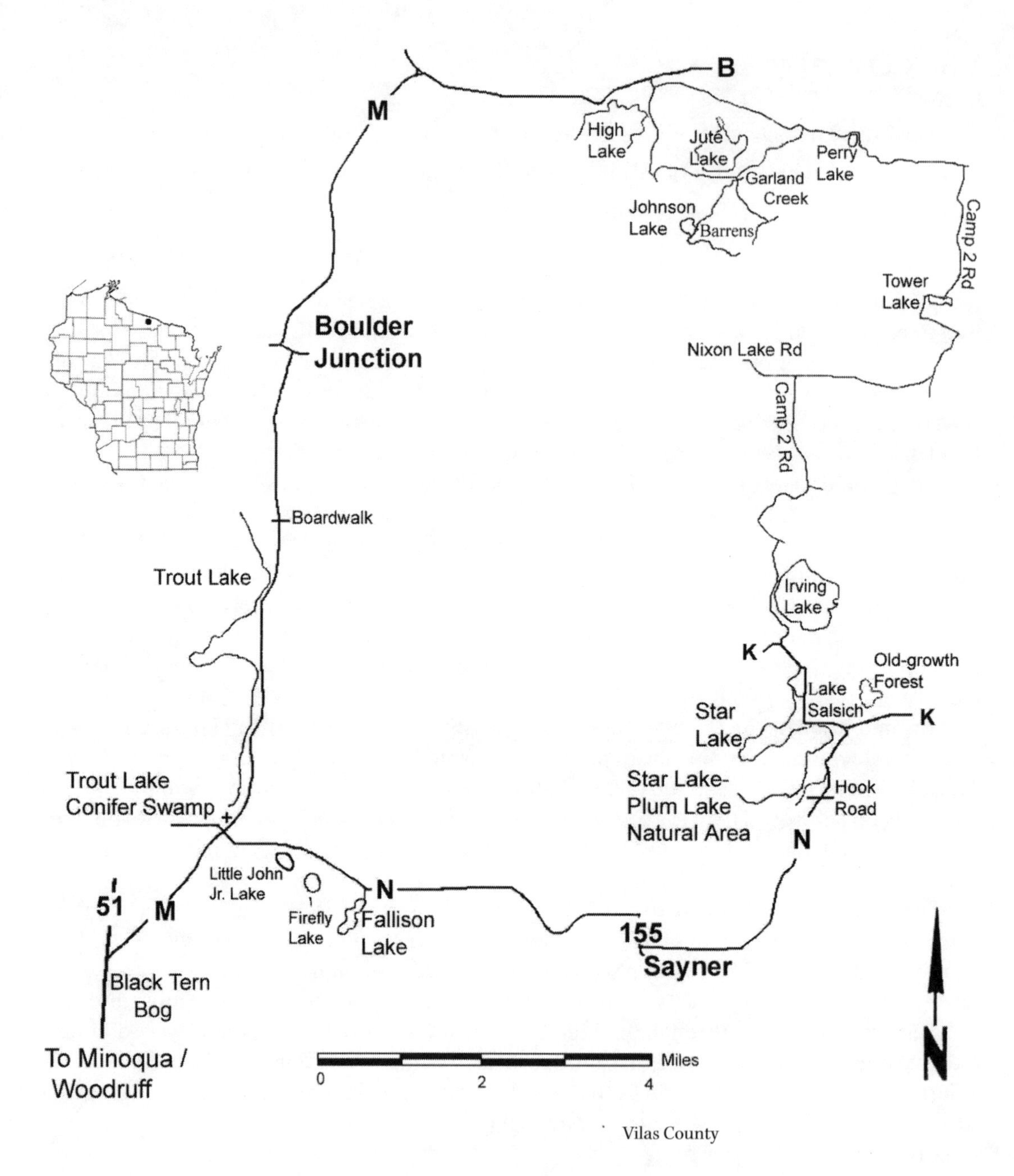

Vilas County

a sedge area, neither of which species was known to frequent the area but may have been there all along. This just might be an excellent area for discovery.

Continue the trip by following the forest roads east of Jute Lake. These poorly marked roads change direction often, but keep generally heading east. Jute Lake Road merges with old County Highway B. At this point you're still heading east. At

Lungworts

Old northern hardwood–hemlock forests may be the only places to appreciate fully arboreal lichens because the trees live long enough to allow the lichens to grow large. Nearly every tree in Wisconsin has lichens growing on it. The tree serves a supporting role, providing the lichen with a stable base. Several lichen species are found only on certain tree species. Others are found only after the trees reach a certain age. The large lichens can reach one to four inches in size and develop into a frill-like covering on tree trunks.

The lungwort lichen *Lobaria pulmunaria* is the most easily identifiable of the large lichens (see color insert). This species can grow to four inches in size but withers into flat brown structures that lie flat on the trunk. When it rains, however, the lichen absorbs water, becomes green, and expands into its large identifiable shape. At close range, the underside has pockets that are lung-like in appearance. This species only grows on hardwoods, especially maples in mature to old-growth forests.

the intersection with Tower Lake Road, turn south. Continue on this road to Camp 2 Road, then follow this south until the intersection with County Highway K just west of Star Lake. This eastern part of the trip is in a real backwoods area. East of Star Lake take County Highway N south about a mile to Hock Road, which leads to the Star Lake–Plum Lake Natural Area. The area between these lakes has an excellent hemlock–hardwood forest more than 250 years old with only minor disturbances.

Follow County N west through Sayner to the next stop, the nature trail at Fallison Lake. The well-marked trail traverses the hardwood forest surrounding this undeveloped lake. A different flora covers the ground, but the real treat is on the lake's sandy bottom.

Excellent sterile rosette floras are submerged in Fallison, Firefly, and Little John Junior Lakes. Several unusual rosette-forming plants occur in these very soft lakes; quillworts and plantain shoreweed are especially interesting because their geographical range centers in Wisconsin. In addition, there are good populations of small purple bladderwort.

From here it's just a short distance west to complete the circuit. By following this route, you can survey most of the Vilas County communities. This route does not cover all of the wonders of this vast area, but it does give a good sample.

Specialties

Plants

algal-leaved pondweed
blunt-lobed grape fern
Canadian black currant
Canadian ricegrass
daisy-leaved grape fern
dog lichen
dragon's mouth
least moonwort
lenticular sedge
lungwort lichen
plantain shoreweed
purple bladderwort
rose acacia
small purple bladderwort
twin-stemmed bladderwort

Animals

Connecticut warbler
laurentian skipper (butterfly)
West Virginia white (butterfly)

Uncommon Species

Plants

blunt-leaved orchid
bog St. John's wort
brownish sedge
Canada St. John's wort
dog violet
dwarf watermilfoil
fringed polygala
golden corydalis
heartleaf twayblade
Hooker's orchid
horned bladderwort
lake quillwort
marsh cinquefoil
northern bog sedge
northern cranesbill
northern sedge
purple virgin's bower
red honeysuckle
Robbin's spike-rush
round-leaved orchid
rusty cotton-grass
sheathed sedge
showy lady's slipper
smooth white violet
snake-mouth orchid
spiny-spored quillwort
stalked sedge
tall northern bog orchid
tesselated rattlesnake plantain
white beak-rush
woodland horsetail
wool grass
yellow monkey flower

Lichens

bug-on-a-stick
dog lichen
leather lichen

Insects

arctic skipper
Boll's grasshopper
bracken borer
brown elfin
cracker grasshopper
roadside skipper

Reptiles and Amphibians

spotted salamander

Birds (Breeding)

black-throated blue warbler
northern goshawk
olive-sided flycatcher

Birds (Winter)

Bohemian waxwing
pine grosbeak
red crossbill
white-winged crossbill

Mammals

American pygmy shrew
badger
black bear
bobcat
fisher
gray fox
hoary bat
lemming mouse
meadow jumping mouse
northern flying squirrel
otter
porcupine
snowshoe hare
southern saddle-back shrew
water shrew
woodland deer mouse

Whitefish Dunes State Park

Communities: White Pine–Hardwood Forest, Northern Hardwood–Hemlock Forest, Lake Dune, Lake Beach, Cliffs

Whitefish Dunes State Park is a focal point for exploration of the natural communities adjacent to Lake Michigan. Two distinct areas—dunes to the south and cliffs to the north—are within walking distance of each other. Although these areas touch, there is a definite demarcation. The dunes exemplify various stages of stabilization, from open wind-blown dunes to older dunes which grow beech forests. Largest among Wisconsin's dunes, Whitefish shows all stages of succession. There are some areas, mostly near Clark Lake, that have wetland affinities with some bog and sedge meadow species.

To the northeast of the dunes is an area of more resistant dolomite. The change can be readily seen at the lakeshore, which goes abruptly from sand beach to dolomite ledge. Cave Point is a very popular place because of the scenery reminiscent of the Maine coast. There are areas of wave-cut caverns, cliffs, small points, and, if the waves are big enough, some places with wave surge blow holes. Beyond the obvious shoreline amenities, there is a very nice beach forest inland from shore. A road traverses the park, but the only way to get an understanding of the natural communities and species is by hiking the trails.

Dune Specialists

The dune thistle, a federally threatened species, is protected at Whitefish Dunes, the largest dune system in the state. This thistle grows on the foredunes. Wind from the lake causes blowing sand, and the first series of dunes always has shifting sands. The dune thistle thrives on those shifting sands in more open areas among the beach grass. No wonder the plant is rare: the natural community is very limited, the habitat requirements within the dunes are also limited, and it's a favorite place of people for congregating.

These dunes are also home to the Lake Huron locust. This species is only found on dunes along the upper Great Lakes. The locusts inhabit the open sandy areas in the foredunes and feed on the sparse vegetation. This species is rare because of its limited habitat and because of the presence of people on the dunes that trample their food plants.

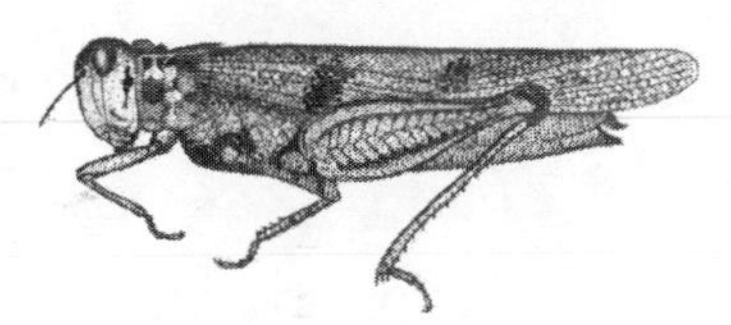

Lake Huron locust (illustration adapted by author)

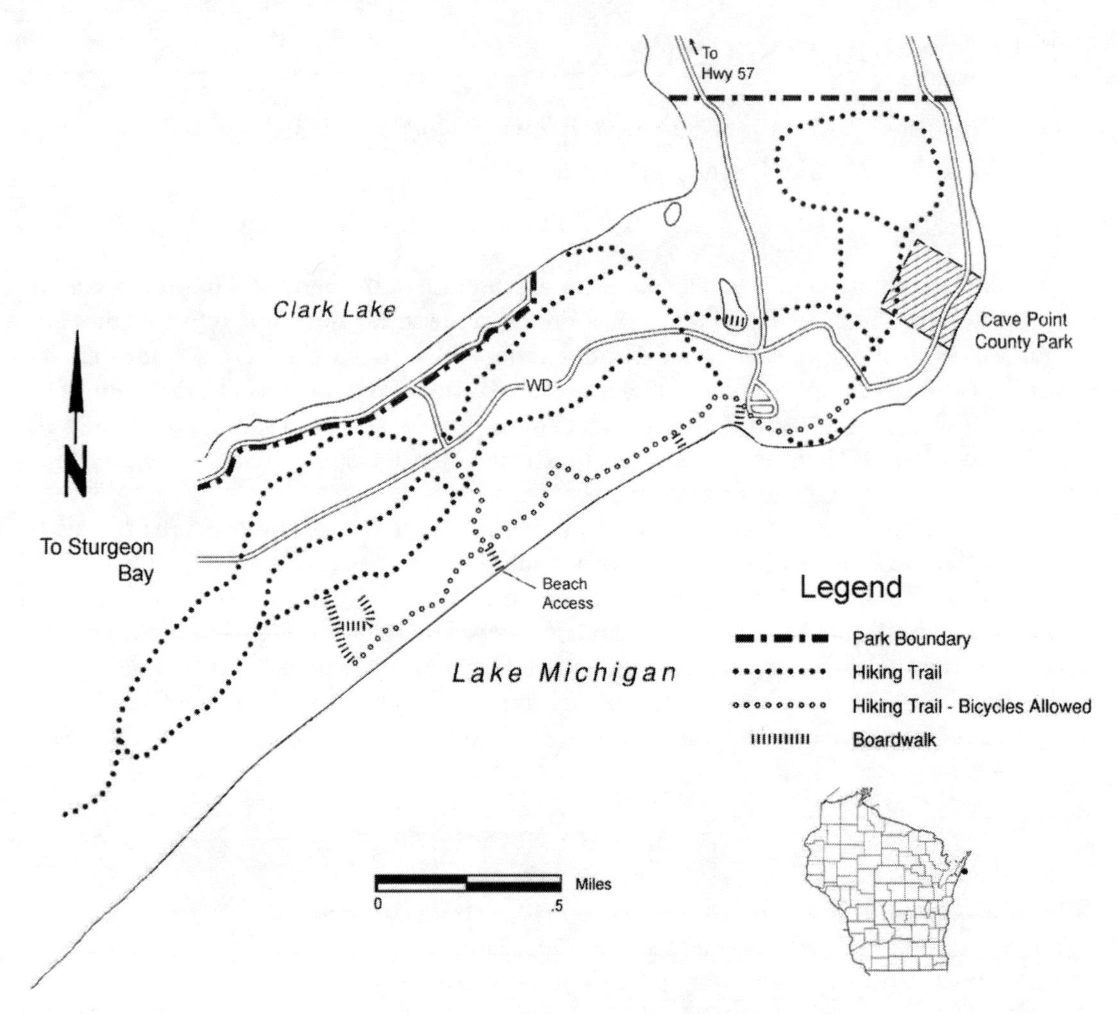

Whitefish Dunes State Park

Specialties

Plants
- cord moss
- dune goldenrod
- dune thistle
- long-spurred violet
- Ohio goldenrod
- sand tickseed
- thickspike wheatgrass
- white adder's mouth

Animals
- dainty tiger beetle
- Lake Huron locust

Uncommon Species

Plants
- beech drops
- bottlebrush sedge
- broad-leaved toothwort
- brownish sedge
- bur-reed sedge
- common bog arrow-grass
- common scouring-rush
- ditch stonecrop
- Dutchman's breeches
- flowering rush
- Garber's sedge
- golden sedge
- hair beak-rush
- hoary willow
- inflated sedge
- long-beaked sedge
- marsh cress
- prickly lettuce
- purple false oats
- rough cinquefoil
- sand-reed
- seaside spurge
- sessile bellwort
- shining willow
- small seed false-flax
- sweet William
- thimbleberry
- thyme-leaved speedwell
- toad rush
- twig rush
- upland wild timothy
- variegated scouring-rush
- white bear sedge

Mushrooms
- bluing bolete

Insects
- narrow-winged grasshopper

Birds (Breeding)
- field sparrow
- mourning warbler
- whip-poor-will

Birds (Migrants)
- black scoter
- Bonaparte's gull
- long-tailed duck
- ruddy turnstone
- surf scoter
- white-winged scoter

Mammals
- gray fox
- long-tailed weasel
- northern flying squirrel

Southeast Wisconsin

Chiwaukee Prairie

Communities: Dry-Mesic Prairie, Wet-Mesic Prairie, Wet Prairie, Bur Oak Opening, Lake Dune

If ever there was a textbook area for studying the conflicts over land use and planning, Chiwaukee would be the place. Monumental battles occurred over the "right" to develop and the "responsibility" to maintain vestiges of our natural heritage. The battles have been ongoing since the 1920s when developers subdivided the prairie.

In the 1960s, the Nature Conservancy began buying land for preservation. This slow intrusion into the developers' domain began to fester until the mid-1980s when a land-use plan was developed for the area. It identified 900 acres of the 1,660 acres of prairie to be preserved as open space. The remainder was available for continued development. Since then, prairie protection lot by lot has been about 70 percent successful, although building continues on prime prairie land.

Chiwaukee Prairie's location poses the biggest threat. It lies on Lake Michigan within an urban corridor, which makes it extraordinarily attractive to developers who have a waiting list of people who will pay almost anything to live on the lake. The combination of attractive real estate and a high-demand market imposes a costly burden on those who wish to protect the prairie. The slow methods of acquiring land one-quarter acre at a time may not ensure long-term protection of the species living there. For example, the federally endangered prairie white-fringed orchid depends on sphinx moths for pollination. If habitat continues to be lost or adjacent land-use practices decimate the sphinx moth populations, then the prairie white-fringed orchid's future becomes bleak.

Chiwaukee Prairie is a type of prairie known as a lake plain prairie. The nearness of a

Thismia

Sometimes the view of the natural world is very fleeting. On a prairie on the south side of Chicago in 1913 and 1914, a young botanist made the discovery of her lifetime. This very observant scientist found a small button of blue-green hiding under dense vegetation among some mosses. The mass was not much more than the size of a small puffball mushroom, but it was a flowering plant in full bloom. She gave the plant the scientific name *Thismia americana*. This plant of the lakeplain prairies has not been seen since it was discovered, and it is presumed to have become extinct within a few years of its initial sighting (Mohlenbrock 1983). This story illustrates the need to observe our natural world carefully. Since Chiwaukee Prairie is one of the last prairies of this type, who knows what surprises await the careful observer.

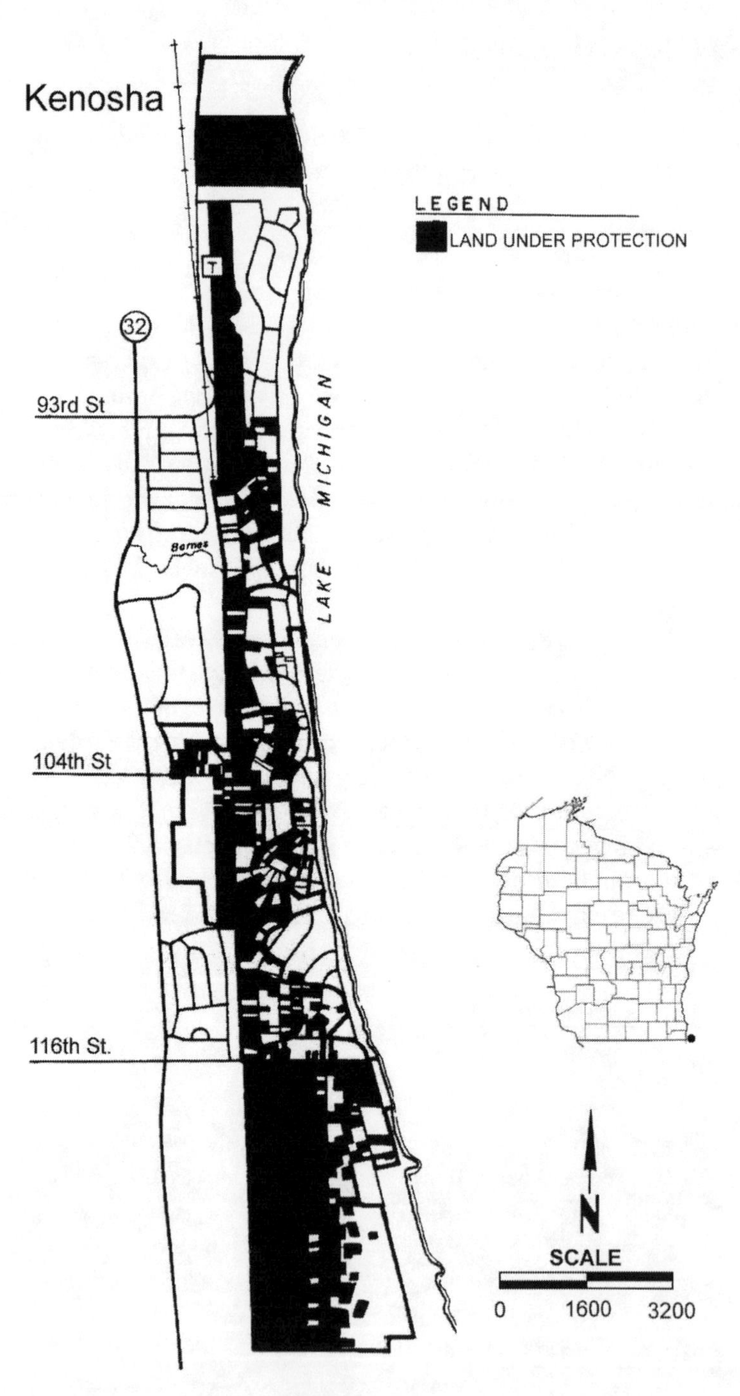

Chiwaukee Prairie

large lake was enough to influence prairie composition, by cooling the spring air and providing constant water only a few feet down, making it unique in Wisconsin. Downtown Chicago was the center of the lake plain prairie, which ran eastward along the lake to Indiana Dunes and northward to Chiwaukee Prairie. Chiwaukee and Illinois Beach State Park may be the only viable patches left. To many, the wonders of Chiwaukee Prairie belong to all the residents of Wisconsin, whose rights by far supersede the rights of a few to live by the water.

Ridges and swales are the defining landforms of Chiwaukee. Undulating two or three feet from high point to low, these landforms have dramatic differences in vegetation and animals. The sandy ridges have a dry sandy prairie flora, while a few feet away, a swale can be dominated by wet prairie, fen, or sedge meadow vegetation.

The best way to visit is by driving the roads of Carol Beach looking for State Natural Area signs or walking the trails near the Nature Conservancy sign.

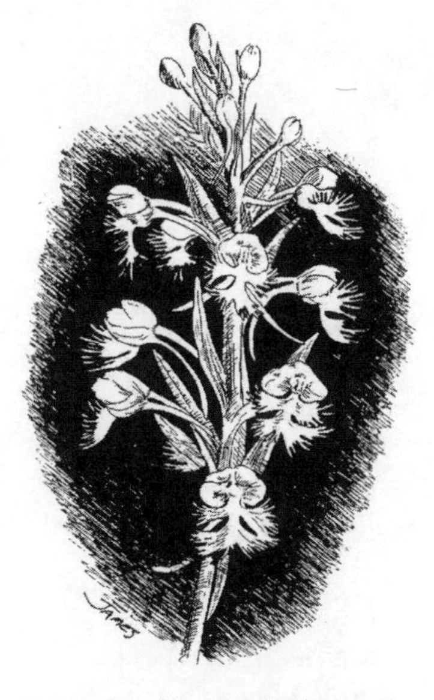

Prairie white-fringed orchid (illustration by Jim McEvoy; courtesy of Wisconsin Department of Natural Resources)

Specialties

Plants

- chestnut sedge
- false asphodel
- low nut-rush
- marsh blazing star
- Ohio goldenrod
- pale false foxglove
- pink milkwort
- prairie Indian plantain
- prairie milkweed
- prairie white-fringed orchid
- purple milkweed
- round-stemmed false foxglove
- sand-reed
- sand tickseed
- smooth phlox
- tall nut-rush
- white lady's slipper

Animals

- Blanding's turtle
- Butler's garter snake
- dion skipper (butterfly)
- Henslow's sparrow
- king rail
- liatris borer moth
- mulberry wing (butterfly)
- northern harrier
- silphium borer moth
- smoky eyed brown (butterfly)

Uncommon Species

Plants

- autumn sedge
- awl sedge
- Bebb's sedge
- Bicknell's sedge
- bog goldenrod
- bog lobelia
- bottlebrush sedge
- bulbous cress
- Canada milk-vetch
- Canadian rush
- cardinal flower
- clammy ground cherry
- clustered beak-rush
- colicroot
- common blue-eyed grass
- common bog arrow-grass
- cream wild indigo
- cursed buttercup
- ditch stonecrop
- downy gentian
- ear-leaved brome
- false sunflower
- fescue sedge
- field milkwort
- fox sedge
- fringed gentian
- glaucous white lettuce
- grass pink
- groundnut
- hair beak-rush
- hairgrass
- hairy beard-tongue
- Kalm's St. John's wort
- lance-leaved ground cherry

large-flowered beard-tongue
leafy satingrass
low calamint
lupine
marsh cress
mat panic grass
Mead's sedge
monkey flower
nodding ladies' tresses
nodding wild onion
pale beard-tongue
panicled sedge
prairie alum-root
prairie blue-eyed grass
prairie brome
prairie lily
prairie panic grass
prairie sedge
prairie violet
purple milkwort
Richardson's sedge
Riddell's goldenrod
Robin's fleabane
rough false foxglove
sand milkweed
sand primrose
Seneca snakeroot
short green milkweed
smooth false foxglove
starry campion
swamp thistle
swan sedge
sweet grass
tall green milkweed
turtlehead
twig rush
waxy meadow-rue
white wild indigo
wild onion
woolly sedge
yellow pimpernel
yellow stargrass
yellow water crowfoot

Insects

Acadian hairstreak
Atlantic grasshopper
cloudless sulphur
Culver's root borer moth
fiery skipper
goatweed butterfly
hoary elfin
lined ruby tiger moth
meadow-rue borer moth
northern flower moth
pandorus sphinx
pink-streak
salt marsh moth
seaside grasshopper
sedge skipper
short-lined chocolate
sleepy orange
turtlehead borer moth
wrinkled grasshopper

Reptiles and Amphibians

Fowler's toad (probable)

Mammals

hoary bat
least weasel
silver-haired bat

Harrington Beach State Park

Communities: Northern Hardwood–Hemlock Forest, Forested Swamps

Harrington Beach is a small park, just over 600 acres, but this speck of land is a meeting place of sand and surf. The water–land interface is obvious, but botanically, this is the tension zone, where many species reach the limits of their range. Northern species and southern species meet here, and this is most obvious in the ash–cedar forest. You can detect the meeting place by looking at the vegetation: the smatterings of yellow birch, red maple, and white birch, plus many herbaceous species with northern affinities mixed with beech–sugar maple forest, signify that neither the northern forest nor the southern forest will predominate. Another meeting place, a geological tension zone, is underground. Devonian rocks, with their great assortment of fossils including brachiopods, cephalopods, gastropods, bryozoans, trilobites, crinoids, coral, and fish plates, are exposed at the quarry. Just a few miles away Silurian rocks with fewer fossils overlay these older Devonian rocks.

These areas of flux seem to translate into highly varied animal usage. Harrington Beach is famous in birding circles for its value as a migrant trap, a combination of landforms and prevailing winds that concentrate birds in certain locations. Both small migrant birds and hawks tend to concentrate at Harrington Beach when westerly winds force birds to the lake, over which they are reluctant to cross. These concentrations are especially evident here because birds have to backtrack to stay over land. By going directly south from Harrington Beach, a bird would not make landfall again until it reached central Racine County. Water covers a great westward bend of the shore of Lake Michigan along Milwaukee and most of Ozaukee Counties. The tired migrants must now fly back into the prevailing winds, which causes huge concentrations of birds.

Inland from the lakeside forest is more than 400 acres of oldfield grassland that supports habitat for grass-loving birds. Foremost among these is the short-eared owl. If there can be a place called consistent for this highly erratic species, this would be it. These owls are regular visitors in late fall and winter.

Sea Ducks

The offshore area produces large numbers of scoters and grebes when very few birds can be found elsewhere along the lake. These "sea ducks" (white-winged scoter, surf scoter, black scoter, long-tailed duck, and greater scaup) concentrate off Harrington Beach in October and November in great flocks called rafts.

These birds are all divers and can reach great depths in their search for favorite foods. Long-tailed ducks have been caught in gill nets set at 240 feet below the surface (Bellrose 1976). They prefer feeding near the bottom for food like mussels, snails, crustaceans, and aquatic insect larvae. The appearance of these sea ducks year after year in the same location must attest to a rich food source. Occasionally other rare seabirds such as harlequin ducks, king eiders, brants, and Pacific loons have been seen in these large rafts.

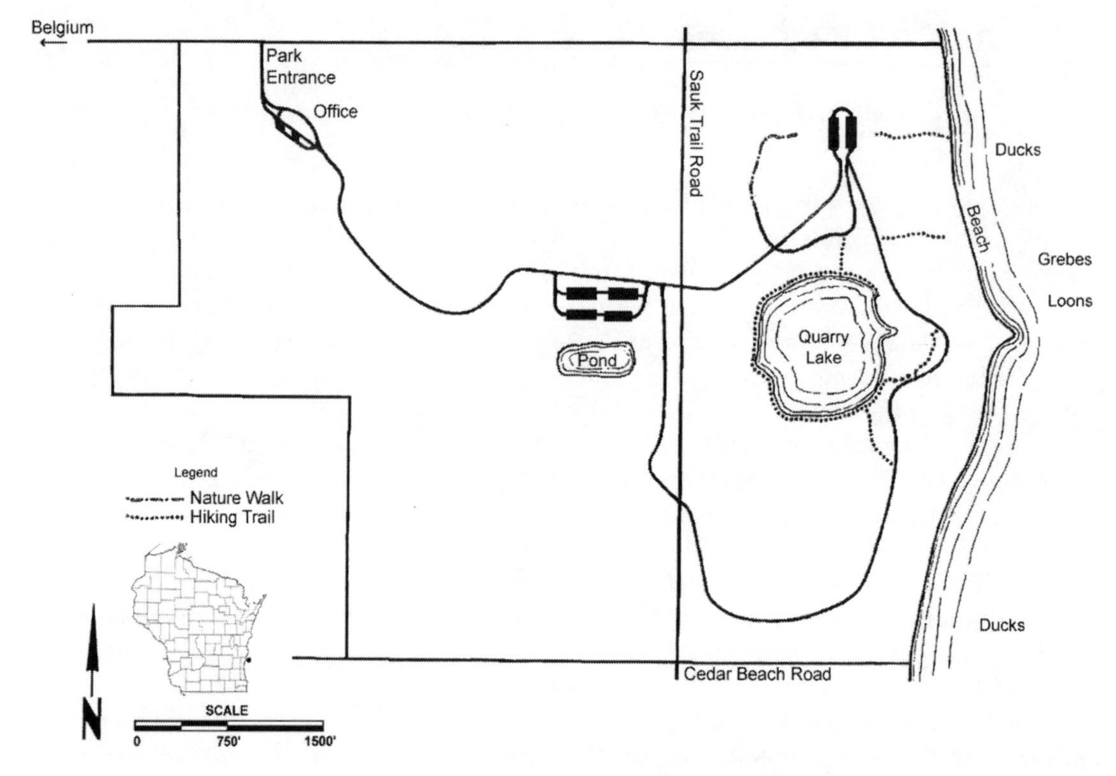

Harrington Beach State Park

Specialties

Plants

hairy beard-tongue
helleborine
sea rocket
seaside spurge
slender bog arrow-grass

Animals

migrating diving ducks
migrating grebes
migrating hawks
migrating loons
Reakirt's blue butterfly

Other Features

Fossiliferous Devonian outcrops

Uncommon Species

Plants

arrow-leaved tearthumb
baltic rush
beech drops
bur-cucumber
cardinal flower
columbine
common bugseed
common mallow
creeping snowberry
cup plant
dwarf ginseng
early coralroot
green dragon
groundnut
highbush cranberry
Indian pipe
joint rush
large-leaved shinleaf
marsh cress
moccasin flower
moonseed
northern bugleweed
partridgeberry
purple giant hyssop
purple spring cress
round-leaved shinleaf
sand tickseed
scarlet pimpernel
sessile bellwort
side-flowering aster
silverweed
smartweed
smooth scouring-rush
striped coralroot
swamp false Solomon's seal
tall buttercup

tall lettuce
tall northern bog orchid
wild golden glow
winterberry
wintergreen
witch-hazel
woodland sedge
wood nettle
yellow wood violet

Insects

black dash
Boll's grasshopper
checkered white
northwest red-winged grasshopper
puzzling dagger
robust cone-head
three-spotted nola

Birds (Breeding)

eastern bluebird
field sparrow
upland sandpiper

Birds (Migrants)

bald eagle
bay-breasted warbler
blackpoll warbler
black scoter
Bonaparte's gull
brant
bufflehead
canvasback
Caspian tern
common goldeneye
common loon
common tern
double-crested cormorant
dunlin
Forster's tern
fox sparrow
Franklin's gull
golden-crowned kinglet
gray-cheeked thrush
greater scaup
Harris' sparrow
hooded merganser
horned grebe
lapland longspur
lesser scaup
Lincoln's sparrow
little gull
merlin
mourning warbler
northern goshawk
orange-crowned warbler
osprey
Pacific loon
peregrine falcon
Philadelphia vireo
red-breasted merganser
red crossbill
redhead
red knot
red-necked grebe
red-shouldered hawk
red-throated loon
ruddy turnstone
rusty blackbird
sanderling
saw-whet owl
semipalmated plover
short-eared owl
snowy owl
surf scoter
tundra swan
water pipit
western grebe
white-crowned sparrow
white-winged scoter
Wilson's warbler
yellow-bellied flycatcher

Birds (Winter)

bufflehead
glaucous gull
greater scaup
harlequin duck
long-tailed duck
northern shrike
short-eared owl
snowy owl

Horicon Marsh

Communities: Southern Sedge Meadow, Cattail Marsh

Horicon Marsh is under dual ownership: the State of Wisconsin owns and manages the southern portion, which consists of around 11,000 acres, and the U.S. Fish and Wildlife Service administers the northern portion as a national wildlife refuge, which covers approximately 21,000 acres. Access is much different for both sections. The southern portion is mostly open to the public. Canoeing is the best way of exploring, and you'll have many memories if you spend a day on the marsh. Unfortunately, most of the prime wildlife concentrations are on the federal portion where access is restricted. Here, access points are two dead-end roads used for fishing, a series of trails that barely touch the northwest corner of the marsh proper, and the crowded and sometimes dangerous Highway 49 that bisects the marsh at the Dodge–Fond du Lac county line.

Horicon Marsh is a magnificent area for concentrations of wildlife. Highway 49 east of Waupun is the best known and the most used observation area. From this road, you can see hundreds of thousands of geese, ducks, and other marsh birds. Tremendous numbers of wildlife watchers congregate around the marsh in October and early November to view the huge waterfowl flocks.

On the west edge of the refuge and south of Highway 49 a short spur road leads to the wildlife trails. These trails traverse mostly oldfield habitat, but a boardwalk and observation platform enters a portion of marsh. During years when southern herons disperse, this area has been truly exciting. Other access points are dead-end fishing access roads on the refuge's eastern side. These can be reached by going south of Highway 49 on County Highway Z. The first dead end is Ledge Road, the second road south. This road enters the marsh for a short distance, affording views of the less well known but common marsh inhabitants. You may want to stop at the refuge headquarters (look for the sign along County Highway Z) to obtain information or a checklist or register your approval for more access.

By continuing on County Highway Z then turning south on County Highway Y into Kekos-

Migrant Herons

Late fall brings migrating northern ducks and geese, which tend to attract the majority of wildlife watchers to Horicon Marsh. Remarkably, few people realize that migrating southern species show up in late summer, from late July through mid-September. At this time of the year, southern herons show up on a regular basis somewhere on the marsh. Cattle egrets and snowy egrets can be found nearly every year. Little blue herons are almost annual. Tricolored herons are more sporadic but usually are found every third or fourth year. In addition to the herons, Horicon seems like a magnet for other southern birds. Several records exist for American avocet, black-necked stilt, glossy or white-faced ibis, black-bellied whistling duck, and Egyptian goose. The last bird was obviously an escapee from a zoo . . . or was it?

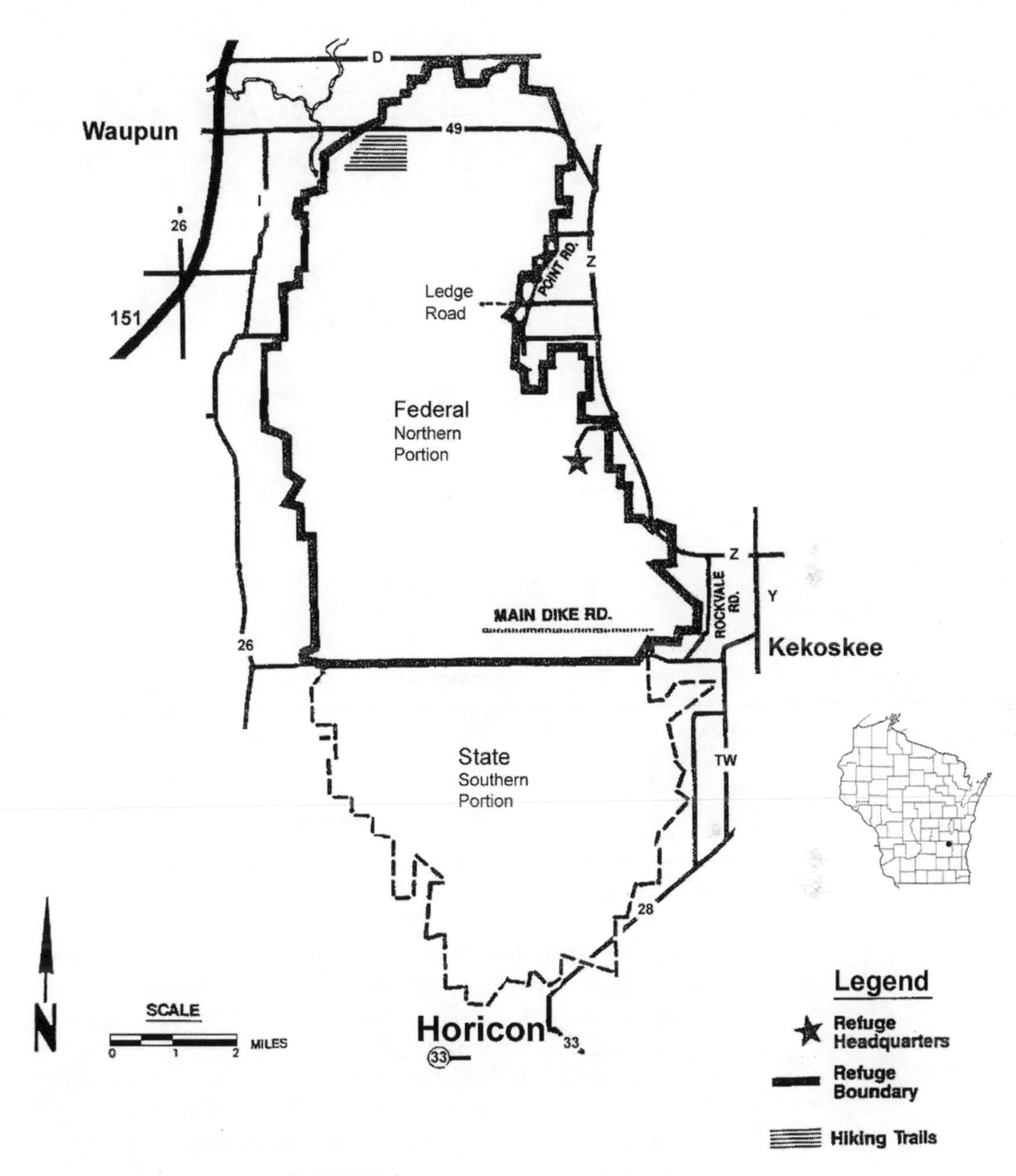

Horicon Marsh

kee, you can turn west onto County Highway TW, which leads to the Main Dike Road. This road is only open from May 15 to September 15 and can be traveled only as far as the dam. While it goes deep into the marsh, there is virtually nowhere to park, although by driving slowly along the road, you can see most of the animals.

To the west of Horicon Marsh is an area of farmland that

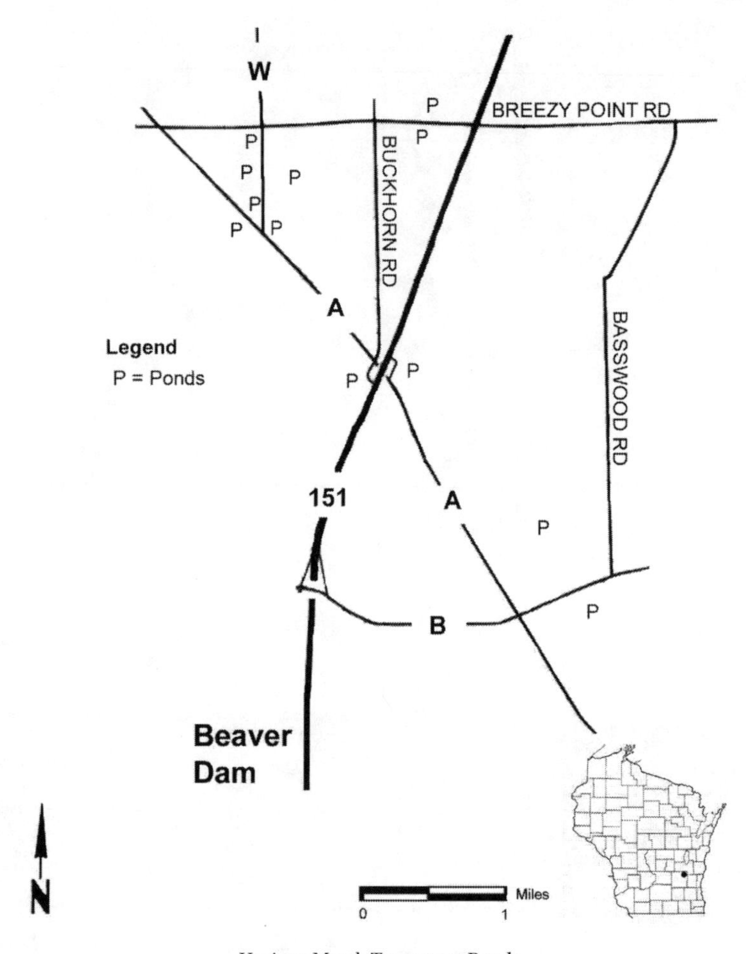

Horicon Marsh Temporary Ponds

floods regularly to form a very unusual and little understood habitat, that of the temporary pond. It is north of Beaver Dam and bounded by county roads A and W and Breezy Point Road. The temporary ponds here are products of flat former prairie lands and shallow flooded basins. They form in spring with the snow melt or the annual flood and are dry by fall in most years because of the evaporation that occurs during summer. These areas have an unusual invertebrate fauna with some species that only live in temporary pools, such as snow melt mosquitoes and some species of bloodworms. They are staging areas for waterfowl and especially shorebirds, which historically have used these environments because there is abundant invertebrate food that they use for refueling on their epic migrations.

The plowing of this area has served these migrants well. Plant life is sparse, but the invertebrate food is abundant, and in most years there is enough water to produce mud flats. Us-

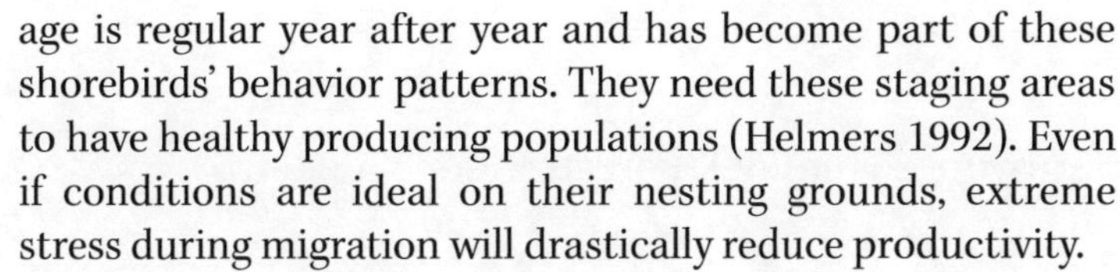

age is regular year after year and has become part of these shorebirds' behavior patterns. They need these staging areas to have healthy producing populations (Helmers 1992). Even if conditions are ideal on their nesting grounds, extreme stress during migration will drastically reduce productivity.

Because this habitat is temporary in nature and serves only passage species, it is never purposefully managed. The opposite is what usually happens. When an excellent shorebird area is purchased, the public agency usually manages it but not for shorebirds. The mud flats revert to grassland (many times to reed canary grass) or are planted to dense nesting cover for purposes of raising a duck or two. There is nothing wrong with this practice, but the decision-making process usually involves consideration for the nesters and not the migrants. Continued loss of these valuable temporary ponds will have a deleterious effect upon shorebirds in the future. Even with ideal nesting and wintering areas, loss of migratory staging areas will result in fewer shore birds.

Specialties

Birds

- cattle egret
- little blue heron
- snowy egret
- tricolored heron
- waterfowl concentrations

Insects

- cattail borer
- oblong sedge borer

Uncommon Species

Plants

- fringed brome
- golden alexanders
- marsh purslane
- slender false foxglove
- stiff water crowfoot
- sweet flag
- yellow water crowfoot

Birds (Breeding)

- common moorhen
- Forster's tern
- gadwall
- king rail
- least bittern
- northern pintail
- upland sandpiper
- Virginia rail
- Wilson's phalarope

Birds (Migrants)

- American avocet
- buff-breasted sandpiper
- dunlin
- hudsonian godwit
- long-billed dowitcher
- marbled godwit
- red knot
- red-necked grebe
- red-necked phalarope
- short-billed dowitcher
- stilt sandpiper
- western sandpiper
- white pelican
- willet
- yellow-crowned night heron

Kohler–Andrae State Park

Communities: Northern Hardwood–Hemlock Forest, Alder Thicket, Lake Dune, Lake Beach

Kohler–Andrae State Park covers 760 acres that feature lake dunes in various vegetative states. The two trails of the park can be taken to view the different dune vegetation. The Indian Pipe Trail at the south end near the campground goes through old, well-stabilized forest. This is a transition area between northern forest and southern forest. Many northern and southern species form a unique mix here, growing side by side in an uncommon setting.

The other trail is the Dunes Cordwalk Trail, next to the nature center. Here the dunes are more open with grasses and flowers holding the sands in place. These are the largest open dunes in Wisconsin and have species restricted to dunes growing on them. The dominant species is beach grass. Along the dunes trail, some areas are beginning to succeed into forest. Some spots have an abundant groundlayer of creeping juniper. The dunes proceed through open pine stands and pole timber forest to mature forest. This is the cycle of dune succession.

There are some low areas within the dunes that reach nearly to the water table. These interdunal wetlands have a much different species composition. Baltic rush grows here in almost pure mats. But other species do find a home here, especially ladies' tresses orchids, which brighten any August day. Farther inland and running parallel to the dunes is an almost impenetrable thicket of alder, which is the best developed and largest in the region.

Fish-burying beetles (illustration by author)

Beach Bugs

People love to spend time on the beach, although most don't notice the natural world around them while there. At Kohler–Andrae keep an eye out for a fast-flying little bug that never lets anyone get too close. These are tiger beetles, Wisconsin's most efficient beach predators that eat other insects.

For the truly inquisitive, the beach offers an exceptional opportunity to view how death becomes life. It takes patience, a strong stomach, and a weak nose. First of all, find a dead fish. Not a real fresh one but something the gulls pass by. Sit down beside the fish and observe. You'll probably see large black-and-orange bugs crawling all over the fish. These critters are dead fish-burying beetles. They work together to dig the soil from under the fish and bury it. The beetles will then lay their eggs in the rotting flesh so their young will have something tasty to eat when they hatch.

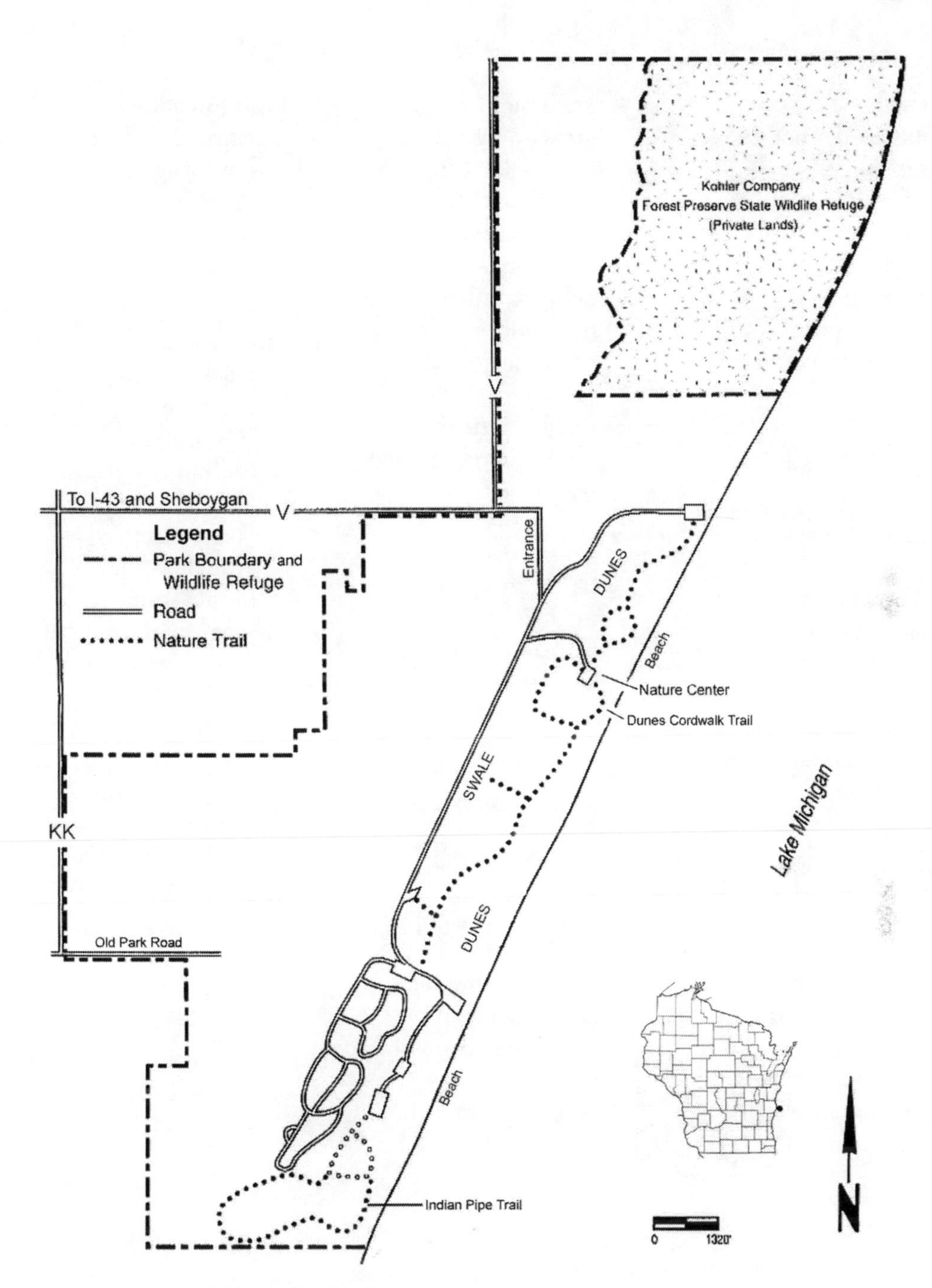

Kohler–Andrae State Park

Specialties

Plants

clustered broomrape
dune thistle
dune willow
prairie moonwort
sand-reed
sand tickseed
seaside spurge
thickspike wheatgrass

Animals

deep water migrants
fringed dart (moth)
inordinate apamea (moth)
Manitoba dart (moth)
pale lichen moth
phyllira tiger moth
rosewing (moth)
ruddy quaker (moth)
seaside grasshopper
rubbed dart (moth)

Uncommon Species

Plants

beech drops
blanketflower
bristly crowfoot
Canada wild rye
cinnamon willow-herb
common bugseed
early coralroot
grass pink
great St. John's wort
hair beak-rush
hairy beard-tongue
leather-leaved grape fern
long-beaked sedge
nodding ladies' tresses
pearlwort
purple false oats
ragged fringed orchid
Richardson's sedge
round-leaved sundew
sessile bellwort
snowberry
sweet grass
tall mannagrass
turtlehead
variegated scouring-rush
winterberry

Beetles

dainty tiger beetle

Insects

Atlantic grasshopper
black dash
Boll's grasshopper
checkered white
large-headed grasshopper
narrow-winged grasshopper
robust conehead

Birds (Breeding)

clay-colored sparrow
field sparrow
green heron
red crossbill
Virginia rail

Birds (Migrants)

black-bellied plover
black-throated blue warbler
Cape May warbler
Caspian tern
Forster's tern
Franklin's gull
horned grebe
long-eared owl
long-tailed duck
merlin
northern goshawk
red knot
ruddy turnstone
short-eared owl
water pipit
white-winged scoter
willet

Mammals

gray fox
meadow jumping mouse
red-backed vole

Maribel Caves

Communities: Sugar Maple–Basswood Forest, Northern Hardwood–Hemlock Forest, Forested Swamps, River, Cliffs

Maribel Caves is a quaint little county park along County Highway R, formerly Highway 141, north of Maribel in Manitowoc County. Sugar maple–beech forest covers the land on top of the ledge, and white cedar dominates along the cliff edge and near the river. There is a 50-foot-high cliff running the length of the park. Along this cliff are small solution caves indented into the cliff face. The cliffs have an abundance of ferns, and in some places the walking fern is as thick as anywhere known in the state. The flora is very rich, with snow trillium found in the vicinity. This entire Twin River valley is an excellent spot for this rare plant, with additional colonies along Highway 147 and farther south near County YY. Just to the north is another rare plant, the Christmas fern. One good location to see this plant is two miles north of the Kewaunee County line and Highway Q. This plant grows on private land requiring permission to enter, but it also should occur elsewhere in the vicinity.

The Maribel Caves area is small—only a few dozen acres—but it harbors a legacy of the area's natural history. Many similar areas in eastern Wisconsin are small, but they, too, may be refuges for rare plants and invertebrates. Most of these plant and invertebrate habitats are on private land, and most often the landowners do not know what's on their land. I hope that as landowners become educated, they will identify and preserve these small holdings of exceptional quality.

Walking Fern

As you are walking the trails at Maribel Caves, notice any rock. Chances are very good you will see one of the strangest ferns in the state growing on the rock's surface. The walking fern grows abundantly on the rocks at this park. It only grows on limestone, so you won't find these ferns throughout much of the state. This fern is different than most ferns because the narrow tapering frond can grow a new plant from its tip. The frond arches toward the rock's surface. When the tip reaches the rock it establishes a foothold, and a new walking fern frond begins to grow.

Walking fern (illustration by Jim McEvoy; courtesy of Wisconsin Department of Natural Resources)

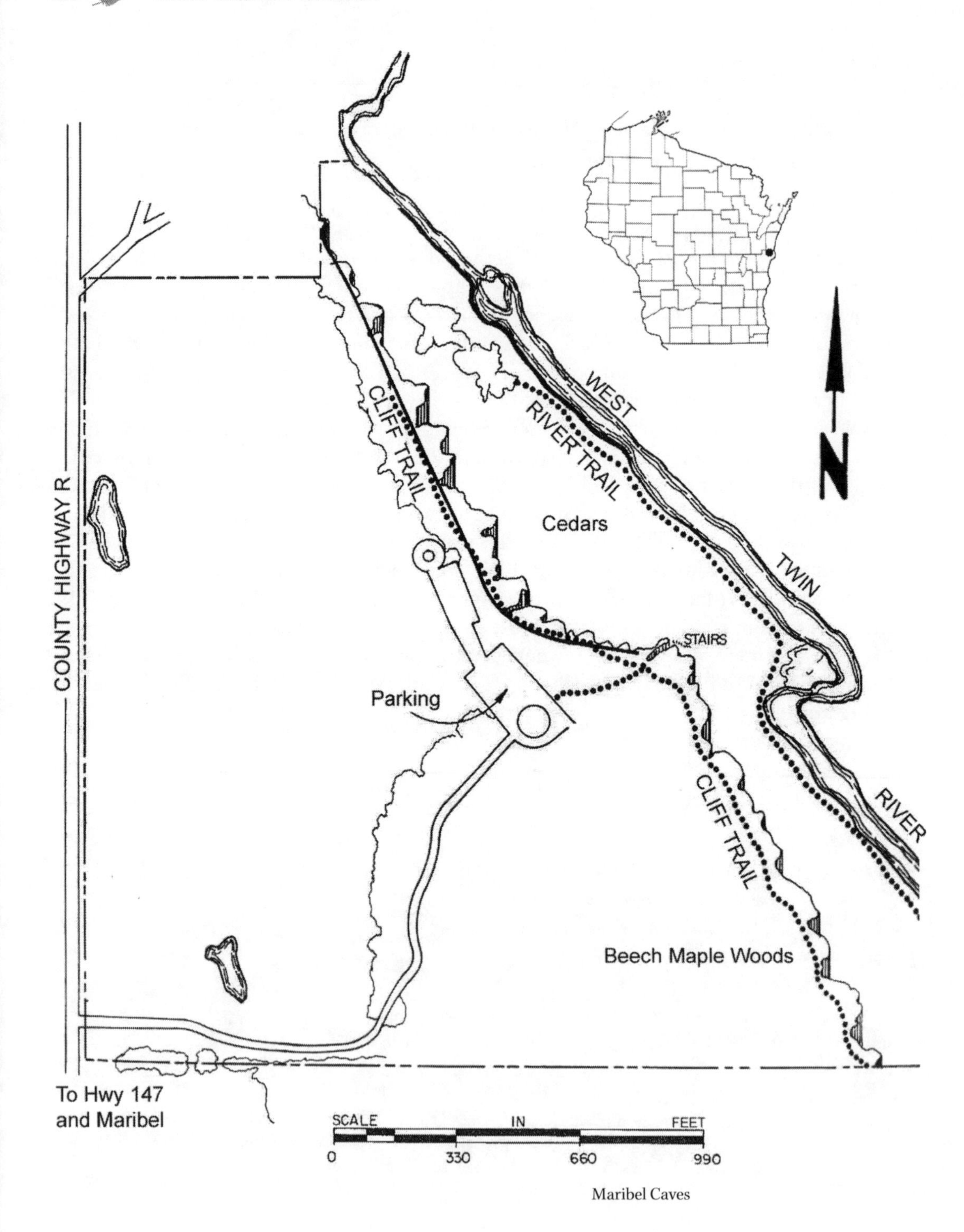

Maribel Caves

Specialties

Plants

Allegheny vine
Christmas fern
snow trillium
twinleaf

Animals

golden redhorse (fish)

Other Features

caves in the glaciated area

Uncommon Species

Plants

bladdernut
bloodroot
bristly sarsaparilla
Carolina spring beauty
clearweed
cliff goldenrod
Dutchman's breeches
early meadow-rue
false Solomon's seal
golden ragwort
great St. John's wort
hairy goldenrod
large yellow lady's slipper
late figwort
maidenhair fern
marsh speedwell
moonseed
mountain maple
nodding trillium
ostrich fern
pellitory
rattlesnake fern
red elderberry
rock cress
rock spikemoss
round-leaved serviceberry
round-lobed hepatica
side-flowering aster
silvery glade fern
smooth cress
smooth serviceberry
smooth sweet cicely
snowberry
squirrel corn
swamp false Solomon's seal
tall white violet
Virginia waterleaf
walking fern
water forget-me-not
water pennywort
white snakeroot
wild ginger
wild leek
winterberry
wiregrass sedge
yellow jewelweed
yellow trout lily

Birds

mourning warbler
whip-poor-will
yellow-bellied flycatcher

Birds (Migrants)

Connecticut warbler
white-crowned sparrow
Wilson's warbler

Birds (Winter)

evening grosbeak
red crossbill
white-winged crossbill

Mammals

gray fox

North Kettle Moraine

Communities: Southern Red Oak–Mixed Forest, Sugar Maple–Basswood Forest, Floodplain Forest, Forested Swamps, Shrub-Carr, Shallow Marsh, Lakes, Rivers

The north unit of the Kettle Moraine State Forest is a vast forest land for southern Wisconsin, covering about 26,000 acres. Because of its location near population centers, this unit has considerable recreational use. If you have an interest in glacial geology, you'll find this state forest very accommodating, with many trails and auto tours that lead to the glacial features of the forest. The better places to view glacial features are the Parnell Esker at the Butler wayside, the Ice Age Trail, and Dundee Mountain. You can get information on exact locations and interpretation of glacial geological features by visiting the Ice Age Visitors Center just south of Dundee. The best place to observe a drumlin, which is an oval-shaped hill, is northwest of Campbellsport. To find these features take County V north to County Y, then proceed west into the heart of the drumlins.

With such a large area, the most obvious question is: "Where do I go to see the natural areas?" You can see a very good cross section of these areas by hiking the existing trails. The

Esker (Endangered Resource file photograph; courtesy of Wisconsin Department of Natural Resources)

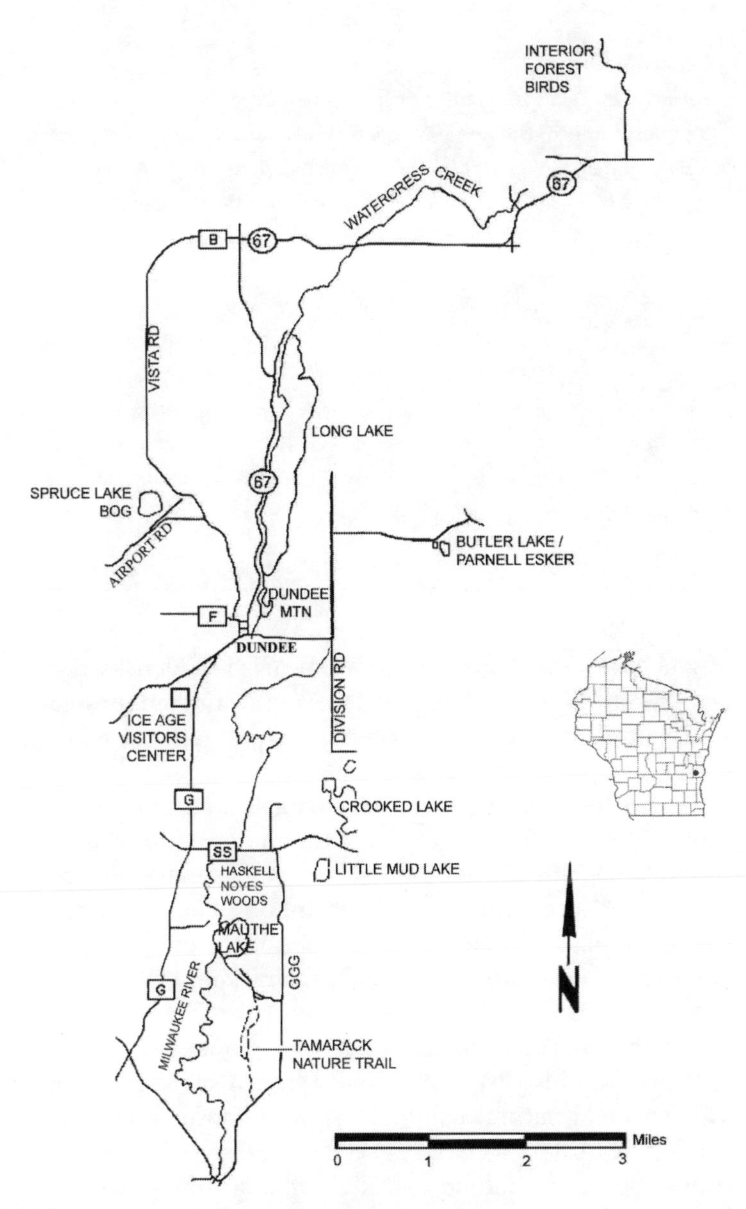

North Kettle Moraine

Glacial Trail is the longest, winding its way through the park. There's a nature trail at the Long Lake recreation area and one that circles Mauthe Lake called the Tamarack Nature Trail. The Zillmer hiking and skiing trails near the Ice Age Visitors Center also offer a pleasant hike. If hiking is not available, then a good overview can be had by driving the well-marked Kettle Moraine scenic drive.

The above gives the amateur naturalist a good sampling of the Kettle Moraine, but the outstanding natural communities of the forest remain in scattered pockets. Since these areas are

Hooded Warbler

Several neotropical migrant bird species require large blocks of forest. The north unit of the Kettle Moraine State Forest offers enough large forested areas to attract significant populations of these habitat-restricted birds. The biggest block of forest is near the Greenbush Trails. This block is designated as a Habitat Protection Area for these forest interior birds.

The hooded warbler is a species that exemplifies the importance of the interior forest. This bird likes shrubby areas with dense saplings and especially brambles. We have plenty of this type of habitat in the state, so why is this a species of concern? Hooded warblers also need the edges of the interior of the forest—sort of like a species that will not live on the exterior of a block of Swiss cheese but will live around the holes in the middle. Therefore, hooded warblers need large blocks of forest with young patches embedded in it.

fragmented, a field-trip format will be used to identify these areas. Many of the best natural areas are away from the heavily used trails, and only the experienced naturalist should attempt to explore them.

A trip could start east of Armstrong on Highway 67 approximately one-half mile east of the intersection where 67 turns 90 degrees to the south. There you'll find Watercress Creek. This stream flows south into Long Lake. Along the banks are places where springs seep into the creek. Cattails and sedge with scattered areas of tamarack and white cedar cover these seeps.

Spruce Lake bog, an area of national importance, is to the southwest and is very easy to reach from Watercress Creek. This national natural landmark is a 2½-foot-deep seepage kettle lake. A narrow band of sedge bog then a much wider conifer bog with abundant poison sumac surround the small lake. These bog communities have many plants with northern affinities, several at or near their southern limit. To view the exceptional natural features of this area, take the boardwalk deep into the bog.

To reach the boardwalk, continue south on Highway 67 to Dundee. At Dundee, turn west on County F, then quickly turn north onto Vista Road. Follow Vista Road to Airport Road. Take Airport Road west for about 0.2 miles to a parking area, and from there a trail leads to the boardwalk.

Returning to Dundee, proceed south on Highway 67 to County Highway G, then take G south to County Highway SS. Drive east on SS through New Prospect to County Highway

GGG and Haskell Noyes Woods. This is an old-growth sugar maple–red oak woods located south of SS at the intersection with GGG. There are trails leading into the area. This woods is on the moraine. It has a sparse shrub layer due to intense shading, but the spring flora is fantastic.

Continue east of Haskell Noyes on County SS to two quite different areas. By taking the glacial hiking trail north, you will skirt Crooked Lake. On the west edge of the lake is an extensive area of the shrub-carr community. Little Mud Lake is to the south and west of the glacial hiking trail. This is a small muck-bottomed lake surrounded by cranberry meadow. The area holds great interest for the herpetologist because four kinds of salamanders breed in this lake.

Resuming the trip, return to County Highway GGG and head south. About 0.8 miles south on GGG is the road to Mauthe Lake. The hardwood and conifer swamp along the Milwaukee River, which flows south out of Mauthe Lake, is particularly interesting. These areas are almost impenetrable. To best explore them, go downstream from the Tamarack Nature Trail.

Specialties

Plants

arrow-arum
showy lady's slipper
small yellow lady's slipper

Animals

Acadian flycatcher
cerulean warbler
hooded warbler
northern ribbon snake
red-shouldered hawk
swamp metalmark (butterfly)
western ribbon snake

Other Features

drumlins
eskers
interlobate moraines
kames
kettles
outwash plains

Uncommon Species

Plants

blue phlox
bog rosemary
Carolina cranesbill
chara
Dewey's sedge
fern pondweed
Goldie's fern
goldthread
heartleaf twayblade
hooded ladies' tresses
hooked crowfoot
marsh cinquefoil
marsh St. John's wort
pipewort
purple fringed orchid
purple spring cress
purple twayblade
putty root
showy orchis
silky wild rye
silvery glade fern
turtlehead
white bear sedge
wild calla
wild ginger
witch-hazel
yellow trout lily

Mushrooms

tree volvariella

Insects

northern eudeilinea
rosy maple moth
stripped sedge grasshopper

Reptiles and Amphibians

Butler's garter snake
four-toed salamander

Birds (Breeding)

- blue-winged warbler
- Brewster's (hybrid) warbler
- golden-winged warbler
- Lawrence's (hybrid) warbler
- mourning warbler
- pine warbler

Mammals

- badger
- gray fox
- meadow jumping mouse

Point Beach State Forest

Communities: Northern Hardwood–Hemlock Forest, Forested Swamps, Alder Thicket, Lake Dune, Lake Beach

Alternating dune ridges and swales are the main features of this state forest. Point Beach has 17 of these landforms throughout the forest. The ridges formed when lake levels dropped. The closer to the beach, the younger the ridges. Each ridge has a little different vegetation than any other ridge. To observe these ridges and swales, hike the Ridges or Red Pine Trails. These trails will bring you close to most of the forest's features.

Dune willows at Point Beach (illustration by Jim McEvoy; courtesy of Wisconsin Department of Natural Resources)

The First Dune

The most interesting portion of the dune–swale complex is the area from the beach through the first two dunes and swales. It is here that the rarest plant species are located at Point Beach. The beach itself varies in width depending on the lake's water level. At high water the beach is almost nonexistent, with the waves lapping the more stabilized dunes. During low levels, the beach extends 100 or more feet toward the open lake. Sea rocket, a plant of the upper beach, thrives during the low lake levels and is rarely found during high lake levels.

The first dune back from the beach harbors rare flora, including dune thistle, thickspike wheatgrass, and sand reed. This area is also one of a handful of places in the state where you can see the dune willow. This species, which grows to three or four feet high, is able to tolerate surface temperatures on the sand of 140 degrees one day and be shrouded in 60-degree fog the next. Although a true survivor of adversity, the plant cannot tolerate competition from other woody plants and dies out quickly if the dunes become forested.

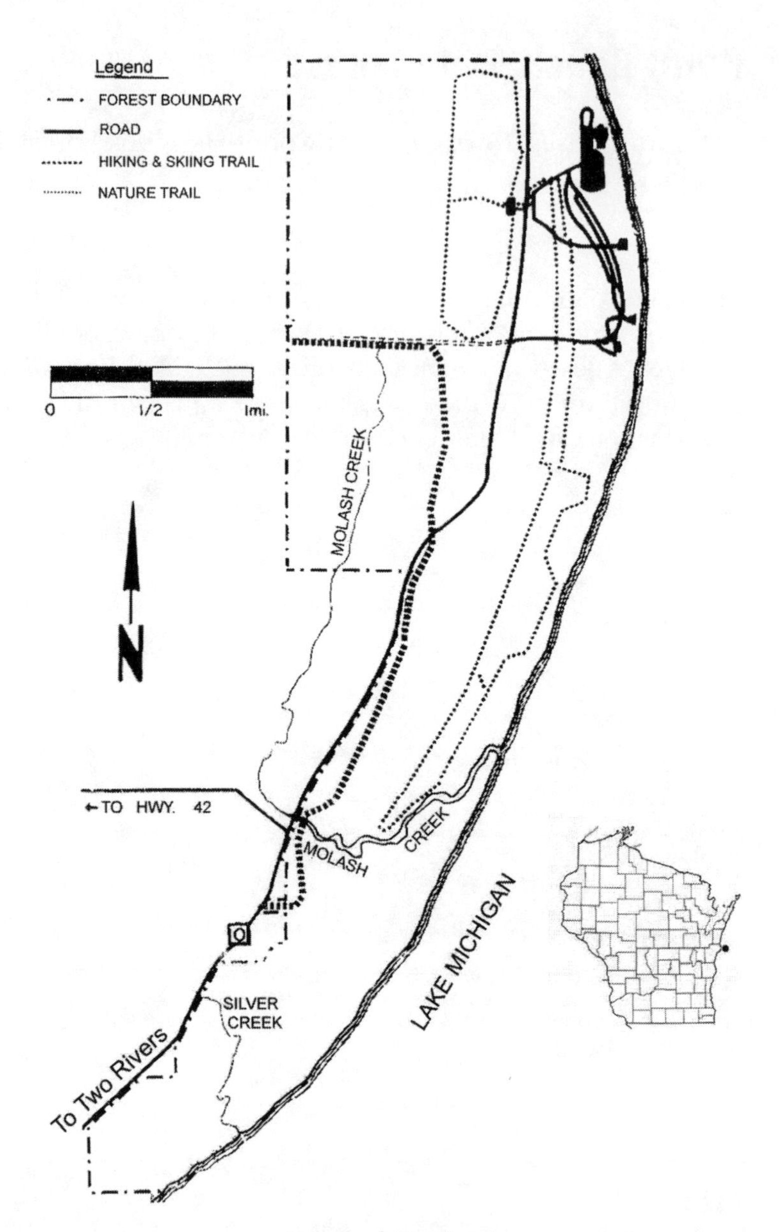

Point Beach State Forest

Aside from hiking the trails, you can walk the six miles of beach, although on most weekends the beaches are crowded. You can easily see here the succession process of beach to dune to forest. To the east is the open lake whose bottom is strewn with some 26 vessels that were stranded before the lighthouse was built. This light is the most powerful one on the Great Lakes.

Most visitors do not use the western portion of the forest, especially during summer. The cross-country ski trails go through an immense area of black ash swamp, but you can only use these trails during the winter months. White cedar makes up a significant portion of the canopy. This area has had little biological exploration, so surprises are surely in store. The northern waterthrush and the Canada warbler appear regularly in June.

Specialties

Plants

dune thistle
dune willow
Indian cucumber-root
sand tickseed
seaside spurge
thickspike wheatgrass

Animals

eastern box turtle

Uncommon Species

Plants

beaked sedge
beech drops
bog lobelia
creeping snowberry
cuckoo flower
cursed crowfoot
Dewey's sedge
early coralroot
fringed loosestrife
fringed polygala
graceful sedge
grass pink
heartleaf twayblade
hooked sedge
marsh cress
marsh horsetail
moccasin flower
one-sided shinleaf
pale sedge
pink shinleaf
pitcher plant
purple false oats
sand dropseed
Schweinitz's cyperus
smooth false foxglove
snowberry
striped coralroot
tall white violet
upland wild timothy
variegated scouring-rush
witch-hazel
yellow sedge

Insects

black dash
Boll's grasshopper
green marvel
large-headed grasshopper
narrow-winged grasshopper
snowberry clearwing
striped hairstreak

Birds (Breeding)

black-crowned night heron
common tern
green heron
least bittern
little gull
mourning warbler
whip-poor-will

Birds (Migrants)

black-bellied plover
horned grebe
long-eared owl
northern goshawk
red knot
red-throated loon
ruddy turnstone
sanderling
saw-whet owl
whimbrel
white-winged scoter
willet

Birds (Winter)

glaucous gull
Iceland gull
long-tailed duck
snowy owl
Thayer's gull

Mammals

gray fox
red-backed vole

South Kettle Moraine

Communities: Southern Oak Forest, Southern Red Oak–Mixed Forest, Dry Prairie, Wet-Mesic Prairie, Shrub-Carr, Calcareous Fen, Bur Oak Opening, Shallow Marsh, Lakes, Rivers

The south unit of the Kettle Moraine State Forest is much smaller than the north unit, and it also lacks many highly distinctive glacial features. The lack of geological features, other than kettle moraines, is more than compensated for by biological ones. The forest contains rare prairie, savanna, and fen communities. As with the north unit, the south unit's best representative habitats are near the center of the forest boundaries. The main areas of interest are north and west of Eagle.

Near the headquarters, located west of Eagle on Highway 59, is the Wood Duck Kettle Trail. This short nature trail traverses the wooded undulating hills and kettles. It also goes through an area of restored oak opening in which many species are reestablishing themselves. This is the most readily observable example of this extremely rare community. There are better examples farther afield, next to the Ice Age Trail a few miles southwest.

The wet-mesic prairie is another rare natural community. There are several areas within

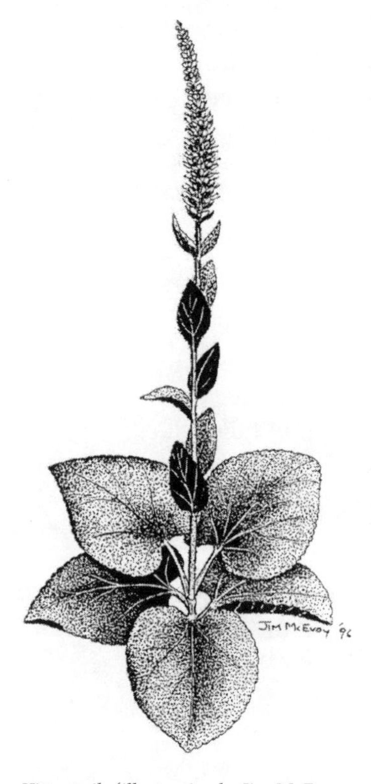

Kittentails (illustration by Jim McEvoy; courtesy of Wisconsin Department of Natural Resources)

Kittentails

Some species have very localized ranges. A good example is kittentails. This plant has a single heart-shaped basal leaf that grows for several years before flowering. A single flower stalk shoots up in May and has yellowish flowers that somewhat resemble common plantain.

Kittentails need filtered sunlight to do best. They can grow in open sun, but other species out-compete the flattened leaf for available sunlight. The dappled sunlight of oak savannas seems ideal for kittentails. The plant gets needed sunlight, but the shading prevents too much competition. Therefore, this species is endemic to the midwestern oak openings found only in Minnesota, Iowa, and Illinois, in addition to Wisconsin. Surprisingly, populations in the other states are extremely low, and without the Wisconsin populations, the plant would be listed as federally threatened. The south unit of the Kettle Moraine State Forest is the center of the world's population of kittentails.

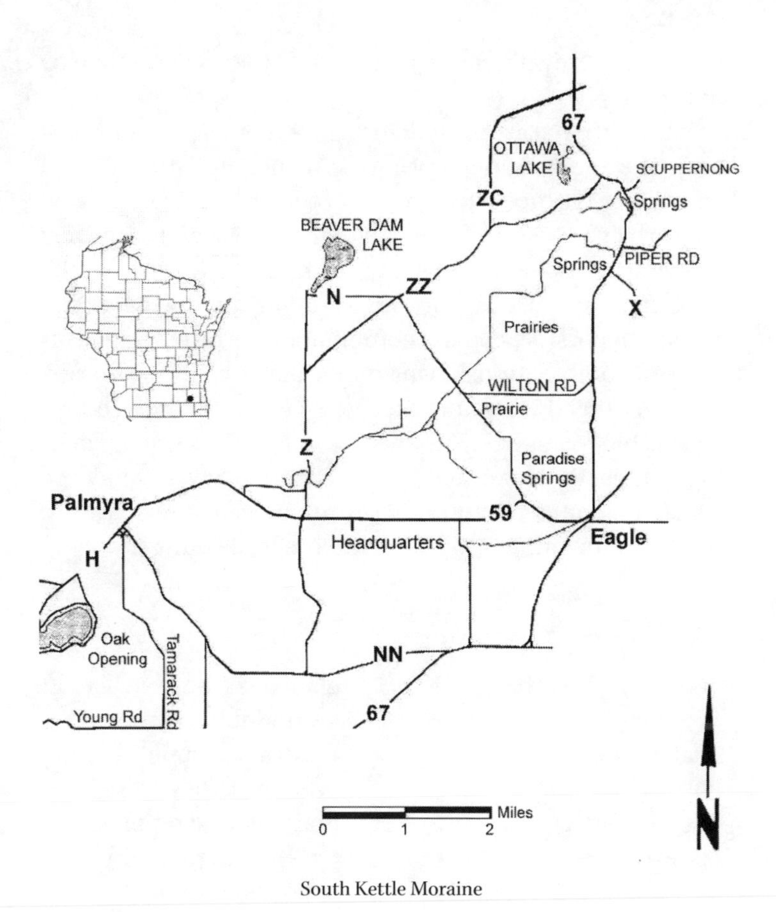

South Kettle Moraine

the Scuppernong basin that contain examples of this natural community. The best remaining wet-mesic prairie is northwest of Eagle at the intersection of County Highway N and Wilton Road. Here are 25 acres of remnant wet-mesic prairie with a presumed complete representation of the historic flora, including many rare plants and animals.

Traveling north on Highway 67 you can find a series of very hard water springs emanating from the base of the moraine. These springs are so high in calcium carbonate that the algae (chara) extracts it. Marl develops from centuries-old decay of this chara. These springs (the best examples are Paradise Springs and Scuppernong Springs) have specialized floras, plus there is always the possibility of finding the rare queen snake.

When these springs percolate through the soil, another highly unusual community forms—the marl fen. These areas are highly alkaline with very distinct and unusual floras. Ottawa Lake is an excellent location to study this community. There are trails around the lake that touch most habitats, but you will need a rowboat or canoe to explore the quality fen.

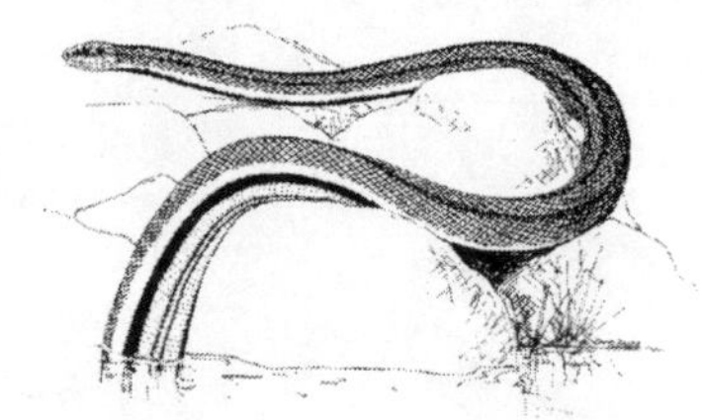

Queen snake (illustration by Georgine Price; courtesy of Wisconsin Department of Natural Resources)

There is a boardwalk into the marl flats that is accessible by canoe.

One additional area should be carefully explored for birds. This is the Scuppernong Springs Trail, next to Ottawa Lake. This trail is well known by birders for the high incidence of hybrid warblers (Brewster's and Lawrence's). The cross between blue-winged and golden-winged warblers, followed by the next generation mix of blue-winged and golden-winged, can produce 16 possible genetic combinations. The Lawrence's is the rarest combination, having only a 1 in 16 chance of occurring. The other 15 variants are Brewster's type, nine combinations; blue-winged type, three combinations; or golden-winged type, three combinations. Theoretically, it would be possible for a mated pair of these crossbreeds to raise within one nest four completely different plumaged young.

Specialties

Plants

adder's tongue fern
beaked spike-rush
cream gentian
eared false foxglove
false asphodel
few-flowered spike-rush
kittentails
prairie Indian plantain
prairie milkweed
prairie white-fringed orchid
purple milkweed
small fringed gentian
tufted hairgrass
white lady's slipper

Animals

Acadian flycatcher
Brewster's warbler (hybrid)
broad-winged skipper (butterfly)
cerulean warbler
dog-face butterfly
hooded warbler
Lawrence's warbler (hybrid)
liatris borer moth
little glasswing (butterfly)
powesheik skipper (butterfly)
queen snake
red-tailed leafhopper
silphium borer moth
silver spotted skipper (butterfly)
swamp metalmark (butterfly)

Uncommon Species

Plants

American gromwell
biennial gaura
bracted orchid
Canada milk-vetch
cream wild indigo
downy gentian
Dutchman's breeches
early horse gentian
fragrant white water lily
fringed gentian
glaucous white lettuce
groundnut
Illinois tick-trefoil
nodding ladies' tresses
nodding wild onion
pale Indian plantain
pale-spike lobelia
poison hemlock
prairie lily
prairie sunflower
purple fringed orchid
rattlesnake master
Richardson's sedge
Riddell's goldenrod
Seneca snakeroot
showy goldenrod
slender beard-tongue
slender false foxglove
small yellow lady's slipper
swamp betony
swamp saxifrage
swamp thistle
sweet grass
sweet Indian plantain
tall green milkweed
tall nut-rush
upland boneset
valerian

white camas
yellow stargrass

Insects

agile meadow grasshopper
Georgian prominent
inconsolable underwing
Linneaus 17-year cicada
red-spotted purple
silvery blue
striped hairstreak
Uhler's katydid
wingless prairie grasshopper

Fish

blacknose shiner
lake chubsucker
pugnose shiner
slender madtom

Birds (Breeding)

least bittern
orchard oriole
red-shouldered hawk

Mammals

ermine
long-tailed weasel
prairie vole
red bat
southern flying squirrel
woodland vole

Upper Mukwonago River

Communities: Southern Red Oak–Mixed Forest, Floodplain Forest, Shrub-Carr, Calcareous Fen, Southern Sedge Meadow, Shallow Marsh, Rivers

The sites described in this section are two small locations that are found in a highly fragmented part of the state along with the Mukwonago River watershed. Both of these areas are found in landscapes that look similar to most of southeastern Wisconsin. Most natural areas are small, but fortunately several protected sites exist in this area. In this quarter of the state, private holdings may contain some of the rarest plants we have. These plants are rare because the communities they live in are rare. The rarest communities in this part of the state are old-growth forests, especially those with an abundance of sugar maples that have had little past disturbance. The other critical community is deep soil prairie–savanna. Most remaining areas of this community are along railroad rights-of-way, but a few may exist along streams or lakes where soil moisture has prevented plowing. These remaining areas within this quadrant of Wisconsin, unless found and protected within the next 10 years, will most likely be lost forever.

The Lulu Lake State Natural Area is the exceptional site in southeastern Wisconsin. Fourteen natural communities prosper in and around the 84-acre lake and its watershed. Some of these, including calcareous fen and oak opening (see color insert), are the rarest natural communities in Wisconsin. The rapid change from one community to another, combined with a few northern relict species, makes Lulu Lake one of the most significant biological areas in southeastern Wisconsin. Public access to Lulu Lake is not easy. A parking area on County Highway J leads to an old farm road that serves as a walk-in access. The Nature Conservancy permits access to its members on the western side of the lake. This place offers the last refuge for many species. Its value will increase dramatically as time goes on.

Another superb but small natural area is Muskego hardwood forest. The natural area is a part of the greater Muskego Park that is part of the Waukesha County Park system, affording

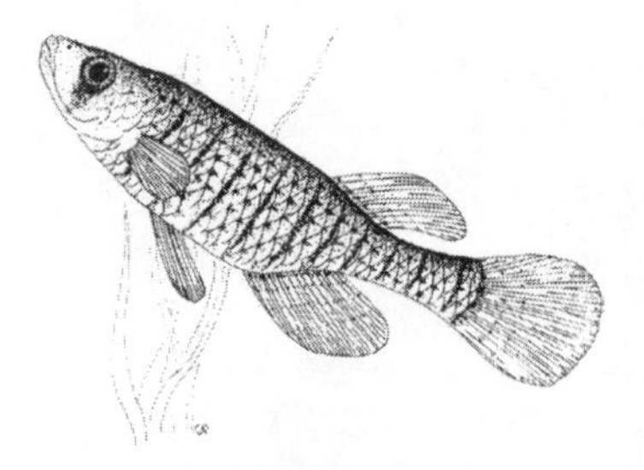

Starhead topminnow (illustration by Georgine Price; courtesy of Wisconsin Department of Natural Resources)

Mukwonago River

The Mukwonago River contains the highest known diversity of river fishes, freshwater mussels, and aquatic insects in southeastern Wisconsin. Nutrient-rich waters supplied by the numerous springs and fens along the stream and the gravel and medium-sized rocks deposited by the glacier provide the conditions needed for many species. Several species of darters live in the riffle areas where they find ideal conditions among the gravel and rocks. Little-known fish, like the starhead topminnow, forage in the backwaters and lakes where abundant vegetation grows. In most areas of the state, it would be acceptable for nature lovers to capture specimens for study, but this resource is so rare and tenuous in this part of Wisconsin that we should simply be content knowing they are there and leave them alone.

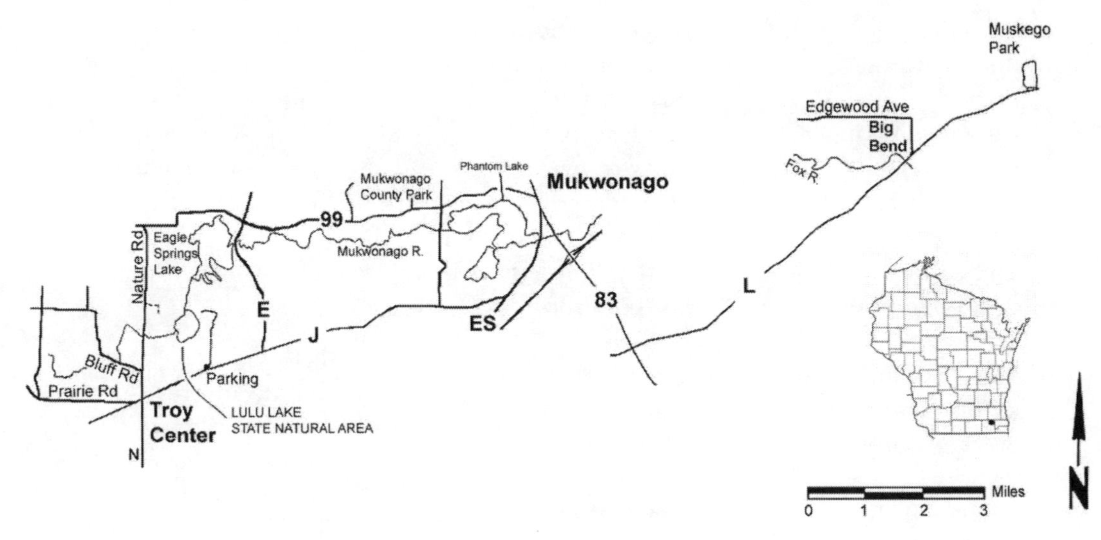

Upper Mukwonago River

it protection. There is a small entrance fee, but it is well worth it to see remnant old-growth forests in southeast Wisconsin. Muskego is an upland forest ranging from relatively dry to wet conditions with an excellent herbaceous layer. The featured species in the park is the very rare blue ash, easily the largest rare plant in the state. These trees can be separated from other ashes because they have square tips to their branches.

The western Mukwonago River watershed contains scattered natural communities. The obvious problem of accessing private land can be resolved by canoeing the river from County N in Walworth County downstream to Phantom Lake in Mukwonago. You will find many communities along the shores in almost unheard of condition for this part of the state. The aquatics of the river system are good, with a few endangered species still thriving. The shoreline has excellent areas of fen, sedge meadow, marly springs, and even some oak savanna.

Specialties

Plants

adder's tongue fern
American gromwell
beaked spike-rush
blue ash
common bog arrow-grass
crow-spur sedge
downy willow-herb
false asphodel
forked aster
hemlock parsley
Kentucky coffee tree
kittentails
many-headed sedge
Ohio goldenrod
red trillium
small fringed gentian
swamp agrimony
swamp rose mallow
tufted bulrush
tufted hairgrass
white lady's slipper

Animals

Acadian flycatcher
black dash (butterfly)

Blanding's turtle
Brewster's warbler (hybrid)
broad-winged skipper (butterfly)
cerulean warbler
creek heelsplitter (mussel)
dion skipper (butterfly)
ellipse (mussel)
hooded warbler
lake chubsucker (fish)
Lawrence's warbler (hybrid)
least darter (fish)
lilypad forktail (dragonfly)
longear sunfish
mulberry wing (butterfly)
pugnose shiner (fish)
rainbow mussel
round pigtoe (mussel)
slipper shell (mussel)
starhead topminnow (fish)

Other Features

trilobite fossil

Uncommon Species

Plants

bog goldenrod
bog lobelia
bog rosemary
butterfly weed
butternut
crested sedge
ditch-grass
downy willow-herb
dragonhead
dragon's mouth
Dutchman's breeches
false mermaid
flax-leaved aster
grass-of-parnassus
green dragon
large-flowered trillium
large yellow lady's slipper
late coralroot
leatherleaf
low nut-rush
mare's tail
marsh blazing star
marsh St. John's wort
marsh valerian
nodding ladies' tresses
Ohio buckeye
pale-spike lobelia
prairie blazing star
prairie dropseed
putty root
Riddell's goldenrod
rosinweed
showy orchis
spotted St. John's wort
swamp saxifrage
swamp thistle
sweet black-eyed Susan
tall agrimony
waxy meadow-rue
western sunflower
white beak-rush
wild leek
wild onion
yellow monkey flower
yellow stargrass

Insects

afflicted dagger
coral hairstreak
grizzly grasshopper
harnessed moth
Juvenal's dusky wing
mustard white
orange halomelina
small-eyed sphinx
snout butterfly
tawny-edge skipper
twin-spotted sphinx

Fish

banded darter
banded killifish
blackside darter
blackstripe topminnow
large scale stoneroller
mottled sculpin
orange-spotted sunfish
rainbow darter
rosyface shiner
stonecat
suckermouth minnow
tadpole madtom

Reptiles and Amphibians

blue-spotted salamander
brown snake
central newt
eastern milk snake
pickerel frog
smooth green snake
stinkpot

Birds (Breeding)

common moorhen
field sparrow
mourning warbler
red-shouldered hawk
tufted titmouse
Virginia rail

South Central Wisconsin

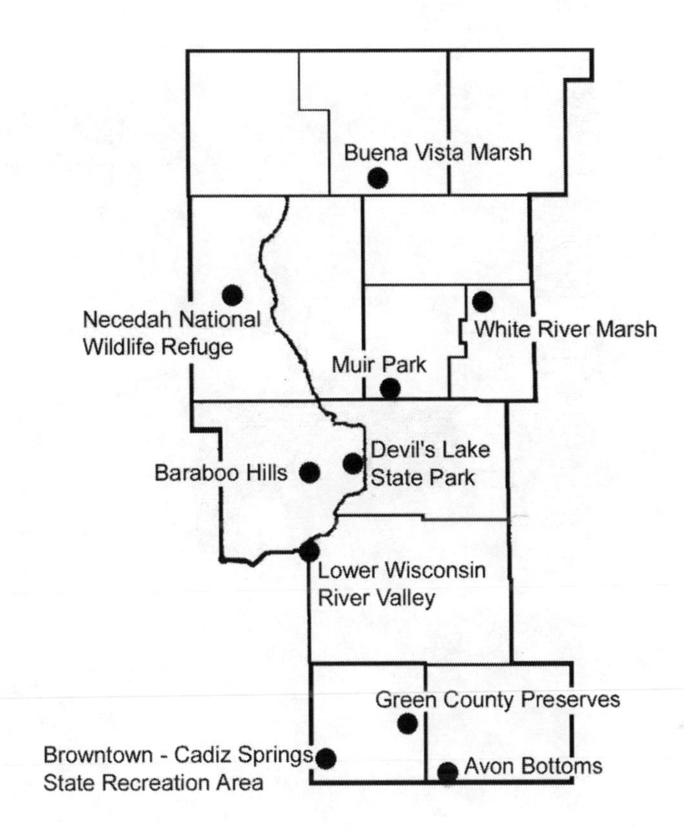

Avon Bottoms

Communities: Floodplain Forest, Wet-Mesic Prairie, Bur Oak Opening, Rivers

Southwestern Rock County lies in an area that was missed by the Wisconsin glacier, but older glaciers covered the land. The terminal moraine from the Wisconsin glacier lies to the north, and a great outwash plain is east of the Rock River. Originally this portion of Rock County was almost entirely grassland, with forest thriving only along rivers. The grassy areas were either thinly wooded oak savanna or pure prairie. Today these prairie areas are no longer vast, and even the floodplain forest is a mere fraction of its past acreage. Remaining portions of the natural habitat are widely scattered and greatly reduced in size.

Natural areas become more important to us as the land around them becomes more altered. In addition, the presence of rare species because of range limitations can elevate conservation status of a site. When few natural areas remain and they harbor rare species, we can say the sites are critical resources. That can be said of the few remaining natural areas in southwestern Rock County and adjacent southeastern Green County.

We will never again see large areas of intact prairies in the Avon area. But a tour of the place will allow you to see, feel, touch, and appreciate its remaining natural heritage. I like to start and finish in Brodhead, but you could join the route anywhere and continue the circuit or just search one of the remnants.

From Brodhead, go south and west on Highway 11 to the intersection with Highway 81. Turn east on Highway 81 and travel one-half mile. Just south of the road lies an area where nat-

Discovery

The presence of sycamore trees at the Avon Wildlife Area led me to make some assumptions that eventually resulted in the discovery of a rare bird in the area. Yellow-throated warblers nest primarily in pine forests of the southern United States, but a subspecies in the center of the country also uses sycamores as habitat. With that information, I surmised that if sycamores are in the state, then maybe yellow-throated warblers could be in the state, too.

In the early 1980s, I went to the University of Wisconsin Herbarium to search for records of naturally occurring populations of sycamore in the state. The best-known location was along the Sugar River in Rock County. In addition, through my birding contacts, I knew that yellow-throated warblers had been observed in June north of Rockford, Illinois. With this information in hand, I set out to see if my theory was correct. In 1986, I spent two days venturing into the Avon Bottoms. Upon discovering the first grove of sycamores, I discovered yellow-throated warblers. In more extensive forays I have located six clumps of sycamores, each used by yellow-throated warblers.

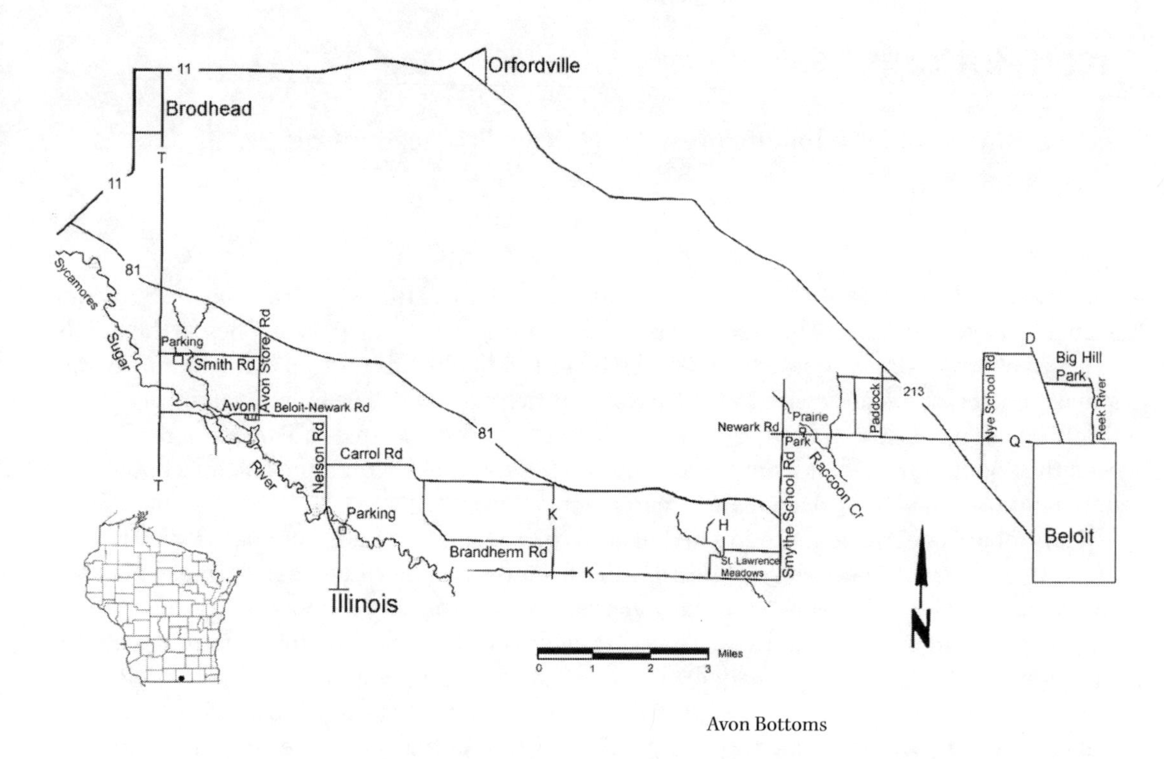

Avon Bottoms

ural sycamores grow at the northwest limit of their range. The area is public and a part of the Avon Wildlife Area.

Continue east on Highway 81 to County Highway T. This road bisects Rock and Green Counties. Take County Highway T south to Smith Road, the first intersection, then continue east one-quarter mile to a parking area. The area south of Smith Road is also part of Avon Wildlife Area. This is an area of narrow floodplain forest with oak woods on higher ground but also some prairie and a small area of oak savanna. The significant prairie–savanna area is found just southeast of the parking area. In the attempt to locate the prairie–savanna area, you can sharpen your community identification skills.

To get to the best floodplain forest, go east to Avon Store Road, then south to Beloit-Newark Road, east to Nelson Road, then south to the bridge crossing the Sugar River. Just north and just south of the bridge are parking areas from which you can begin exploring. This area is very exciting in both flora and fauna. Several plants reach their northern limits here, including hop tree, scarlet oak, American chervil, beak grass, and tall melic grass. The fauna is equally impressive. Check the oxbow lakes for reptiles and amphibians and the trees for southern warblers, including the recently discovered yellow-throated warbler. Also, if you do discover the hop trees, look at them closely because this is the sole food of the brown-bordered ermine moth. Discoveries like these give credence to the assumption that by saving one threatened species another species can coincidentally be saved.

Another rewarding way to observe the Avon Bottoms area is by canoe. The river is difficult to canoe, with many downed trees and log jams, but the effort can be highly rewarding.

To the east of Avon is an area of sedge meadows, low prairies, and some oak woods, known as the St. Lawrence Meadows. The degraded area still holds many species, including pokeweed and cleft phlox. St. Lawrence Meadows can be reached by taking Carrol Road just north of the Avon Bottoms and proceeding east to the intersection with County Highway K and turning south for 1½ miles, where the road turns east. This road has three names: St. Lawrence Road, County Highway K, and County Highway H. Disregard everything and continue straight east for 3½ miles to the meadows on the north side of the road.

From here, continue east to the first intersection, Smythe School Road, and turn north for 2½ miles. You will intersect with Beloit-Newark Road. Go east for one-half mile to Newark Road Prairie. This is an excellent remnant prairie, one of the

Prairie milkweed (Endangered Resource file photograph; courtesy of Wisconsin Department of Natural Resources)

best remaining in the state. The prairie varies in composition depending on moisture, proceeding from mesic prairie on the upland sites to sedge meadow in the wettest area. Extreme care must be taken in observing this prairie because many species, such as the prairie milkweed and white lady's slipper, are very susceptible to trampling.

Big Hill Park is at the north edge of the city of Beloit. It is a well-used city park with some unusual species recorded. Most exciting is the report of tinted spurge at its only known Wisconsin location.

To get to Big Hill Park, continue directly east on Beloit-Newark Road to the intersection with Highway 213. Proceed directly east now on County Q to the intersection with County D. Turn north on D and go one mile to Big Hill Road. After exploring Big Hill Park, return to Highway 213 and proceed northwest to Orfordville, where you should go west on Highway 11 to Brodhead, but make sure you observe the area between the railroad tracks and the highway west of Orfordville. There are many areas of remnant prairie in the first three miles.

Specialties

Plants

- American gromwell
- beak grass
- black haw
- bluets
- blunt-leaved spurge
- cleft phlox
- cream gentian
- fragrant sumac
- glade mallow
- hop tree
- kittentails
- marbleseed
- pale purple coneflower
- prairie Indian plantain
- prairie milkweed
- prairie parsley
- prairie straw sedge
- prairie white-fringed orchid
- purple milkweed
- Richardson's sedge
- round-fruited St. John's wort
- scarlet oak
- small fringed gentian
- snowy campion
- sycamore
- tall melic grass
- tinted spurge
- white lady's slipper
- wild chervil
- wild petunia
- yellow giant hyssop

Animals

- Acadian flycatcher
- Blanchard's cricket frog
- Blanding's turtle
- brown-bordered ermine moth
- cerulean warbler
- dark rubyspot (dragonfly)
- gravel chub (fish)
- lark sparrow
- ornate box turtle
- riverine clubtail (dragonfly)
- russet-tipped clubtail (dragonfly)
- starhead topminnow (fish)
- weed shiner (fish)
- yellow-breasted chat
- yellow-crowned night heron
- yellow-throated warbler

Uncommon Species

Plants

- biennial gaura
- Canada milk-vetch
- Canadian St. John's wort
- cardinal flower

climbing bittersweet
cream false indigo
downy gentian
dragonhead
ear-leaved brome
ebony sleenwort
false mermaid
false pimpernel
fringed gentian
glaucous white lettuce
heart-leaved alexanders
heart-leaved aster
large-flowered beard-tongue
large yellow lady's slipper
meadow parsnip
nodding wild onion
Ohio buckeye
one-flowered broomrape
pale Indian plantain
pokeweed
scarlet paintbrush
showy blazing star
slender beard-tongue
slender false foxglove
small-flowered cranesbill
small sundrops
small yellow lady's slipper
smooth false foxglove
smooth rose mallow
starry campion
sweet Indian plantain
tall nut-rush
toothed cress
valerian
weak St. John's wort
wild senna
woodland boneset
yellow stargrass

Insects

ash-gray leaf bug
delightful dagger
eight spot
fiery skipper
giant cicada killer
grizzly grasshopper
gulf fritillary
ottoe skipper
snout butterfly
sycamore lace bug
wingless prairie grasshopper

Fish

American eel
blackside darter
blackstripe topminnow
northern brook lamprey
river redhorse
sand shiner
shorthead redhorse
silver chub
silver redhorse
slenderhead darter
stonecat

Birds (Breeding)

Carolina wren (irregular)
prothonotary warbler
white-eyed vireo

Birds (Migrants)

American golden plover
short-billed dowitcher
white-crowned sparrow

Birds (Winter)

field sparrow
white-crowned sparrow
yellow-rumped warbler

Mammals

least shrew
southern flying squirrel

Baraboo Hills

Communities: Southern Red Oak–Mixed Forest, White Pine–Hardwood Forest, Cliffs, Rivers

What a welcome sight to a migrating forest bird—big woods, the first seen after crossing the Illinois prairies. The Baraboo Hills are the first such place they see since the hills of southern Illinois or Indiana. The only "forests" to the south are scattered islets of trees in a sea of farmland. A satellite view of the landscape confirms this conclusion; the Baraboo Hills are an island of forest in a sea of agriculture. These hills are so important biologically that they are covered in two chapters. This chapter describes the area west of Highway 12, and the Devil's Lake chapter describes the area east of Highway 12.

Various conservation organizations, primarily the Nature Conservancy, now protect many portions of this area in perpetuity. Efforts to maintain such a large tract of woodland this far south in Wisconsin have been monumental. Several years ago it would have been unthinkable to mention protection for entire watersheds, but the unthinkable has been accomplished. This project integrates protection activities between conservation organizations and private landowners.

Baxter's Hollow, a cool ravine descending south out of the hills, is an important biological site in the Baraboo Hills because it encompasses an entire forested watershed. To reach the

Southern Birds

The yearly presence of warblers that reach their absolute northern limit here in southern Wisconsin is an exciting event for Wisconsin bird watchers. The abundance of Louisiana waterthrushes and cerulean warblers would rightfully make the area worth visiting. While the additional bonus of finding Kentucky warblers would make a trip outstanding, the yearly presence (from May to July) of hooded and worm-eating warblers (see color insert) could make this a phenomenal trip for a birder.

Hooded warblers live in the big blocks of woods out of the valley. They inhabit oak woods with dense shrubs or areas where old logging activity allowed young saplings to grow. This type of logging has helped the hooded warblers in the sugar maple–basswood forest because it promotes dense understory development while still maintaining the large block of forest needed. It is not dissimilar to openings created by dead trees.

The other southern bird, the worm-eating warbler, is much more difficult to find, although it is present yearly. The nature of the bird is secretive; it feeds on the ground or low in the bushes. But when it sings, which is rarely, it perches in midtree and stays put. Only patient and persistent observation can bring one into view. Be careful, though, because the valleys have many chipping sparrows whose songs are very similar.

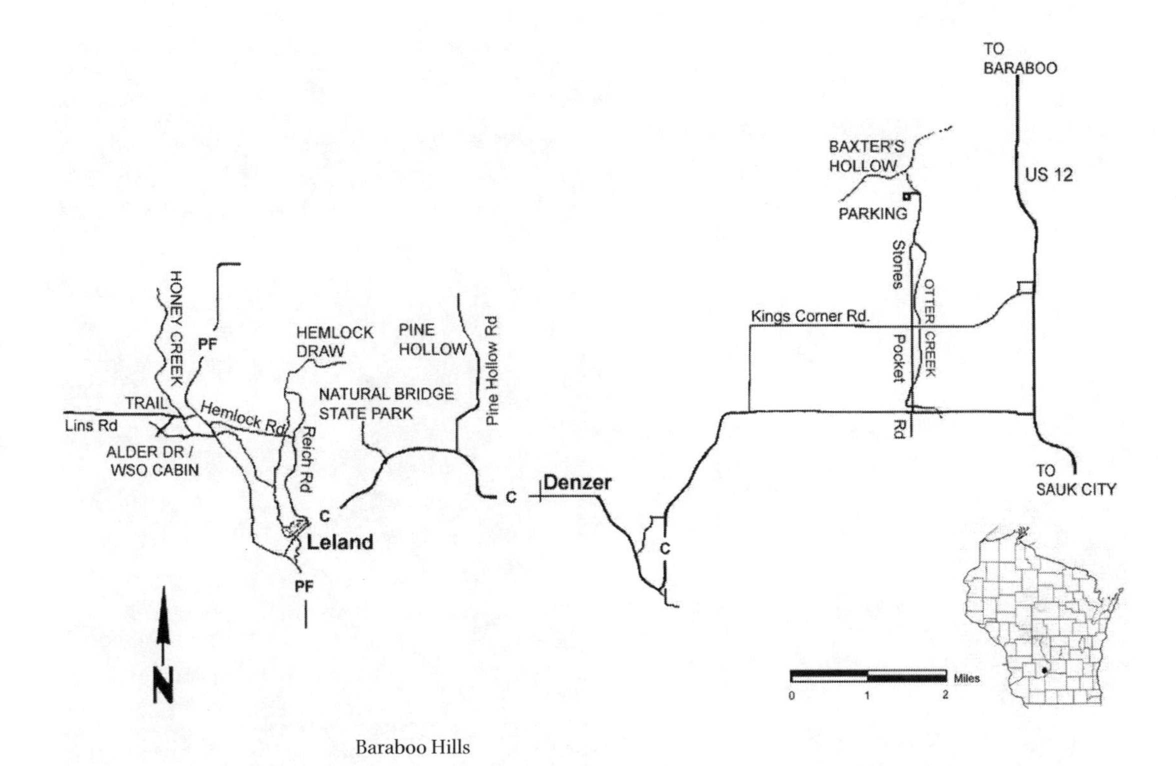

Baraboo Hills

hollow, take County Highway C west of Highway 12 to Stones Pocket Road, then go north on this road into Baxter's Hollow. This place, as with other hollows and draws in the "hills," is a blend of northern and southern species. Each area has its own relict species from the north, and most have hemlock as the dominant tree, but here at Baxter's Hollow are towering white pines. There, of course, is the complementary northern pine herbaceous layer that goes along with the pine relict canopy. Even more remarkable is a relict fauna of many northern birds that still use these relict forests. Many warbler species occur here with the Canada, blackburnian, black-throated green, and magnolia varieties present nearly every year.

Baxter's Hollow was formerly the site of a sometimes noisy campground and much traffic. Most users today are nature observers and pursuers of trout in the exceptionally clean Otter Creek. Otter Creek is the home of a great multitude of caddisflies, some of which are quite unusual. It is also one of the few places in the state where large water starwort can be found. The flora varies, with northern and southern species growing nearly side by side. You might find the one-flowered broomrape growing in the stream-swept gravel or see a beautiful purple fringed orchid or a cardinal flower growing within feet of each other.

Nodding Pogonia

The nodding pogonia, or three-bird's orchid, is a species of moist forests in southern Wisconsin. It can be found in sugar maple–basswood forest or maple–red oak forest. The delicate orchid appears for a week or two in late July. This species can spend up to 10 years underground, feeding off the roots of other plants. Above-ground populations can vary greatly from year to year. I experienced this first-hand on a 40-acre preserve in the western Baraboo Hills. In the dry summer of 1989, I counted seven flowering nodding pogonias, while in the wet summer of 1990, I counted over 1,000 plants of the same species in the same woods.

Natural bridge (Endangered Resource file photograph; courtesy of Wisconsin Department of Natural Resources)

The other areas of the Baraboo Hills are just as exciting and have their own rewards. To reach these, continue west on County Highway C through Denzer to Pine Hollow Road. This is the first road west of Denzer. This road leads to Pine Hollow, with its cool hemlocks and rock shelters. Beyond Pine Hollow is Natural Bridge State Park. County Highway C traverses the park. This 530-acre park features a 25-foot-high and 35-foot-wide natural bridge, the largest of its kind in Wisconsin. At the base of the bridge is a 60-foot by 30-foot rock shelter once used by Native Americans. Excavations here have found bones of elk, mountain lion, timber wolf, fisher, marten, and passenger pigeon. If you follow the self-guided nature trail to the rock shelter, it will be the closest you will ever get to a passenger pigeon. You might stand over a passenger pigeon bone lying a few feet under the ground.

To reach Hemlock Draw, proceed west on County Highway C to the east edge of Leland, turn north on Hemlock Road, and stay right onto Reich Road. The cliffs here have many hemlocks, but the unique feature of this draw is the length of time the ice hangs on—during some seasons until mid-June. Species preferring these cool microenvironments thrive here.

Returning on Reich Road this time, take Hemlock Road west to County Highway PF. Make a short jog north on PF, then west on Lins Road to Alder Drive and the Wisconsin Society for Ornithology property. Although well known for its birds, Honey Creek properties have had more than 500 plant species recorded. The area has been subject to degradation in the past, but the cliff and alder–tamarack bog communities are well intact. An unusual plant found in the alder–tamarack swamp is the bog bluegrass, a species that is rare throughout its range and usually found in cool, mossy springy areas much farther north.

Specialties

Plants

blunt-lobed grape fern
bog bluegrass
cliff goldenrod
cliff saxifrage
drooping sedge
erect bur-head
great Indian plantain
great white lettuce
large water starwort
late coralroot
narrow-leaved dayflower
nodding pogonia
one-flowered broomrape
purple cliff-brake
purple milkweed
rock clubmoss
Rocky Mountain sedge
round-stemmed false foxglove
slender bush clover
sullivantia
woodland boneset
wooly milkweed
yellow giant hyssop

Animals

Acadian flycatcher
blackburnian warbler
black-throated green warbler
Brewster's warbler (hybrid)
Canada warbler
cerulean warbler
Henslow's sparrow
hooded warbler
Kentucky warbler
western ribbon snake
worm-eating warbler

Other Features

natural bridge
quartz crystals

Uncommon Species

Plants

aromatic aster
bladdernut
bracted orchid
butternut
Canada yew
cardinal flower
dry woods sedge
Goldie's fern
hooked crowfoot
leafcup
long-beaked sedge
panicled tick-trefoil
partridgeberry
prickly pear cactus
purple fringed orchid
purple giant hyssop
purple virgin's bower
putty root
sand milkweed
showy orchid
sicklepod
stiff sandwort
walking fern
wild yam
wintergreen
yellow trout lily

Insects

arrowhead spiketail
caddisflies (six rare species)
clamp-tailed emerald
predacious diving beetle
speared dagger
stygian shadowfly

Reptiles and Amphibians

four-toed salamander
timber rattlesnake

Birds (Breeding)

golden-winged warbler
mourning warbler
northern goshawk

tufted titmouse
whip-poor-will

Birds (Winter)

common redpoll
northern shrike
red crossbill

Mammals

gray fox
least weasel
meadow jumping mouse
otter
red bat
southern flying squirrel

Buena Vista Marsh

Communities: Shrub-Carr

Buena Vista Marsh is now a marsh in name only. Formerly a great expanse of tamarack and sedge meadow, agricultural promotion encouraged farmers to drain the area. However, most farms failed through a combination of poor soils and July frosts. After abandonment, most farmland turned into oldfield grasslands, which, when interspersed with continuing farms, proved to be an ideal situation for prairie chickens.

Prairie chickens need areas of grass, cover, solitude, and a place to nest. Equally important is their need for secluded booming grounds. In Wisconsin, these birds seem to do best when grassland is interspersed with agriculture areas, which they use as forage areas. Prairie chickens thrive in the surrogate habitat offered by small family farms.

After the depression era, Wisconsin's agriculture became big business, with movement toward larger farms and row crop production. These changes severely reduced habitat for prairie chickens. In the 1950s and 1960s, several citizen groups formed to preserve prairie chicken habitat. Preservation of 11,000 acres occurred throughout the marsh, although in scattered locations. A careful observer can usually find a few flying prairie chickens, but to observe the

Gyrfalcon Encounter

One of my greatest experiences as an observer of nature occurred at a greater prairie chicken blind on a frosty early April morning. The cold chill of a blind is an experience most people do not consider important in their lives, but I do. Even with my outdoor experience and knowledge of frosty temperatures, I often do not dress warmly enough. Such was the case one morning when the temperature dipped into the low 20s, and I was shivering at dawn. But shivering or not, I could not leave the blind until at least 8 A.M. when the males begin to leave the lekking grounds.

Suddenly, the ooohing sound of the males ceased, and I became alert. Looking out one of several openings in the blind to see the cause of the silence, I happened to be at the right window at the right time. A gyrfalcon rocketed by and hit a male chicken with such force that I knew it was instantly dead. The bird had just begun to fly and was no more than two feet off the ground when the collision occurred. The gyrfalcon looped back and landed next to the chicken. Because my line of sight was poor, I could not see how the falcon grasped the chicken, but after several minutes the gyrfalcon flew off with a fine breakfast.

With no males around, it was an easy decision to leave the blind. An inspection of the kill site revealed an abundance of feathers scattered over a large area. The pattern was similar to a debris trail left by a sinking ship. The closer I came to the place where the bird died, the more feathers were on the ground. Later that day, I recalled in hindsight that I was not cold the rest of the day.

Male prairie chicken (illustration by Jim McEvoy; courtesy of Wisconsin Department of Natural Resources)

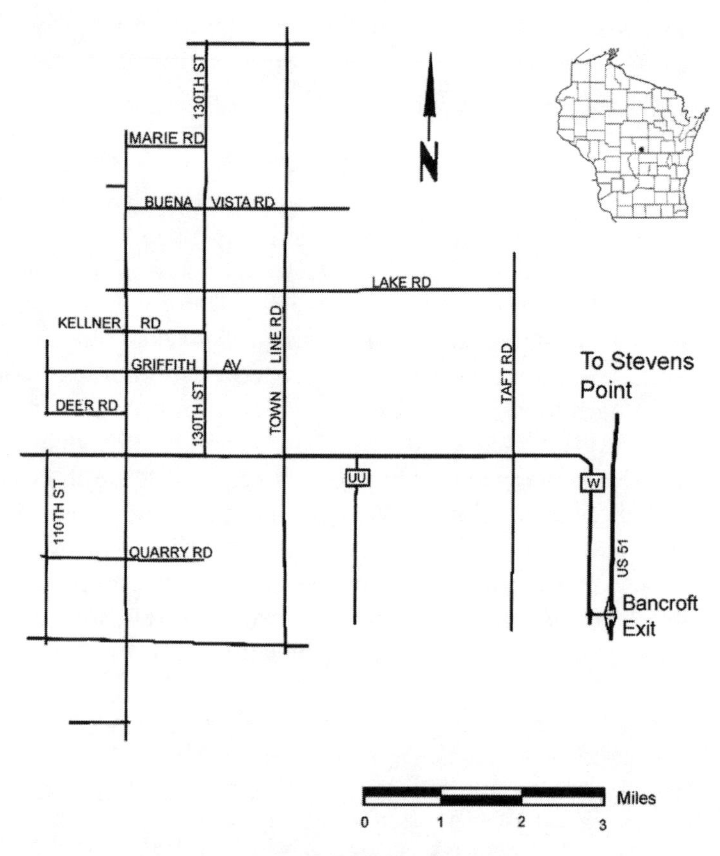

Buena Vista Marsh

chickens booming, you almost have to use blinds. Information about when and how to reserve blinds can be made by inquiring at the Department of Natural Resources office in Friendship, Wisconsin.

Buena Vista is an anomaly to naturalists. Management keeps continually cutting, mowing, burning, and grazing as a part of a new disturbance regime, thus perpetuating a highly unnatural system for the benefit of a formerly abundant resident. We are forming a new restructured grassland community, highly foreign with few native species of plants but still meeting the requirements of our grassland birds.

Another phenomenal discovery has been made recently. One of our rarest butterflies, the regal fritillary, has been observed on the grasslands in large numbers. The butterfly's discovery came when two experts in butterfly habitat sensed that Buena Vista could harbor regals. Their suspicion was confirmed when they discovered a large population of this rare butterfly. Normally a prairie species requiring prairie violets and forb nectar sources, the regals at Buena Vista utilize

Regal fritillary (illustration by Jim McEvoy; courtesy of Wisconsin Department of Natural Resources)

woodland violets that grow in the open as their larval food host. The openness of the area and the European forbs that serve as a nectar source provide the habitat they need to thrive.

These new grasslands are much more economic than restoring original prairie. By doing nothing, a grassland with nonnative cool-season grasses will develop and will be regularly used as a surrogate for prairie by grassland birds. To perpetuate these areas, woody vegetation must be removed by burning, cutting, and applying herbicides. By preserving and perpetuating large areas for the benefit of grassland animals, we are in essence producing a new type of grassland that has yet to stabilize.

Specialties

Plants
- adder's tongue fern

Animals
- greater prairie chicken
- gyrfalcon (winter)
- Henslow's sparrow
- regal fritillary (butterfly)
- short-eared owl
- upland sandpiper

Birds (Breeding)
- clay-colored sparrow
- dickcissel
- sandhill crane

Birds (Migrants)
- American golden plover
- lapland longspur
- rusty blackbird
- tundra swan
- water pipit
- Wilson's phalarope

Birds (Winter)
- northern shrike
- snow bunting
- snowy owl

Mammals
- badger
- prairie deer mouse
- prairie vole

Uncommon Species

Insects
- unarmed wainscot

Cadiz Springs State Recreation Area

Communities: Southern Oak Forest, Southern Red Oak–Mixed Forest, Shrub-Carr, Shallow Marsh

Cadiz Springs State Recreation Area is in southwestern Green County approximately six miles west of Monroe on Highway 11 or one mile east of Browntown on the same road, just a few miles north of Illinois. Many of Wisconsin's rare features tend to lie near the state's borders, and Cadiz Springs is no exception. The area is small because agricultural practices have placed huge demands on the land in this region of the state. An 80-acre portion of the park contains an excellent southern oak forest.

The Browntown oak forest is the natural area of the park. This is an area of old-growth (150 years) red oak forest. Soil types differ between the ridge top and the lower slope, resulting in different species composition. This oak woods is on a north-facing slope, as are most forest remnants from the prairie age. Red oak dominates the upper slope, and farther down slope, black and white oak become more abundant. Slippery elm, sugar maple, ironwood, bitternut hickory, basswood, black walnut, and black cherry are also present in the canopy. With communities that range from southern oak forest on the summit to southern red oak–mixed forest down slope, the area should be monitored long-term because the understory contains many seedlings and saplings of sugar maple, indicating a shift in the future to a more mesic-type forest.

Most of the surrounding land is in succession forest, offering various combinations of pioneering species, such as red-osier dogwood, box elder, and nannyberry. The exemplary plant

Edge of Range

Some species are rare because of their low populations and restricted habitat. Other rare species have a low population but a large range. And some species have large, dense populations but very restricted habitat and therefore are vulnerable to extirpation if the habitat disappears. Still other species are considered rare because they are at the edges of their range. If a political boundary, especially an international or state boundary, falls near the edge of the range of a species, the species is considered rare in the particular country or state.

Such is the case with two species at Cadiz Springs. The yellow-breasted chat and white-eyed vireo are birds of scrub, second-growth forest, and fencerows in the southern United States. Their ranges just reach into Wisconsin. Sightings of these species warrant an automatic inclusion on the state's birding hotline. The same two species in Alabama or Tennessee would never be mentioned as important. These edge-of-range species do merit protection. Many times individuals at the edge of their range have different genetic material than center-of-range individuals and therefore are of significant conservation value.

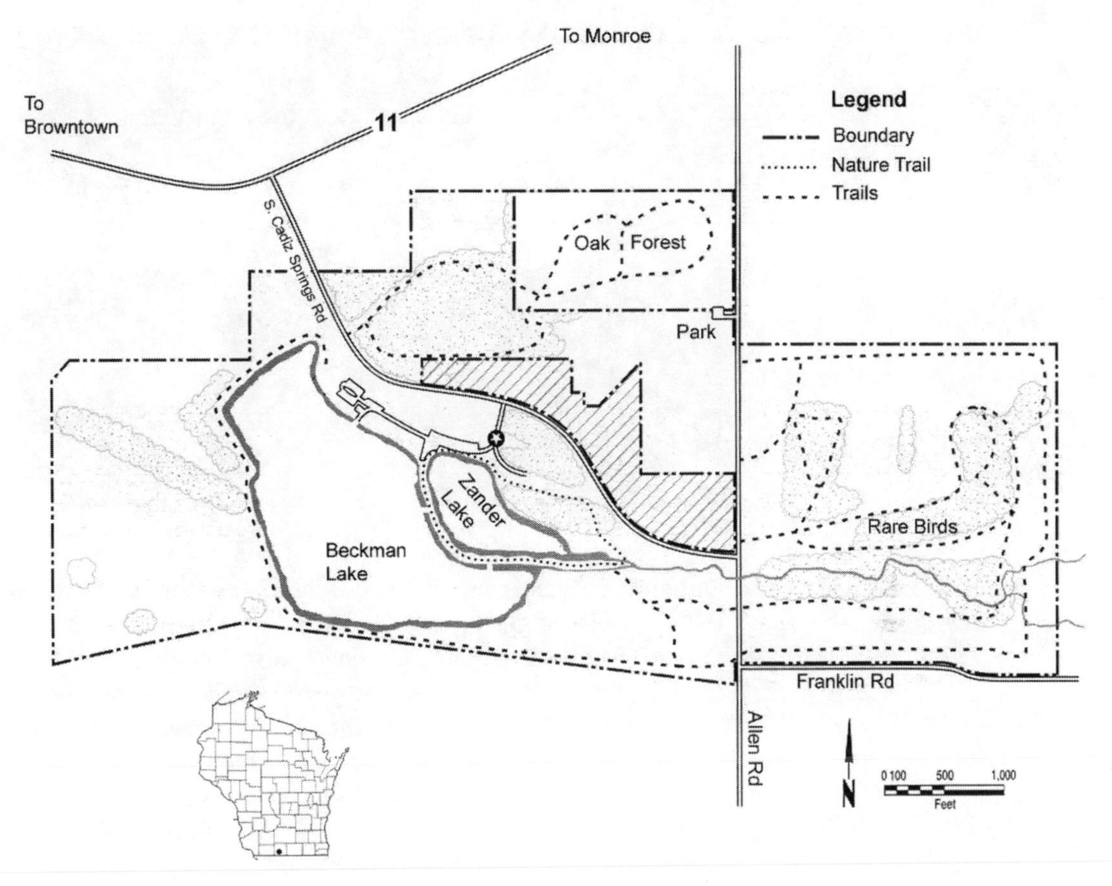

Cadiz Springs State Recreation Area

communities and rare plant species occur in the old-growth oak forest and other woodlands elsewhere on private land in the Browntown area. One such rare species is the red trillium, a species that is at the limit of its range in the southern tier of Wisconsin counties, but is abundant throughout the Browntown area. The increasingly uncommon golden seal is also in the area; thankfully it is in a well-protected area. Other private lands contain this species, but it is rare.

The forest is without a doubt the most important feature of the park scientifically, but it receives minimal attention from park visitors. The attractions drawing most visitors are Zander and Beckman Lakes, which were formed by damming streams. While neither prime nor ideal habitats, the lakes' location in the state can be attractive to species which have wandered too far north. There are trails around both lakes, including an interpretive nature trail around Zander Lake.

East of the lakes is an area of varied habitat. The terrain is undulating with a variety of fields, forest, scrubby areas, streams, meadows, and alder thickets. The diverse habitat

Red trillium (photograph by Thomas A. Meyer; used by permission)

along with the nearness to the border makes this spot attractive to southern bird species. The shrubby places along the stream east of Allen Road have been the most consistent spot to view the white-eyed vireo and yellow-breasted chat. These birds, though very common in their main range, are difficult to observe at their periphery range, which coincides with southern Wisconsin. Both species have been very dependable in this area, with as many as three or four white-eyed vireos and upwards of six chats seen by birders in June. The uncommon Bell's vireo prefers similar shrubby–brushy habitats but is more of an upland bird.

No uncommon or unusual invertebrate species are known to be in this area, though due to its nearness to the southern border it may harbor some butterflies or moths that drift north each fall. A careful examination of blooming fields might reveal some surprises.

Specialties

Plants

- golden seal
- red trillium
- snowy campion

Animals

- Bell's vireo
- Juvenal's dusky wing (butterfly)
- white-eyed vireo
- yellow-breasted chat

Uncommon Species

Plants

- American gromwell
- bracted orchid
- Culver's root
- hooked crowfoot
- large yellow lady's slipper
- leafcup
- purple giant hyssop
- Robin's fleabane
- Short's aster
- wood sedge
- yellow pimpernel

Insects

zabulon skipper

Reptiles and Amphibians

bullfrog
fox snake
gray treefrog
northern brown snake
northern water snake
pickerel frog
red-bellied snake
smooth softshelled turtle
tiger salamander
wood frog

Birds (Breeding)

dickcissel
field sparrow
gray partridge
green heron
Henslow's sparrow
northern bobwhite

Birds (Migrants)

bald eagle
common loon
great egret

Birds (Winter)

golden-crowned kinglet

Mammals

badger
gray fox
woodland vole

Devil's Lake State Park

Communities: Southern Oak Forest, Southern Red Oak–Mixed Forest, Lake, Cliffs

Devil's Lake State Park, our largest state park, covers more than 7,000 acres. It is the most visited park between the Appalachians and the Rocky Mountains, prized both for its natural beauty and for interesting geological and biological history. Geologists have studied this area for over a century, and educators have held field trips here for decades. Scientists consider the park a mecca for its biological diversity. Rock climbers visit here so much that published rock-climbing guides are available. Most people, however, come to view the scenery, and justly so because what great scenery it is!

Your first stop should be at the park headquarters to get a detailed map of the park's nearly 20 miles of trails and natural wonders and to meet the full-time naturalist who is on duty to answer questions or help in identifying plants. You may choose to explore the deep, clear Devil's Lake itself or wander up the mountainous bluffs.

Devil's Lake was originally part of the Wisconsin River, but with the onset of the last glacier, the flow of the river was diverted. The glacier advanced to the edge of Devil's Lake, depositing mounds of rock and debris called a terminal moraine on both ends of the gap, which allowed Devil's Lake to form. If you would like to know more about the natural history of Devil's Lake State Park, I recommend *Ancient Rocks and Vanished Glaciers,* by former park naturalist Ken Lange, which provides a wealth of information on the natural history of the area.

The bluffs surrounding Devil's Lake are 500 feet high. If you start at a bluff base and climb to the summit, you will notice changes in the vegetation from the bottom of the bluff to the top. Many northern Wisconsin species, such as Allegheny vine and leafcup, live at the bottom

Pygmy forest (photograph by Eric Epstein; used by permission)

Pygmy Forest

Most of the Devil's Lake State Park shows typical southern Wisconsin forest habitats. At the bluff tops, though, the habitat becomes uncharacteristically dry for Wisconsin, and the forest community changes to dry, un-Wisconsin-like growth. These bluff tops, with their thin soils, extra sunlight, and drying winds, have plants that are much more typical west and south of here, including many dry prairie species.

Even more unusual is a stunted oak woods growing on the east bluff summit. This "pygmy" forest grows on very thin soils over the top of extremely hard rock. White oak, the most common tree species, grows to a height of only 15 to 20 feet. In rich soils these oaks can grow to well over 100 feet tall. The gnarly branches and stunted growth make the woods appear similar to the oak woods in the deserts of the Southwest or the glades of the Ozarks. Sedges and savanna plants make up the groundlayer in this forest.

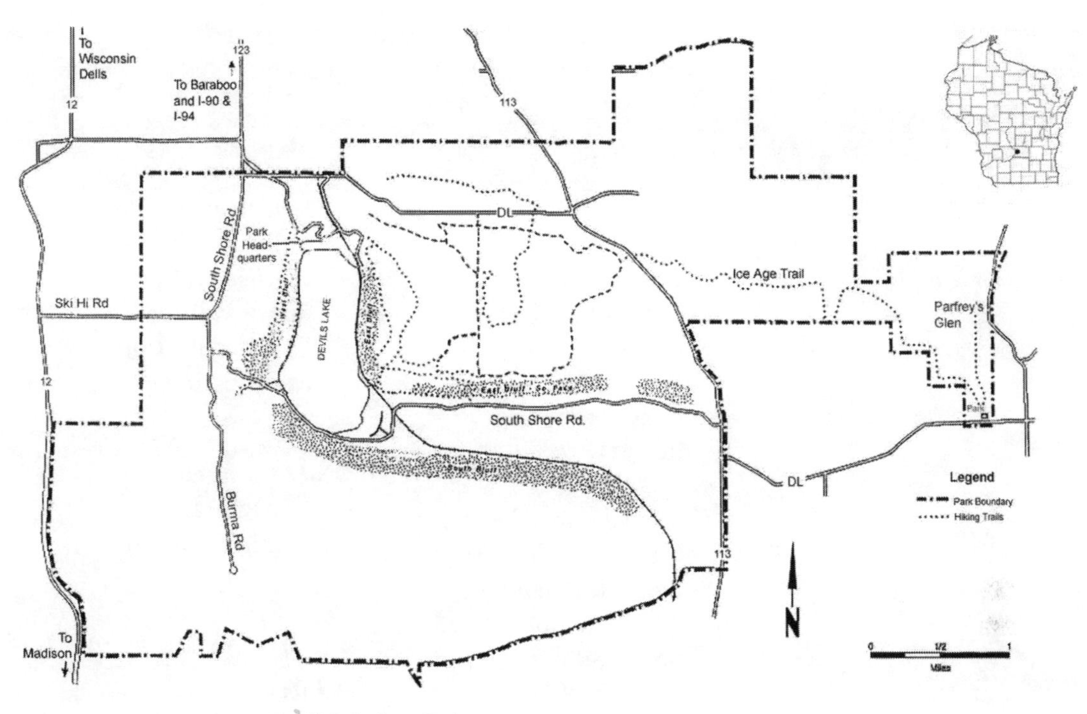

Devil's Lake State Park

of the slope because of dense shading and the cool air that settles at the bottom of the bluff. These species are relicts of a bygone age when the entire Wisconsin landscape harbored northern plants and animals. The grotto trail at the base of the east bluff is an exceptional area for viewing many of these northern species.

Parfrey's Glen, a small canyon to the east of Devil's Lake and part of the park, is another superb area to investigate because the cool climate provides conditions ideal for many rare species. Parfrey's Glen is a narrow gorge that was cut into sandstone conglomerate by Parfrey's Glen Creek. The small stream harbors distinctive rare caddisfly larvae. The gorge walls, however, are the real treat. Water seepage over the rock and cool air drainage allows many highly unusual northern species, such as rock clubmoss, to live on the walls.

To reach Parfrey's Glen, take County Highway DL east from Highway 113 just south of Devil's Lake State Park. Follow DL to the Parfrey's Glen signs at the parking area. Because Parfrey's Glen has a history of visitors trampling the gorge tops and the stream bed, a graveled trail and boardwalks into the gorge were constructed to alleviate the abuse. Please stay on the trail to ensure that many future generations can enjoy this wonderful area.

Northern monkshood (illustration by Jim McEvoy; courtesy of Wisconsin Department of Natural Resources)

Devil's Lake State Park and Parfrey's Glen host nearly one-third of all ferns and flowering plants recorded in Wisconsin.

Specialties

Plants

- broad beech fern
- cliff saxifrage
- drooping sedge
- dry woods sedge
- Hooker's orchid
- large water starwort
- lichen (*Parmelia substygia*)
- maidenhair spleenwort
- Missouri rock cress
- nodding pogonia
- northern monkshood
- poverty grass
- prairie parsley
- purple milkweed
- rock clubmoss
- Rocky Mountain sedge
- round-stemmed false foxglove
- slender bush clover
- sullivantia
- tubercled orchid
- tufted hairgrass
- Vasey's pondweed
- violet bush clover
- woodland boneset

Animals

- arrowhead spiketail (dragonfly)
- black rat snake
- Blanding's turtle
- bobcat
- cerulean warbler
- freshwater jellyfish
- hooded warbler
- Kentucky warbler
- long-eared bat
- rare caddisflies
- rare diving beetles
- timber rattlesnake
- Townsend's solitaire
- turkey vulture concentrations
- worm-eating warbler

Other Features

- quartzite talus
- terminal moraines

Uncommon Species

Plants

- Allegheny vine
- arrow-leaved violet
- bearberry
- bracted orchid
- broad-leaved goldenrod
- cliff cudweed
- cliff goldenrod
- common rockrose
- creeping rattlesnake plantain
- Culver's root
- Drummond's rock cress
- early coralroot
- false pimpernel
- five-parted toothwort
- graceful sedge
- great-spurred violet
- hooded ladies' tresses
- hooked crowfoot
- large yellow lady's slipper
- long-leaved bluets
- lupine
- pale-leaved sunflower
- pale-spike lobelia
- prairie blue-eyed grass
- prairie dandelion

purple fringed orchid
purple giant hyssop
purple twayblade
putty root
showy orchis
smooth false foxglove
swamp betony
twinleaf
western sunflower
wild ginger
woodland bluegrass
woodland brome
wood sandwort
yellow false foxglove

Ferns

blunt-lobed cliff-fern
broad beech fern
bulblet fragile fern
Christmas fern
dissected grape fern
dwarf scouring-rush
flat-branched ground-pine
Goldie's fern
hairy lip-fern
lake quillwort
leather-leaved grape fern
maidenhair spleenwort
northern fragile fern
rock-cap fern
rock spikemoss
rusty cliff-fern
shining clubmoss
slender cliff-brake
slender lip-fern
smooth cliff-brake
smooth scouring-rush
triangle grape fern
walking fern
water horsetail

Parfrey's Glen Special Plants

fir clubmoss
mountain ash
narrow-leaved dayflower
partridgeberry
small enchanter's nightshade
white moss
witch-hazel
yellow birch

Insects

Acadian hairstreak
Baltimore
banded hairstreak
black dash
checkered skipper
convergent lady beetle
coral hairstreak
Edward's hairstreak
giant swallowtail
hickory hairstreak
hoary edge
little wood satyr
mulberry wing
pine elfin
snowy tree cricket
striped hairstreak
tawny-edged skipper
tawny emperor
13-year cicada
two-spotted skipper

Fish

bluntnose minnow
burbot
spottail shiner

Reptiles and Amphibians

blue racer
brown snake
four-toed salamander
pickerel frog
red-bellied snake

Birds (Breeding)

black and white warbler
blackburnian warbler
black-throated green warbler
brown creeper
Canada warbler
Louisiana waterthrush
mourning warbler
northern goshawk
peregrine falcon
saw-whet owl
tufted titmouse
winter wren

Birds (Migrants)

Bell's vireo
Bewick's wren
Carolina wren
eared grebe
surf scoter
yellow-bellied flycatcher

Birds (Winter)

evening grosbeak
red crossbill

Mammals

big brown bat
bobcat
eastern pipistrelle
gray fox
least weasel
long-tailed weasel
prairie vole
southern flying squirrel

Green County Preserves

Communities: Sugar Maple–Basswood Forest, Dry Prairie

Central Green County contains some very small but well-known preserves located a few miles southwest of Albany. Abraham's Woods, Oliver Prairie, and Muralt Bluff Prairie, all located within a few miles of each other, are remnants of the original prairie savanna and sugar maple–basswood forest in Green County. There are other small remnants of prairies and maple woods in central Green County, but those are mostly on private land. It is a shame that only postage-stamp-sized remnants remain, but at least a few of the remnants are preserved, and just as with stamps, they increase in value as other areas succumb to human progress.

Lying on a limestone ridge, Muralt Bluff Prairie at 62 acres is the largest of the three preserved areas. Part of the prairie is healthy, but other parts have been damaged over the years by landowners pasturing cattle on the land. Fortunately, a large component of the prairie flora and fauna can still be found. The remaining prairie flora includes a magnificent spring flower display of pasque flowers, bird's-foot violets, and shooting stars. Green County now owns this area.

Abraham's Woods is 40 acres of old-growth sugar maple–basswood forest. The hills to the south and west formed a barrier against prevailing winds and fire, which allowed the maple–basswood forest to develop. One-half mile to the west is Oliver Prairie, a four-acre dry-prairie remnant that has felt the full force of past prairie fires. Both preserves are owned by the University of Wisconsin Arboretum. Restoring the intervening agricultural lands between Oliver Prairie and Abraham's Woods would provide an unparalleled opportunity for the prairie–forest continuum to be demonstrated in the field. It would be an invaluable teaching and research tool for professionals and amateurs alike. A visit to the interior *must* be coordinated with the Arboretum. There are no plants or animals found here that cannot be found in other areas, so you need not be tempted to enter without permission.

Whitney's underwing (photograph by Les Ferge; used by permission)

Prairie Moths

Many naturalists associate moths with woods, but moths are found in nearly every natural community in the state. A number of very interesting moths appear on the prairies of the state, including a few species that are active during the day. Many times I have flushed a strikingly patterned haploa moth from the prairie edge or observed a nectar-feeding hummingbird moth foraging in a manner similar to hummingbirds.

One group of moths depends on leadplant, a prairie species, for nearly their entire existence. The larvae (caterpillars) of two large and colorful moth species, the abbreviated underwing and Whitney's underwing, feed exclusively on leadplant. The caterpillars are almost impossible to find because they hide in dead plant material on the ground during the day, but they ascend the leadplant each evening for a night of feasting. Each morning before dawn they descend back into their hiding places. The best way to observe these leadplant moths is with black-light equipment.

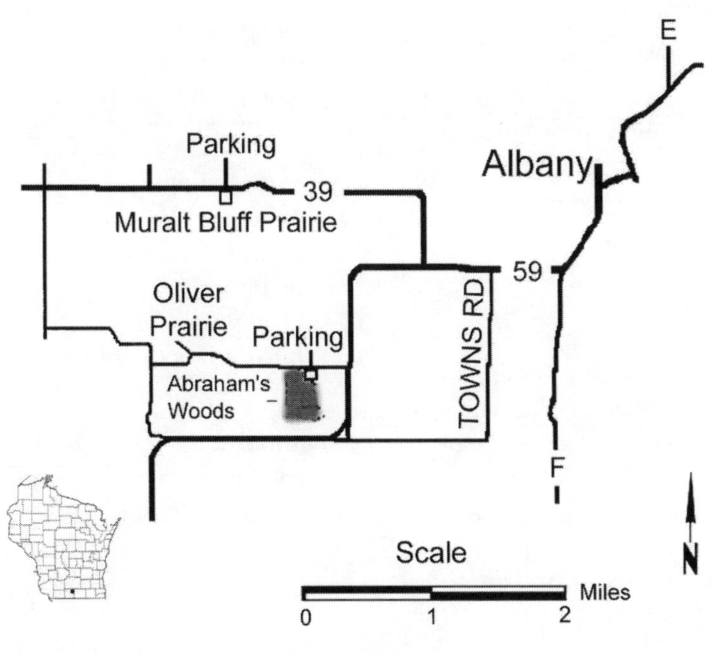

Green County Preserves

Specialties

Plants

- kittentails
- prairie dandelion
- prairie Indian plantain
- prairie satingrass
- prairie thistle
- round-stemmed false foxglove
- yellow giant hyssop

Animals

- abbreviated underwing (moth)
- Bell's vireo
- grasshopper sparrow
- ottoe skipper (butterfly)
- pawnee skipper (butterfly)
- regal fritillary (butterfly)
- Whitney's underwing (moth)
- yellow-breasted chat

Uncommon Species

Plants

ABRAHAM'S WOODS

- arrow-leaved aster
- bladdernut
- blue beech
- blue cohosh
- bristly greenbriar
- broad-leaved toothwort
- Dutchman's breeches
- early coralroot
- false mermaid
- Goldie's fern
- Hitchcock's sedge
- hooked crowfoot
- James' sedge
- moonseed
- putty root
- Short's aster
- showy orchis
- small-flowered buttercup
- spring beauty
- white baneberry
- white bear sedge
- wild yam
- wood phlox
- yellow jewelweed

MURALT BLUFF AND OLIVER PRAIRIES

- aromatic aster
- bird's-foot violet
- butterfly weed
- clammy ground cherry
- cream wild indigo
- crooked aster
- downy gentian
- glaucous white lettuce
- grooved yellow flax
- hairy aster
- hoary puccoon
- Illinois tick-trefoil
- long-leaved bluets
- New Jersey tea
- one-flowered broomrape

pale-spike lobelia
porcupine grass
prairie alum-root
prairie blue-eyed grass
prairie brome
prairie dropseed
prairie phlox
prairie smoke
prairie violet
purple milkwort
purple twayblade
rattlesnake master
Robin's fleabane
Seneca snakeroot
shooting star
short green milkweed
sky blue aster
spreading dogbane
stiff aster
stiff gentian
sweet black-eyed Susan
violet wood sorrel
western sunflower
yellow stargrass

Insects

capsule moth
common wood nymph
Delaware skipper
Gorgone checkerspot
lead-plant flower moth
LeConte's haploa
meadow fritillary
nessus sphinx moth
olive hairstreak
wingless prairie grasshopper
woody underwing

Birds (Breeding)

Acadian flycatcher
bobolink
dickcissel
loggerhead shrike
northern bobwhite

Mammals

badger
prairie deer mouse
red bat

Lower Wisconsin River Valley

Communities: Floodplain Forest, Dry Prairie, Shrub-Carr, Oak Barrens, Sand Barrens, River, Cliffs

This site is large and diverse, extending from the dam at Sauk City west to Spring Green, and from the bluffs on the south side to the bluffs on the north side. I recommend that you tour the entire area, traveling downstream on the south side of the river, crossing at Spring Green, and returning on the north side. With a few stops of an hour or so, you can get a flavor of the natural amenities of the Lower Wisconsin River Valley in a day. Of course, you may choose to spend an entire day at one spot to really get to know the natural community there.

A good starting point would be County Road Y at Highway 78, south of Sauk City. Traveling west on Y, you will encounter an area of very scrubby-looking oaks. This is the Mazomanie oak barrens, about 100 acres of sand prairie and oak barrens. The state-owned parcel contains most of the species found in these two natural communities. Watch for the Natural Area sign, then a parking area on the left. You can explore the barrens by walking back to the sign. There are no trails or facilities at this site. While exploring this area, be on the lookout for the rare slender glass lizard, a legless lizard that looks like a small brown snake, which you might glimpse if you are exceptionally lucky.

Continue west on County Highway Y to Laws Road, turn right onto Laws Road, then make another right onto Conservation Road, directly into the heart of the Mazomanie bottoms. This well-known area attracts bird watchers, especially those interested in the passerine migrations. The area is also exceptional for reptiles, amphibians, insects, and plants. There was concern over the heavy public use that caused traffic problems, and a new parking area requires a hike of more than a mile to reach the best natural habitat.

Getting back to the tour, continue on Y through Mazomanie, then drive west on 14, south on K, and make a right turn onto Pinnacle Road, so named for a large exposed sandstone pillar known as the pinnacle. The pinnacle is on private land, but the drive gives you a feel for the driftless area landscape. You can also attain an understanding of the importance of visual natural landmarks to the residents of the area.

Returning on either County Road K or County Road HH and H, drive to Arena, then north of town to Helena Road. Just north of town are some old barrens areas on state-owned land where buttonweed grows. No one knows why this plant grows here, but it does and it's found nowhere else in Wisconsin. Continue west on Helena Road; you will soon come to an area where the Helena–Arena railroad tracks run near the road for the next few miles. The road and railroad run parallel, with sand prairie between them. The prairie varies in quality from poor to excellent due to digging and encroachment by exotic species. It is exciting to compare areas and see the differences. Poppy mallow grows in profusion in some of these prairie areas.

This road merges with Highway 14. From here, turn right, cross the Wisconsin River, turn right on Highway 23, then turn right again onto Jones Road. After 1¼ miles, turn north on Angelo Lane to a parking area. You will encounter an area of dry prairies and bluffs. The Nature Conservancy owns a portion of the finest sand prairie remaining in Wisconsin here. This preserve is presently open to the public. These dry, south-facing bluffs have faunas unique to

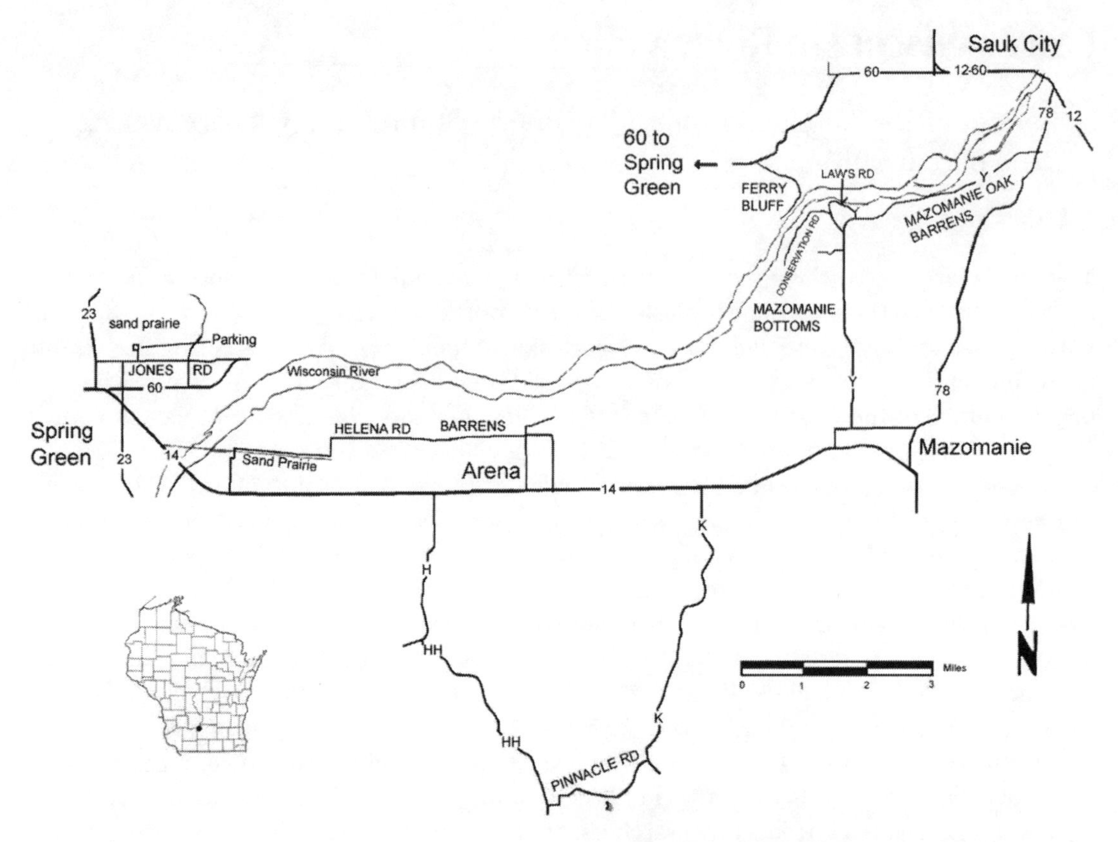

Lower Wisconsin River Valley

Oithona and Phyllira tiger moths (photograph by Les Ferge; used by permission)

Wisconsin, including blue racers (see color insert), timber rattlesnakes, and hundreds of habitat-restricted insects, such as oithona and phyllira tiger moths.

Next take Jones Road east to 60 toward Sauk City, then turn right on Ferry Bluff Road. Ferry Bluff, an important winter roosting site for eagles, lies farther north at the end of this series of bluffs. In the winter, these south-facing bluffs and associated valleys around Ferry Bluff protect the eagles from the prevailing winds. Ferry Bluff is also important for migrating hawks and eagles in the spring. The bluffs warm rapidly, and they send thermals of warm air high into the sky. Under the right conditions in April or May, you might see hundreds of raptors soaring to great heights on these thermals. Ferry Bluff is state-owned and accessible by trail from the end of Ferry Bluff Road.

You can also tour the Lower Wisconsin River Valley by canoe, exploring the seemingly endless variety of sandbars and islands. Each weekend in the summer, there are hundreds of canoes on this portion of the Wisconsin River.

Specialties

Plants

- buttonweed
- cliff goldenrod
- clustered fescue
- late coralroot
- Maryland senna
- narrow-leaved day flower
- northern monkshood
- one-flowered broomrape
- prairie bush clover
- prairie dandelion
- purple cliff-brake
- rough white lettuce
- small forget-me-not
- Wilcox panic grass
- wooly milkweed

Animals

- bald eagle
- Bell's vireo
- black rat snake
- blue-legged grasshopper
- blue racer
- Kentucky warbler
- large-headed grasshopper
- lark sparrow
- oithona tiger moth
- olive hairstreak (butterfly)
- ornate box turtle
- ottoe skipper (butterfly)
- phyllira tiger moth
- prairie polyamia leafhopper
- red-veined leafhopper
- six-lined racerunner (lizard)
- slender glass lizard
- Virginia big-headed tiger beetle
- yellow-breasted chat

Uncommon Species

Plants

FLOODPLAIN FOREST AND RIVER SAND BARS

- downy gentian
- dragonhead
- ear-leaved brome
- false pimpernel
- field milkwort
- field scorpion grass
- green dragon
- groundnut
- heart-leaved aster
- hop sedge
- indigo bush
- Muskingum sedge
- sand milkweed
- short green milkweed
- Tuckerman's sedge

SANDY BARRENS AND BLUFFS

- aromatic aster
- awned cyperus

bearberry
Bicknell's sedge
bracted spiderwort
butterfly weed
clammy ground cherry
cliff goldenrod
cream wild indigo
crooked aster
downy paintbrush
ebony spleenwort
false heather
fringed puccoon
green needlegrass
large cottonweed
lupine
needlegrass
prairie blue-eyed grass
prairie crowfoot
prickly pear cactus
rock jasmine
rough false foxglove
sand croton
sand dropseed
sand-reed
small skullcap
starry campion
stiff gentian
stiff goldenrod
venus looking glass
white camas
white wild indigo
Wilcox panic grass
yellow false foxglove
yellow stargrass

Insects

broad-necked grasshopper
columbine dusky wing
crossline skipper
dainty sulphur
giant cicada killer
gravel grasshopper
Haldeman's grasshopper
larger sand cricket
long-horned grasshopper
mottled dusky wing
mottled sand grasshopper
northern bush katydid
regal fritillary
roadside skipper
sand grasshopper
silvery blue
smaller sand cricket
two-striped mermiria

Fish

brook lamprey
flathead catfish
longnose gar
mooneye
paddlefish
pirate perch
shortnose gar
speckled chub
tadpole madtom

Reptiles and Amphibians

false map turtle
map turtle
Ouachita map turtle
timber rattlesnake
wood turtle

Birds (Breeding)

dickcissel
orchard oriole
prothonotary warbler

Birds (Migrants)

Cape May warbler
Caspian tern
Connecticut warbler
Forster's tern
golden-winged warbler
LeConte's sparrow
merlin
Nelson's sharp-tailed sparrow
peregrine falcon
saw-whet owl
water pipit
white-crowned sparrow

Mammals

badger
gray fox
least shrew
least weasel
long-tailed weasel
otter
prairie deer mouse
prairie vole
southern flying squirrel

Muir Park

Communities: Wet-Mesic Prairie, Shrub-Carr, Calcareous Fen, Shallow Marsh, Lake

Muir Park is another one of those small areas that is so appealing. The site is a county park encompassing 135 acres. Additionally, the Sierra Club, a private conservation organization, has recently purchased an adjoining 27 acres that will be used as a buffer while prairie restoration work goes on. These acres surround a 30-acre lake. Even though the name for this lake on most maps is Ennis Lake, many residents call it Muir Lake. The name that probably should be used is Fountain Lake because that is what the first settlers called it. This is a kettle lake in a ground moraine with both springs and seepage adding to the flow. At 30 feet deep with a marl bottom, this lake is exceptionally clear.

Surrounding the lake is bog, sedge meadow, fen, and wet-mesic prairie. The uplands are oak-dominated woods that are actually overgrown oak openings. Furthermore, this area is the boyhood home of John Muir, the famous naturalist who received his early environmental training right here. The scenes of wonder that inspired Muir are still here. The intricate interactions of water, wetlands, prairie, and savanna are still in place within this park. The western end of the park has a picnic area, but the remainder of the park is natural. A trail circles the lake and provides a good way to view the natural communities. This small place is much grander than the surface acreage would suggest. It is a place of thought, a place of idealism, a place to let your mind contemplate nature, and possibly a place to inspire the next great naturalist.

John Muir

A much better description of this area 150 years ago can be enjoyed by reading John Muir's *The Story of My Boyhood and Youth.* We know this place now as Muir Park, and we can sit by the lake and contemplate or we can walk the trail and perhaps observe a sight that Muir might have seen. Do these thoughts give us insight into his environmental development? The question might never be fully answered, but the natural communities and their associated plants and animals that helped inspire Muir's words are still here. The numerous and intricate connections that fascinated the conservationist can still be found with diligent searching. We need to maintain areas such as this to provide places where the next Muir can be inspired.

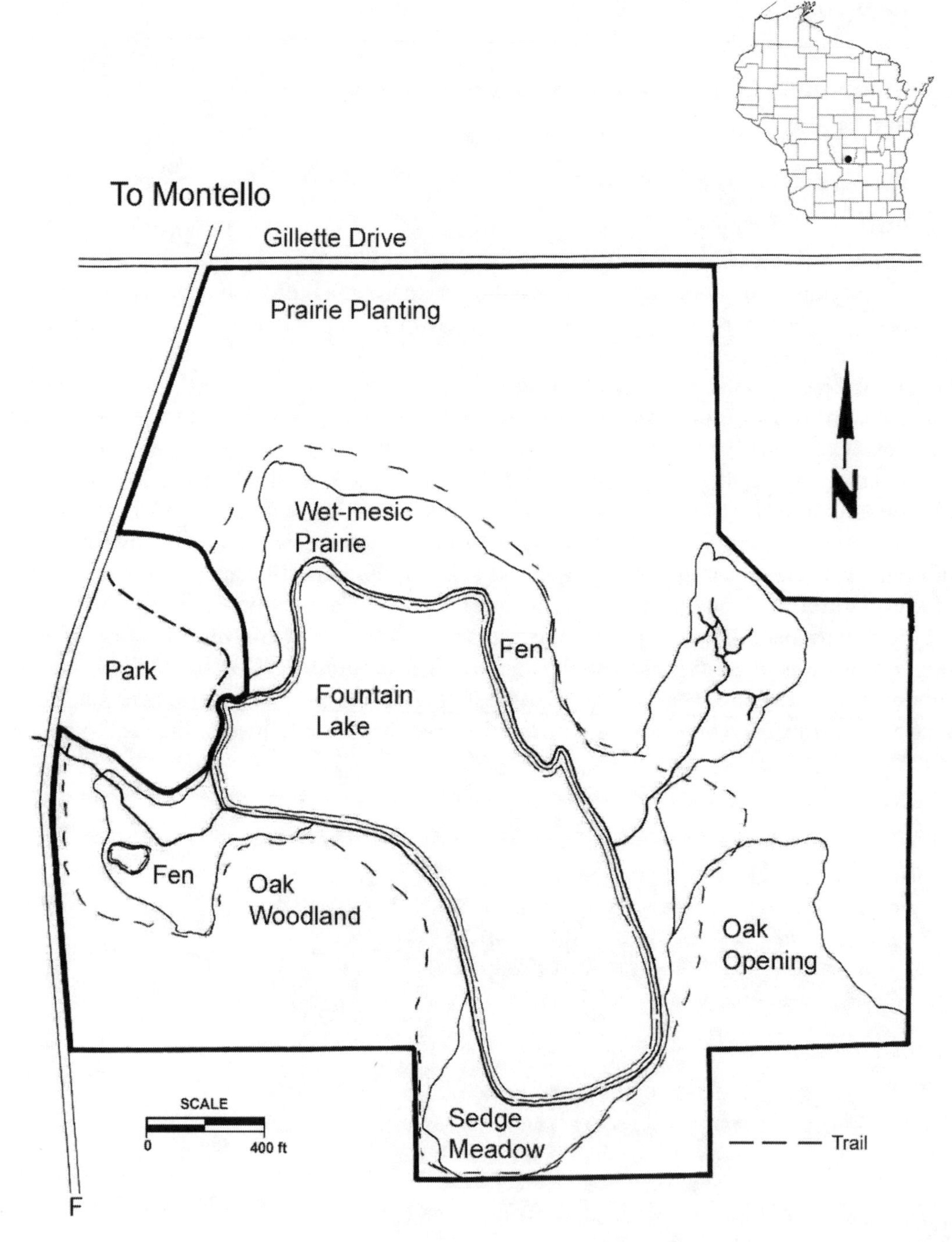
To Montello
Gillette Drive
Prairie Planting
N
Wet-mesic
Prairie
Fen
Park
Fountain
Lake
Fen
Oak
Woodland
Oak
Opening
SCALE
0
400 ft
Sedge
Meadow
Trail
F

Muir Park

Specialties

Plants

false asphodel
small yellow lady's slipper
white lady's slipper

Uncommon Species

Plants

American royal fern
arrow-leaved violet
bog bedstraw
bog goldenrod
bog lobelia
bottle gentian
brook sedge
bulblet water hemlock
clustered beak-rush
crested wood-fern
downy willow-herb
fame flower
few-seeded sedge
flat-leaved bladderwort
flat top aster
golden alexanders
grass pink
great bladderwort
hairy grama
hairy puccoon
Illinois pondweed
inland sedge
jointweed
Kalm's St. John's wort
marsh pea
marsh stitchwort
marsh St. John's wort
marsh wild timothy
New England aster
nodding ladies' tresses
northern bog violet
pasture thistle
pipewort
porcupine grass
Riddell's goldenrod
rough-headed sedge
slender false foxglove
small skullcap
swamp thistle
thin-scale cotton-grass
twig rush
white beak-rush
white prairie clover
wood sandwort
yellow pond-lily

Insects

mulberry wing

Birds (Breeding)

eastern bluebird
field sparrow
orchard oriole

Mammals

badger
gray fox
long-tailed weasel
otter
prairie deer mouse
southern flying squirrel

Necedah National Wildlife Refuge

Communities: Floodplain Forest, Dry Pine Forest, Shrub-Carr, Alder Thicket, Southern Sedge Meadow, Pine Barrens, Oak Barrens, Shallow Marsh

Necedah National Wildlife Refuge is the focus of this chapter, but the same natural features can also be found at Meadow Valley Wildlife Area, Sandhill Demonstration Area, Wood County Wildlife Area, and a section of far eastern Jackson County known as the Bear Bluff Marshes. All of these areas, which encompass approximately 210 square miles, are part of the same system of barrens, scrub oak–pine forest, and wetland resting on acidic sands. There are many uncommon plant species growing in the area, several whose closest population is the Atlantic coastal plain.

Where should you go in such a large expanse? Anywhere is probably a good answer. Travel any of the roads, hike any of the 50 miles of trails. The wildlife refuge proper, however, is a good place to begin. The north sides of Rynearson and Sprague–Mather flowages are managed for extensive savanna communities. Barrens, whether jack pine or black oak, was an important

Karner blue butterfly (photograph by Thomas A. Meyer; used by permission)

Barrens and Karners

Many times, protection or restoration of rare natural communities can have unintended positive consequences for rare species. Conversely, managing conditions to benefit one species usually does not benefit other species. The former case is evident at Necedah in the restoration of barrens communities. The beneficiary is the frosted elfin, a small butterfly that utilizes wild lupine as its larval host. This diminutive butterfly, about one inch in size, flies as an adult for a few weeks in late May. Continued barrens-restoration efforts at Necedah should provide security for this species, as well as for the Karner blue, because land managers are restoring a natural community, not a species habitat.

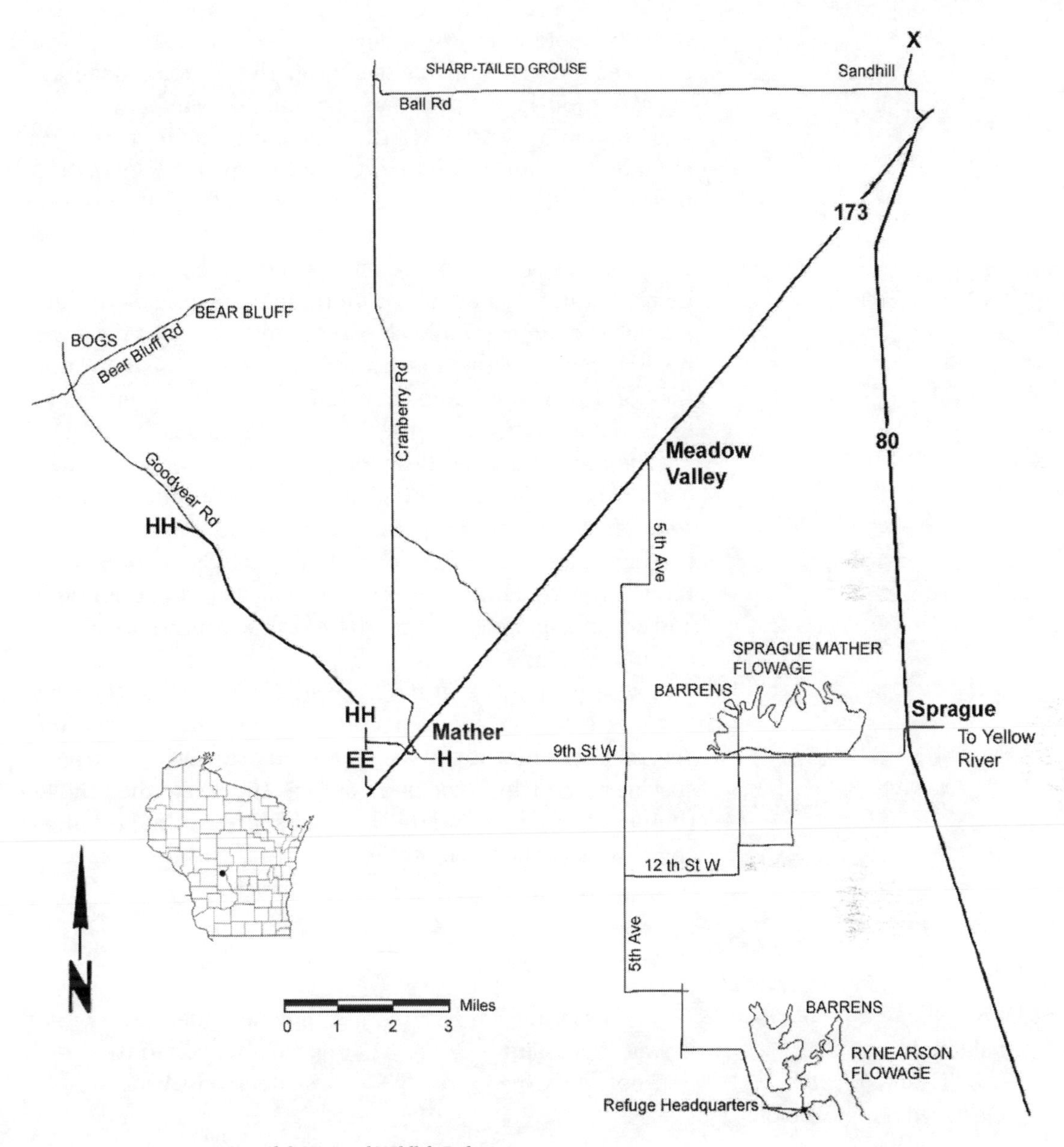

Necedah National Wildlife Refuge

component of the original ecosystem. Several years ago Karner blue butterflies were discovered in areas with remaining barrens communities. This butterfly is a federally endangered species, with the best populations in the world found in Wisconsin. Barrens restoration will dramatically improve the chances for recovery of this species.

The Yellow River bottoms area has an excellent floodplain forest with a good population of eastern massasauga rattlesnakes. The eastern massasaugas are best seen in April or

May and September, when they warm themselves on logs near the river. They can only be seen properly by a long meandering canoe journey. Only a few entry and take-out points exist, and the canoe journey can be frustrating at best because abundant meanders and even more abundant logjams impede travel. A few miles on a map can take all day, if not more, to cover by canoe. However, the intrepid naturalist can find rewards on such a trip. Logs in and near the water serve as highways and homes for an inordinate amount of wildlife. Herons use them as ambush perches to stab fish in deep water. Raccoons use them as highways. Small birds will nest in the flotsam. Snakes, especially water snakes but occasionally eastern massasauga, will use them as warming areas.

The human-altered Bear Bluff marshes have many cranberry production areas and peat-moss operations. However, some areas, mostly on public land, retain the character of this huge bog. Sphagnum-dominated wetlands surround Bear Bluff. These wetlands have water flowing in all four directions and act as a sponge that absorbs and slowly releases water to the adjacent lands.

These wetlands still have a wild aspect and a northern flavor, exemplified by the reappearance of the timber wolf. Several packs now range the Bear Bluff area. Other northern species here include the black spruce, the raven, the yellow-rumped warbler, the white-throated sparrow, the LeConte's sparrow, and the jutta arctic butterfly. All of these species reach their southern range limit in this area.

Specialties

Plants

bald rush
bog St. John's wort
clustered poppy mallow
clustered sedge
cross milkwort
Farwell's water milfoil
long sedge
marbleseed
Massachusetts fern
meadow beauty
oval milkweed
purple bladderwort
screwstem
spotted pondweed
Vasey's pondweed
Vasey's rush
water purslane
wooly milkweed

Animals

bald eagle
buck moth
Connecticut warbler
eastern massasauga
five-lined skink (lizard)
frosted elfin (butterfly)
golden eagle
Henslow's sparrow
Karner blue butterfly
LeConte's sparrow
northern marbled locust
sharp-tailed grouse
slender glass lizard

Other Features

glacial lake bed

Uncommon Species

Plants

American brooklime
broom sedge
butterfly weed
cardinal flower
clammy hedge-hyssop

colicroot
cream wild indigo
crooked aster
dragonhead
glaucous white lettuce
goat's rue
golden pert
indigo bush
large-flowered beard-tongue
lupine
marsh purslane
mermaid weed
netted nut-rush
nodding wild onion
prairie dropseed
purple spring cress
ragged fringed orchid
short green milkweed
slender beard-tongue
slender false foxglove
small fringed gentian
smooth false foxglove
snowy campion
stiff gentian
tall green milkweed
tubercled orchid
yellow-eyed grass
yellow false foxglove

Insects

blue-legged grasshopper
broad-necked grasshopper
coral hairstreak
dion skipper
Henry's elfin
hoary elfin
mottled sand grasshopper
Olympia marble
pine elfin
sand grasshopper
smaller sand cricket
spot-winged grasshopper

Birds (Breeding)

American widgeon
gadwall
golden-winged warbler
green-winged teal
long-eared owl
pine warbler
saw-whet owl
upland sandpiper
whip-poor-will

Birds (Migrants)

American golden plover
least sandpiper
osprey
rusty blackbird
solitary sandpiper
tundra swan

Mammals

badger
bog lemming
Franklin's ground squirrel
harvest mouse
least weasel
long-tailed weasel
meadow jumping mouse
otter
porcupine
prairie vole
red-backed vole
snowshoe hare

White River Marsh

Communities: Floodplain Forest, Wet-Mesic Prairie, Wet Prairie, Shrub-Carr, Southern Sedge Meadow, Shallow Marsh, River

White River Marsh is an extensive wetland complex in northwestern Green Lake County primarily known as a state-owned hunting area. To the naturalist or bird watcher, this is Wisconsin's biggest staging area for sandhill cranes. Several thousand sandhill cranes congregate after the breeding season to prepare for the flight south. The largest numbers are in September and October. White River has other exceptional amenities, too. There are extensive areas of sedge meadow, shrub-carr, cattail marsh, wet-mesic prairie, and oldfield grassland. They all provide excellent nesting areas for most marsh and grassland birds of Wisconsin.

Areas of wet prairie, calcareous fen, tamarack swamp, and floodplain forest are also located within this complex of communities. These native communities have most of the specialty species in them. Some of the wet prairies found along the White and Puchyan Rivers have huge populations of prairie parsley. This species seems to require occasional disturbance to thrive. It depends on seeds dispersing over an area. The seeds must be able to find mineral soil to germinate, then they grow as rosettes for many years until conditions are right

Large Grasslands

An evaluation of important grassland areas for birds rated the White River Marsh area as the most important wet grassland area in southern Wisconsin. In *Managing Habitat for Grassland Birds*, a Wisconsin Department of Natural Resources guidebook (Sample and Mossman 1997), this site was identified as offering the best opportunity in the state for managing wet grassland birds because of the size of the area, its potential for expansion, and the extensive bird use.

White River Marsh contains more than 4,000 acres of open sedge meadow, wet prairie, and shrub communities. If patches of forest and especially fencerows were removed from within the marsh, it would expand the grasslands even more. Grassland raptors, such as the northern harrier and the short-eared owl, need large areas to forage for their food. They cannot catch enough grassland mice and voles on a few acres to feed themselves and their families. If an area provides enough space for the widest-ranging grassland species, then those with smaller habitat requirements, such as Henslow's sparrow and the sedge wren, should find homes, also.

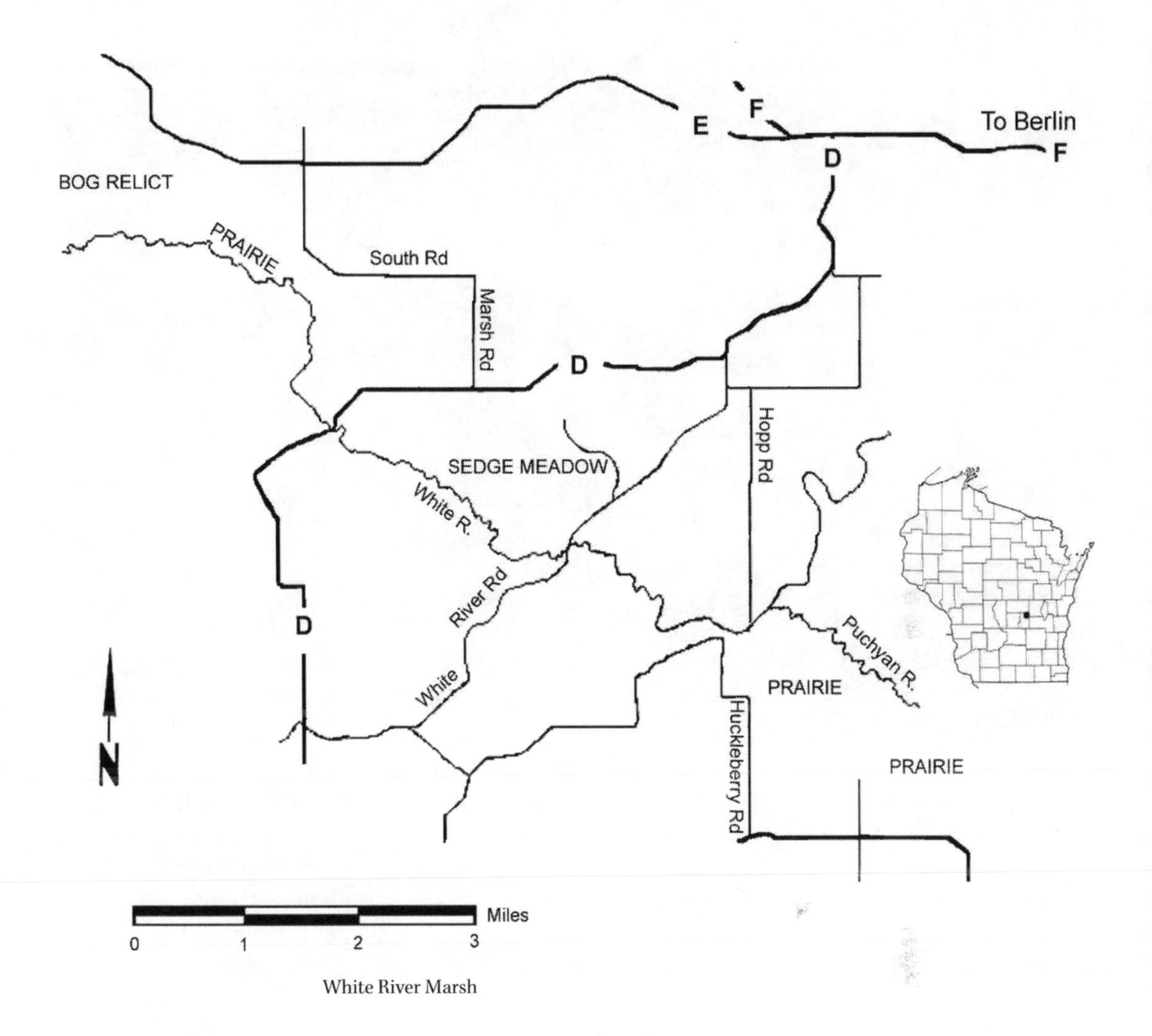

White River Marsh

for flowering, and then they die. A wildfire started by a hunter in the early 1990s provided the right conditions for a massive blooming the following summer. If these stimulating burns do not continue to occur, will this species gradually decline in population until it ceases to be among our state's flora?

White River Marsh is entirely accessible as a day-use area. To reach the wetlands, take County Highway F west of Berlin. Turn south onto County Highway D, which leads to the marsh. Two roads, County Highway D and White River Road, bisect the marsh. There are 18 parking areas from which you can set out on foot. Hunters are the primary users of the area in October and November.

Sandhill cranes (courtesy of Wisconsin Department of Natural Resources)

Specialties

Plants

- false asphodel
- mat muhly
- prairie parsley
- tubercled orchid

Animals

- Acadian flycatcher
- Blanding's turtle
- cerulean warbler
- club-horned grasshopper
- Henslow's sparrow
- king rail
- poweshiek skipper (butterfly)
- sandhill crane
- spotted-winged grasshopper

Uncommon Species

Plants

- aromatic aster
- bog lobelia
- bottle gentian
- common mountain mint
- cursed crowfoot
- fringed brome
- golden alexanders
- great angelica
- Illinois tick-trefoil
- inflated sedge
- leafy satingrass
- marsh pea
- prairie Indian plantain
- prairie lily
- showy goldenrod
- shrubby cinquefoil
- slender false foxglove
- snowy campion
- swamp goldenrod
- water plantain
- water speedwell
- wiregrass sedge

Prairie Plants

- aromatic aster
- Bicknell's sedge
- butterfly weed
- Buxbaum's sedge
- clammy false foxglove
- cream wild indigo
- glaucous white lettuce
- golden ragwort
- heath aster
- Kalm's St. John's wort
- New Jersey tea
- pale-spike lobelia
- prairie blazing star
- prairie blue-eyed grass
- prairie brome
- prairie dock
- prairie lily
- Riddell's goldenrod
- scarlet paintbrush
- shooting star
- short green milkweed
- showy goldenrod
- sneezeweed
- swamp milkweed
- swamp thistle
- sweet grass
- valerian

white wild indigo
wild onion
yellow stargrass

Insects

club-horned grasshopper
many-lined wainscot
spotted-winged grasshopper
unarmed wainscot

Birds (Breeding)

Bell's vireo
black-crowned night heron
clay-colored sparrow
common moorhen
field sparrow
gadwall
great egret
green heron
least bittern
lesser scaup
redhead
red-necked grebe
ring-necked duck
Virginia rail
Wilson's phalarope

Mammals

gray fox
otter
prairie deer mouse
prairie vole

Appendix

Glossary

Bibliography

Index

Appendix

Common and Scientific Names of Species Belonging to Groups Not Found in Single Sources

Mosses, Lichens, and Liverworts

British soldiers	*Cladonia cristatella*
Bugs-on-a-stick	*Buxbaumia aphylla*
Byrum, giant	*Rhodobyrum roseum*
Byrum, matted	*Byrum caespiticium*
Byrum, silvery	*Byrum argenteum*
Catharinea, wavy	*Atrichum undulatum*
Cetraria, pitted	*Platismatia tuckermanii*
Cladonia, brown-fruited cup	*Cladonia fimbriata*
Cladonia, cornucopia	*Cladonia pyxidata*
Entodon, flat-stemmed	*Entodon cladorrhizans*
Entodon, round-stemmed	*Entodon seductrix*
Hair cap, awned	*Polytrichum piliferum*
Hair cap, common	*Polytrichum commune*
Hair cap, juniper	*Polytrichum juniperinum*
Knight's plume	*Ptilium crista-castrensis*
Leucodon, northern	*Leucodon brachypus*
Leucodon, southern	*Leucodon julaceus*
Lichen, antler	*Pseudevernia consocians*
Lichen, black and blue	*Toninia caeruleonigricans*
Lichen, blister	*Physcia sp.*
Lichen, blood	*Xanthoria sp.*
Lichen, blue-dot	*Arthonia caesia*
Lichen, blue-isidate	*Parmelia rudecta*
Lichen, boulder	*Parmelia conspera*
Lichen, bristly	*Anaptychia setifera*
Lichen, candlestick	*Cladonia coniocraea*
Lichen, cobra-head	*Physcia adscendens*
Lichen, crisped	*Cladonia crispata*
Lichen, curd	*Lecanora muralis*
Lichen, dog	*Peltigera canina*
Lichen, dotted	*Buellia punctata*
Lichen, elegant	*Xanthoria elegans*
Lichen, empty-cup	*Cladonia cenotea*
Lichen, flabby	*Evernia sp.*
Lichen, forked	*Cladonia furcata*
Lichen, frayed	*Cladonia phyllophora*
Lichen, furrowed-shield	*Parmelia sulcata*
Lichen, goat's horn	*Cladonia cornuta*
Lichen, ladder	*Cladonia verticillata*
Lichen, lateral-fruited	*Cladonia pleurota*
Lichen, limp tufted	*Evernia mesomorpha*
Lichen, little dog	*Peltigera didactyla*
Lichen, lung	*Lobaria sp.*
Lichen, lungwort	*Lobaria pulmonaria*
Lichen, many-fruited dog	*Peltigera polydactyla*
Lichen, match stick	*Cladonia bacillaris*
Lichen, mealy goblet	*Cladonia chlorophaea*
Lichen, miniature tree	*Cladina arbuscula*
Lichen, narrow crowned	*Cladonia rei*
Lichen, orange tree-trunk	*Xanthoria fallax*
Lichen, orange-dot	*Lecanora symmictera*
Lichen, oyster shell	*Hypocenomyce scalaris*
Lichen, plume	*Anaptychia speciosa*
Lichen, puffed shield	*Hypogymnia physodes*
Lichen, reddish dog	*Peltigera rufescens*
Lichen, reindeer	*Cladina rangiferina*
Lichen, rosette	*Physcia aipolia*
Lichen, scalloped	*Lecanora rugosella*
Lichen, script	*Graphis scripta*
Lichen, sieve	*Cladonia multiformis*
Lichen, slender	*Cladonia gracilis*
Lichen, sod	*Acarospora sp.*
Lichen, soft	*Cladina mitis*
Lichen, squamulate	*Cladonia squamosa*
Lichen, starry	*Cladina stellaris*
Lichen, studded leather	*Peltigera leucophlebia*
Lichen, swollen	*Cladonia turgida*
Lichen, tan grape	*Cladonia botrytes*
Lichen, tufted-hair	*Parmotrema crinitum*
Lichen, twig	*Ramalina sp.*
Lichen, wavy-edge	*Physconia detersa*
Lichen, white cup	*Cladonia conista*
Lichen, wrinkled shield	*Pseudoparmelia caperata*
Lichen, yellow-green pored	*Parmelia flaventior*
Lichen, yellow pine	*Cetraria pinastri*
Lichen, yellow wall	*Xanthoria polycarpa*
Lichen, yellow wax	*Candelaria concolor*
Liverwort, leafy	*Frullania sp.*
Liverwort, thallose	*Conocephalum sp.*
Liverwort, xerophytic	*Reboulia hemisphaeria*
Mnium, pointed	*Mnium cuspidatum*
Moss, broom	*Dicranum scoparium*
Moss, common apple	*Philonotis fontana*
Moss, common beaked	*Eurhynchium pulchellum*

Moss, common beard *Grimmia apocarpa*
Moss, common cedar *Hypnum pallescens*
Moss, common feather *Pleurozium scherberi*
Moss, common fern *Thuidium delicatulum*
Moss, common pygmy *Pleuridium subulatum*
Moss, common tree *Climacium dendroides*
Moss, common tree apron *Anomodon attenuatus*
Moss, feather *Hypnum imponens*
Moss, fringe leaf *Hedwigia ciliata*
Moss, giant fountain *Fontinalis hypnoides*
Moss, green-felt *Pogonatum sp.*
Moss, green hair *Dicranella heteromalla*
Moss, homalia *Homalia trichomanoides*
Moss, Iceland *Cetraria islandica*
Moss, knothole *Anacamptodon splachnoides*
Moss, luminous *Schistostega pennata*
Moss, mountain fern *Hylocomium splendens*
Moss, nodding *Pohlia nutans*
Moss, pear-shaped thread *Leptobyrum pyriforme*
Moss, plume *Fissidens sp.*
Moss, powder gun *Diphyscium foliosum*
Moss, purple-horned tooth *Ceratodon purpureus*
Moss, red collar *Splachnum rubrum*
Moss, rivulet cedar *Brachythecium rivulare*
Moss, rock *Andreaea rupestris*
Moss, Schreber's cedar *Calliergon cordifolium*
Moss, slender cedar *Plagiothecium denticulatum*
Moss, spoon-leaved *Bryoandersonia illecebra*
Moss, toothless twisted *Cynodontium sp.*
Moss, torn veil *Rhacomitrium sp.*
Moss, urn *Physcomitrium pyriforme*
Moss, velvet tree apron *Anomodon rostratus*
Moss, water *Dichelyma sp.*
Moss, water measuring cord *Funaria hygrometrica*
Moss, wavy broom *Dicranum undulatum*
Moss, whip fork *Dicranum flagellare*
Moss, white *Leucobyrum glaucum*
Moss, wiry fern *Thuidium abietinum*
Neckera, feathered *Neckera pennata*
Old man's beard *Usnea sp.*
Peat moss *Sphagnum sp.*
Peat moss, reddish *Sphagnum palustre*
Peat moss, spread-leaved *Sphagnum squarrosum*
Riccia, purple-fringed *Ricca sp.*
Rock tripe, blistered *Umbilicaria papulosa*
Rock tripe, fleecy *Umbilicaria vellea*
Rock tripe, smooth *Umbilicaria mammulata*
Spreading leather *Lobaria quercizans*
Wood reveler *Rhytidiadelphus triquetrus*
Yellow eyes *Cyphelium tigillare*

Mollusks and Snails

Amber snail, golden *Succinea ovalis*
Buckhorn *Tritogonia verrucosa*
Bullhead *Plethobasus cyphyus*
Butterfly *Ellipsaria lineolata*
Clam, Asiatic *Corbicula manillensis*
Deer toe *Truncilla truncata*
Disk snail, common *Discus cronkhitei*
Disk snail, minute *Punctum minutissimum*
Disk snail, open *Discus patulus*
Disk snail, parallel *Heliodiscus parallelus*
Disk snail, Singley's *Heliodiscus singleyanus*
Ear snail, American *Pseudosuccinea columella*
Ebony shell *Fusconaia ebenus*
Elephant's ear *Elliptio crassidens*
Elk toe *Alasmidonta marginata*
Ellipse *Venustaconcha ellipsiformis*
Fawn foot *Truncilla donaciformis*
Fingernail clam, grooved *Sphaerium simile*
Fingernail clam, lake *Sphaerium lacustre*
Fingernail clam, long *Sphaerium transversum*
Fingernail clam, pond *Sphaerium securis*
Fingernail clam, rhomboid *Sphaerium rhomboideum*
Fingernail clam, striated *Sphaerium striatinum*
Fingernail clam, swamp *Sphaerium partumeium*
Floater, eastern *Anodonta cataracta*
Floater, paper *Anodonta imbecillis*
Floater, strange *Strophitus undulatus*
Fluted shell *Lasmigona costata*
Forest snail, Foster's *Triodopsis fosteri*
Forest snail, ripe *Mesodon zaletus*
Forest snail, striped *Anguispira alternata*
Forest snail, three-toothed *Triodopsis tridentata*
Forest snail, white-lipped *Triodopsis albolabris*
Fossaria, amphibious *Fossaria parva*
Fossaria, graceful *Fossaria exigua*
Fossaria, modest *Fossaria modicella*
Fossaria, shouldered northern *Fossaria decampi*
Gyraulus, flatly coiled *Gyraulus circumstriatus*
Gyraulus, irregular *Gyraulus deflectus*
Gyraulus, modest *Gyraulus parvus*
Heelsplitter, pink *Proptera alata*
Heelsplitter, white *Lasmigona complanata*

Common name	Scientific name
Hickory nut	*Obovaria olivaria*
Higgins' eye	*Lampsilis higginsii*
Horn snail, flat-sided	*Pleurocera acuta*
Horn snail, Great Lakes	*Goniobases livescens*
Lilliput	*Carunculina parva*
Limpet, dusky lily-pad	*Laevapex fuscus*
Limpet, oval lake	*Ferrissia fragilis*
Limpet, sturdy river	*Ferrissia rivularis*
Looping snail, riverbank	*Pomatiopsis lapidaria*
Maple leaf	*Quadrula quadrula*
Monkey face	*Quadrula metanevra*
Mucket	*Actinonaias ligamentina*
Mucket, fat	*Lampsilis radiata*
Mussel, salamander	*Simpsoniconcha ambigua*
Mussel, zebra†	*Dreissena polymorpha*
Nautilus snail, tiny	*Armiger crista*
Papershell, cylindrical	*Anodontoides ferussa-cianus*
Papershell, fragile	*Leptodea fragilis*
Papershell, pink	*Proptera laevissima*
Pea clam, Adam's	*Pisidium adamsi*
Pea clam, fat	*Pisidium rotundatum*
Pea clam, giant northern	*Pisidium idahoense*
Pea clam, globular	*Pisidium ventricosum*
Pea clam, perforated	*Pisidium punctatum*
Pea clam, quadrangular	*Pisidium milium*
Pea clam, ridge-beak	*Pisidium compressum*
Pea clam, river	*Pisidium fallax*
Pea clam, rusty	*Pisidium ferrigineum*
Pea clam, shiny	*Pisidium nitida*
Pea clam, short-ended	*Pisidium subtruncatum*
Pea clam, tiny	*Pisidium insigne*
Pea clam, triangular	*Pisidium variabile*
Pea clam, ubiquitous	*Pisidium casertanum*
Pea clam, Walker's	*Pisidium walkeri*
Physa, blunt prairie	*Physa jennessi*
Physa, solid lake	*Physa integra*
Pigtoe	*Fusconaia flava*
Pigtoe, Ohio River	*Pleurobema cordatum*
Pill snail, fraternal	*Stenotrema fraternum*
Pill snail, hairy	*Stenotrema hirsutum*
Pill snail, Lea's	*Stenotrema leai*
Pill snail, southern	*Stenotrema stenotrema*
Pimpleback	*Quadrula pustulosa*
Planorbid, Say's toothed	*Planorbula armigera*
Pocketbook	*Lampsilis ventricosa*
Pocketbook, rock	*Arcidens confragosus*
Pond clam, prairie	*Aplexa hypnorum*
Pond snail, great	*Lymaea stagnalis*
Pond snail, showy	*Bulimnea megasoma*
Pond snail, slender	*Acella haldemani*
Promenetus, keeled	*Promenetus exacuous*
Pupa snail, armed	*Gastrocopta armifera*
Pupa snail, brown	*Vertigo ovata*
Pupa snail, five-toothed	*Gastrocopta pentodon*
Pupa snail, Holzinger's	*Gastrocopta holzingeri*
Pupa snail, modest	*Vertigo modesta*
Pupa snail, moss	*Pupilla muscorum*
Pupa snail, toothless	*Columella edentula*
Pupa snail, tree bark	*Gastrocopta corticaria*
Ramshorn, bell-mouthed	*Helisoma campanulata*
Ramshorn, eastern	*Helisoma trivolvis*
Ramshorn, great corinate	*Helisoma pilsbryi*
Ramshorn, two-ridged	*Helisoma anceps*
Retinella, amber	*Retinella electrina*
Sandshell, black	*Ligumia recta*
Sandshell, yellow	*Lampsilis teres*
Shell, tiny star	*Planogyra asteriscus*
Slippershell	*Alasmidonta viridis*
Snaggletooth, wing	*Gastrocopta procera*
Snail, allied labyrinth	*Strobilops affinis*
Snail, apple seed	*Cionella lubrica*
Snail, Iowa pleistocene	*Vertigo iowaensis*
Snail, tiny harp	*Zoogenetes harpa*
Snail, white swamp	*Carychium exiguum*
Snuffbox	*Epioblasma triquetra*
Spectacle case	*Cumberlandia monodonta*
Spike	*Elliptio dilatatus*
Spire snail, campeloma	*Cincinnatia cincinnatiensis*
Spire snail, flat-ended	*Probythinella lacustris*
Spire snail, ordinary	*Amnicola limosa*
Spire snail, small	*Amnicola walkeri*
Stagnicola, common	*Stagnicola elodes*
Stagnicola, lake	*Stagnicola catascopium*
Stagnicola, striped	*Stagnicola reflexa*
Tadpole snail	*Physa gyrina*
Threehorn	*Obliquaria reflexa*
Three ridge	*Amblema plicata*
Vallonia, handsome	*Vallonia pulchella*
Valve snail, ribbed	*Valvata sincera*
Valve snail, three-keeled	*Valvata tricarinata*
Vertigo, Hulbricht's	*Vertigo hulbrichti*
Vertigo, occult	*Vertigo occulta*
Wartyback	*Quadrula nodulata*
Wartyback, purple	*Cyclonaias tuberculata*
Washboard	*Megalonaias gigantea*
Zonite shell, great	*Mesomphix vulgatus*
Zonite shell, shining	*Zonitoides nitidus*

Zonite shell, small	*Striatura exigua*
Zonite shell, tree	*Zonitoides arboreus*
Zonite shell, white	*Hawaiia minuscula*

Insects (Including Beetles)

Alderfly	*Sialis sp.*
Ambush bug, American	*Phymata americana*
Ambush bug, Pennsylvania	*Phymata pennsylvanica*
Ant, Allegheny mound	*Formica exsectoides*
Ant, black carpenter	*Camponotus pennsylvanicus*
Ant, cornfield	*Lasius alienis*
Ant, little black	*Monomorium minimum*
Ant, pavement	*Tetramorium caespitum*
Ant, red carpenter	*Camponotus ferrugineus*
Ant, silky	*Formica fusca*
Ant, thief	*Solenopsis molesta*
Antlion	*Myrmeleon immaculatus*
Aphid, balsam wooly	*Adelges piceae*
Assassin bug, spined	*Sinea diadema*
Backswimmer	*Notonecta undulata*
Bark beetle, eastern spruce	*Dendroctonus piceaperda*
Bark beetle, native elm	*Hylurgopinus rufipes*
Bark beetle, red flat	*Cucujus clavipes*
Bark beetle, two-spotted	*Pisenus humeralis*
Bark beetle, velvety	*Penthe pimelia*
Bark weevil, black elm	*Magdalis barbita*
Bark weevil, red elm	*Magdalis armicollis*
Bee, colletid	*Colletes compactus*
Bee, honey	*Apis mellifera*
Bee, leafcutting	*Megachile latimanus*
Bee, little carpenter	*Xylocopa virginica*
Bee, mason	*Osmia lignaria*
Bee, mining	*Andrena carlini*
Bee, sweat	*Halictus ligatus*
Bee assassin	*Apiomerus crassipes*
Beetle, antelope	*Dorcus parallelus*
Beetle, bluebone	*Necrobia violacea*
Beetle, bombardier	*Brachinus sp.*
Beetle, Brandel's anthill	*Decarthon brandeli*
Beetle, carpet	*Anthrenus scrophularae*
Beetle, carrot	*Bothynus gibbosus*
Beetle, cedar	*Family Rhipiceridae*
Beetle, Colorado potato	*Leptinotarsa decimlineata*
Beetle, dead fish	*Ateuchus histeroides*
Beetle, eastern larch	*Dendroctonus simplex*
Beetle, false longhorn	*Donacia piscatrix*
Beetle, fire	*Dendroides concolor*
Beetle, goldenrod	*Trirhabda canadensis*
Beetle, goldsmith	*Cotalpa lanigera*
Beetle, green dock	*Gastrophysa cyanea*
Beetle, hieroglyph	*Cryptocephalus mutabilis*
Beetle, horse shoe crab	*Limulodes sp.*
Beetle, larder	*Dermestes lardarius*
Beetle, marsh	*Elaphrus ruscarius*
Beetle, May	*Phyllophaga rugosa*
Beetle, minute bog	*Sphaerius sp.*
Beetle, minute marsh-loving	*Limnichus tenuicornis*
Beetle, odor-of-leather	*Osmoderma eremicola*
Beetle, picnic	*Glischrochilus quadrisignatus*
Beetle, pinching bug	*Pseudolucanus capreolus*
Beetle, red milkweed	*Tetraopes tetrophthalmus*
Beetle, red turpentine	*Dendroctonus valens*
Beetle, riffle	*Stenelimis sp.*
Beetle, round sand	*Omophron labiatum*
Beetle, seed corn	*Stenolopus lecontei*
Beetle, skiff	*Hydroscapha natans*
Beetle, small ironclad	*Phellopsis obcordata*
Beetle, snout	*Polydrusus americanus*
Beetle, spotted cucumber	*Diabrotica undecimpunctata*
Beetle, spotted savage	*Omophron tessellatum*
Beetle, stag	*Platycerus virescens*
Beetle, water lily	*Galerucella nymphaeae*
Beetle, water penny	*Psephenus sp.*
Bembid, American	*Bembidion americanum*
Bembid, four-spotted	*Bembidion quadrimaculatum*
Bembid, patruus	*Bembidion patruele*
Billbug, bluegrass	*Sphenophorus parvulus*
Billbug, clay-colored	*Sphenophorus aequalis*
Blister beetle, ash-gray	*Epicauta fabricii*
Blister beetle, black	*Epicauta pennsylvanica*
Blister beetle, clematis	*Epicauta cinerea*
Blister beetle, margined	*Epicauta marginata*
Blister beetle, short-winged	*Meloe angusticollis*
Blister beetle, spotted	*Epicauta maculata*
Blister beetle, striped	*Epicauta vittata*
Borer, black-horned pine	*Callidium antennatum*
Borer, bronze poplar	*Agrilus liragus*
Borer, clover stem	*Languaria mozardi*
Borer, cottonwood	*Plectrodera scalator*
Borer, cylindrical hardwood	*Neoclytus acuminatus*
Borer, dogwood twig	*Oberea tripunctata*
Borer, elm	*Saperda tridentata*
Borer, hemlock	*Melanophila fulvoguttata*
Borer, linden	*Saperda vestita*
Borer, living beech	*Goes pulverulentus*
Borer, locust	*Megacyllene robiniae*

Borer, lurid flat-headed	*Dicerca lurida*	Caddisfly, tube-making	*Family Psychomyiidae*
Borer, oak branch	*Goes debilis*	Calligrapha, elm	*Calligrapha sp.*
Borer, painted-hickory	*Megacyllene caryae*	Carrion beetle, American	*Silpha americana*
Borer, pine heartwood	*Chalcophora virginiensis*	Carrion beetle, garden	*Silpha inaequalis*
Borer, pine stump	*Asemum striatum*	Carrion beetle, red-lined	*Necrodes sp.*
Borer, pole	*Parandra brunnea*	Caterpillar hunter	*Calosoma scrutator*
Borer, poplar	*Saperda calcarata*	Chalcid, golden	*Spilochalis mariae*
Borer, poplar and willow	*Cryptorhynchus lapathi*	Checkered beetle, dubious	*Thanasimus dubius*
Borer, potato stalk	*Trichobaris trinotata*	Checkered beetle, hairy	*Phyllobaenus humeralis*
Borer, raspberry cane	*Oberea bimaculata*	Checkered beetle, red-blue	*Trichodes nutalli*
Borer, red-necked cane	*Agrilus ruficollis*	Cicada, dogday	*Tibicen canicularis*
Borer, red oak	*Romaleum rufulum*	Cicada, Linneaus 17-year	*Magicicada septendecim*
Borer, red pine flat-headed	*Chrysobothris femorata*	Cicada, prairie	*Okanakana ballii*
Borer, ribbed pine	*Rhagium inquisitor*	Cicada, 13-year	*Magicicada tredecim*
Borer, rustic	*Xylotrechus colonus*	Cicada killer, giant	*Sphecius speciosus*
Borer, shot-hole	*Scolytus rugulosus*	Clerid, wasp-like	*Enoclerus nigripes*
Borer, sugar maple	*Glycobius speciosus*	Click beetle, eyed	*Alaus oculatus*
Borer, tan bark	*Phymatodes testaceus*	Click beetle, faded	*Ctenicera hieroglyphica*
Borer, timber	*Orthosoma brunneum*	Click beetle, narrow-necked	*Ctenicera pyrrhos*
Borer, two-lined chestnut	*Agrilus bilineatus*	Click beetle, red-necked	*Ampedus collaris*
Borer, white oak	*Goes tigrinus*	Collops, four-spotted	*Collops quadrimaculatus*
Bruchid, redbud	*Gibbobruchus mimus*	Collops, two-lined	*Collops vittatus*
Bug, ambush	*Family Phymatidae*	Corsair, black	*Melanolestes pictipes*
Bug, anchor	*Stiretus anchorago*	Crane fly	*Family Tipulidae*
Bug, box elder	*Leptocoris trivittatus*	Crane fly, phantom	*Family Ptychopteridae*
Bug, broad-headed	*Family Alydidae*	Curculio, cabbage	*Ceutorhynchus rapae*
Bug, chinch	*Blissus leucopterus*	Damsel bug	*Family Nabidae*
Bug, cranberry toad	*Family Gelastocoridae*	Darkling beetle	*Uloma impressa*
Bug, large milkweed	*Oncopeltus fasciatus*	Darkling beetle, red-neckcd	*Hoplocephala viridipennis*
Bug, leaf-footed	*Acanthocephala terminatus*	Diving beetle, fasciated	*Laccophilus fasciatus*
		Diving beetle, giant	*Cybister fimbriolatus*
Bug, minute pirate	*Orius insidios*	Diving beetle, predacious	*Agabus confinis*
Bug, shore	*Family Saldidae*	Diving beetle, striped	*Agabus disintegratus*
Bug, small milkweed	*Lygaeus kalmii*	Dobsonfly	*Family Corydalidae*
Bug, spined assassin	*Sinea diadema*	Drynid, contort	*Gonatopus contortulus*
Bug, spined soldier	*Podisus maculiventris*	Dung beetle, aphodine	*Aphodius granarius*
Bug, squash	*Anasa tristis*	Dung beetle, fancy	*Ataenius spretulus*
Bug, stilt	*Jalysus spinosus*	Dung beetle, lesser	*Aphodius fimetarius*
Bug, thread-legged	*Emesaya brevipennis*	Dung beetle, maculated	*Aphodius distinctus*
Bug, wheel	*Arilus cristatus*	Engraver, hackberry	*Family Scolytidae*
Bumble bee, American	*Bombus americanorum*	Engraver, pine	*Ips pini*
Bumble bee, yellow	*Bombus pennsylvanicum*	Engraver, six-spined	*Ips calligraphus*
Buprestid, hairy	*Acmaeodera pulchella*	Fiery hunter	*Calosoma calidum*
Burying beetle, gold-necked	*Nicrophorus tomentosus*	Firefly, black	*Lucidota atra*
Caddisfly, finger-net	*Family Philopotamidae*	Firefly, common	*Photinus pyralis*
Caddisfly, large	*Family Phryganeidae*	Firefly, pale	*Photinus scintillans*
Caddisfly, long-horned	*Family Leptoceridae*	Firefly, Pennsylvania	*Photuris pennsylvanicus*
Caddisfly, net-spinning	*Family Hydropsychidae*	Fishfly	*Family Corydalidae*
Caddisfly, northern	*Family Limnephilidae*	Flea beetle, alder	*Altica ambiens*
Caddisfly, snail-case	*Family Helicopsychidae*	Flea beetle, grape	*Altica chalybea*

Flea beetle, sumac	*Blepharida rhois*
Fleahopper, garden	*Halticus bractatus*
Flower beetle, ant-like	*Notoxus talpa*
Flower beetle, bumble bee	*Euphoria inda*
Flower beetle, cloudy	*Anthicus cervinus*
Flower beetle, hermit	*Osmoderma eremicola*
Flower beetle, metallic	*Malachius aeneus*
Flower beetle, pale	*Attalus scincetus*
Flower beetle, rough	*Osmoderma scabra*
Flower beetle, tumbling	*Mordella marginata*
Flower bug, insidious	*Orius insidius*
Flower scarab, bee-like	*Trichiotinus piger*
Fly, bee	*Bombylius major*
Fly, black	*Simulium vittatum*
Fly, black horse	*Tabanus atratus*
Fly, blue bottle	*Calliphora vicina*
Fly, carrot rust	*Psila rosae*
Fly, cluster	*Pollenia rudis*
Fly, currant fruit	*Epochra canadensis*
Fly, deer	*Chrysops callidus*
Fly, drone	*Eristalis tenax*
Fly, face	*Musca autumnalis*
Fly, flesh	*Family sacrophagidae*
Fly, flower	*Metasyrphus americanus*
Fly, house	*Musca domestica*
Fly, humpbacked	*Family Phoridae*
Fly, latrine	*Fannia Scaloris*
Fly, little house	*Fannia canicularis*
Fly, long-legged	*Dolichopod plumpies*
Fly, louse	*Lynchia sp.*
Fly, March	*Bibio albipennis*
Fly, marsh	*Family Sciomyzidae*
Fly, mydas	*Mydas clavatus*
Fly, orchid	*Eurytoma orchidearum*
Fly, robber	*Leptogaster sp.*
Fly, snipe	*Rhagio mystacea*
Fly, soldier	*Family Stratiomyidae*
Fly, stable	*Stomoxys calcitrans*
Fly, tanglewing	*Family Nemestrinidae*
Fly, vinegar	*Drosophila melanogaster*
Fly, walnut husk	*Rhagoletis completa*
Fly, window-pane	*Scenopinus fenestralis*
Fruitworm, eastern raspberry	*Byturus rubi*
Fungus beetle, banded	*Triplax festiva*
Fungus beetle, horned tree	*Cis cornuta*
Fungus beetle, red-shouldered	*Tritoma humeralis*
Fungus beetle, red-spotted	*Megalodacne heros*
Fungus beetle, shiny	*Baeocera falsata*
Fungus beetle, two-banded	*Alphitophagus bifasciatus*
Fungus weevil, checkerboard	*Brachytarsus sticticus*
Fungus weevil, white-banded	*Eurymyeter fasciatus*
Gall, oak apple	*Amphibolips confluenta*
Gnat, eye	*Liohippelats pusio*
Grain beetle, red-horned	*Platydema ruficorne*
Ground beetle, big-headed	*Scarites subterraneus*
Ground beetle, bog	*Elaphrus cicatricosus*
Ground beetle, dingy	*Stenolophus conjunctus*
Ground beetle, European	*Carabus nemoralis*
Ground beetle, flat	*Lebia atriventris*
Ground beetle, fragile	*Badister pulchellus*
Ground beetle, long-necked	*Colliuris pennsylvanicus*
Ground beetle, mud-loving	*Elaphrus ruscarius*
Ground beetle, notch-mouthed	*Galeritula bicolor*
Ground beetle, pedunculate	*Pasimachus depressus*
Ground beetle, Pennsylvania	*Harpalus pennsylvanicus*
Ground beetle, pubescent	*Chlaenius tricolor*
Ground beetle, riverbank	*Geopinus incrassatus*
Ground beetle, seed-eating	*Amara cupreolata*
Ground beetle, slender	*Galeritula janus*
Ground beetle, snail-eating	*Scaphinotus sp.*
Ground beetle, striated	*Lebia scapularis*
Ground beetle, vivid metallic	*Chlaenius sericeus*
Ground beetle, woodland	*Pterostichus adoxus*
Hister beetle, bark	*Platysoma carolinum*
Hister beetle, fancy ant	*Hetaerius sp.*
Hister beetle, Pennsylvania	*Saprinus pennsylvanicus*
Hornet, bald-faced	*Dolichovespula maculata*
Horntail, pigeon	*Tremex columbia*
Ichneumon, black	*Theronia morio*
Ichneumon, giant	*Megarhyssa maerurus*
Ichneumon, white-footed	*Cryptus albitarsis*
June beetle, green	*Cotinis nitida*
June beetle, spotted	*Pelidonta pinctata*
June beetle, ten-lined	*Polyphylla decemlineata*
Lace bug, basswood	*Corythucha tiliae*
Lace bug, chrysanthemum	*Corythucha marmorata*
Lace bug, elm	*Corythucha ulmi*
Lace bug, hackberry	*Corythucha celtidus*
Lace bug, oak	*Corythucha arcuata*
Lace bug, sycamore	*Corythucha ciliata*
Lacewing, green	*Chrysopa oculata*
Lady beetle, American minute	*Scymnus americanus*
Lady beetle, ash-gray	*Olla abdominalis*
Lady beetle, covergent	*Hippodamia convergens*
Lady beetle, maculated	*Coleomegilla maculata*
Lady beetle, milkweed	*Brachycantha ursina*

Lady beetle, nine-spotted	*Coccinella novemnotata*
Lady beetle, parenthesis	*Hippodamis parenthesis*
Lady beetle, thirteen-spotted	*Hippodamis tridecimpuctata*
Lady beetle, three-banded	*Coccinella trifasciata*
Lady beetle, two-spotted	*Adalia bipunctata*
Lady beetle, undulated	*Hyperaspis undulata*
Lake fly	*Family Chironomidae*
Leaf beetle, aspen	*Chrysomela crotchi*
Leaf beetle, cottonwood	*Chrysomela scripta*
Leaf beetle, rose	*Nodonota puncticollis*
Leaf beetle, scriptured	*Pachybrachis tridens*
Leaf beetle, swamp milkweed	*Labidomera clivicollis*
Leaf beetle, willow	*Chrysomela interrupta*
Leaf bug, ash-gray	*Piesma cinera*
Leafhopper, candy-stripe	*Graphocephala coccinea*
Leafhopper, clover	*Aceratagallia sanguinoleta*
Leafhopper, duck-billed	*Attenuipyga vanduzeei*
Leafhopper, European	*Athysanus argentarius*
Leafhopper, grape	*Erythroneura comes*
Leafhopper, painted	*Endria inimica*
Leafhopper, potato	*Empoasca fabae*
Leafhopper, prairie	*Paraphlepsius irroratus*
Leafhopper, prairie polyamia	*Polyamia dilata*
Leafhopper, red-veined	*Aflexia rubranura*
Leafhopper, six-spotted	*Macrosteles fascifrons*
Leather-wing, short-winged	*Trypherus latipennis*
Longhorn, banded	*Typocerus velutinus*
Longhorn, long-necked	*Dorcaschema wildii*
Longhorn, marginated flower	*Leptura emarginata*
Longhorn, red-shouldered	*Purpuricenus humeralis*
Longhorn, stumpy	*Anoplodera cordifera*
Longhorn, two-lined flower	*Acmaeops bivittatus*
Mantidfly	*Family Mantispidea*
Masked hunter	*Reduvius personatus*
May beetle, fuscous	*Phyllophaga fusca*
May beetle, hairy	*Phyllophaga tristis*
Mayfly, burrowing	*Family Ephemeridae*
Mayfly, Pecatonica River	*Acanthametropus pecatonica*
Mayfly, small	*Family Baetidae*
Mayfly, stream	*Family Heptageniidae*
Micro-caddisfly	*Family Hydroptilidae*
Midge, net-winged	*Family Blephariceridae*
Midge, phantom	*Family Chaoboridae*
Mosquito, anaphales	*Anaphales sp.*
Mosquito, pitcher plant	*Wyeomyia sp.*
Mosquito, snow-melt	*Aedes sp.*
Mosquito, tree-hole	*Aedes sp.*
Moss beetle, minute	*Hydraena sp.*
Mud dauber, black and yellow	*Sceliphron caementarium*
Mud dauber, blue	*Chalybion californicum*
Net-wing, banded	*Calopteron reticulatum*
Net-wing, flat	*Calochromus perfectus*
Net-wing, golden	*Dictopterus aurora*
Passolus, horned	*Odontotaenius disjunctus*
Plant bug, big-leaf aster	*Family Miridae*
Plant bug, four-lined	*Poecilocapsus lineatus*
Plant bug, goldenrod	*Lopidea media*
Plant bug, phlox	*Lopidea davisi*
Plant bug, poplar	*Lopidea cuneata*
Plant bug, rapid	*Adelphocoris rapidus*
Plant bug, red	*Coccobaphes sanguinareus*
Plant bug, tarnished	*Lygus lineolaris*
Plant bug, willow	*Lopidea salicis*
Planthopper, corn	*Peregrinus maidus*
Planthopper, huckleberry	*Oliarus cinnamonensis*
Planthopper, meadow	*Scolops sulcipes*
Planthopper, three-ridge	*Stobaera tricarinata*
Planthopper, two-striped	*Acanalonia bivittata*
Prionid, brown	*Orthosoma brunneum*
Prionus, tile-horned	*Prionus imbricornis*
Pruner, oak twig	*Hypermallus villosus*
Ripple bug	*Family Veliidae*
Rootworm, cranberry	*Rhabdopterus picipes*
Rootworm, grape	*Fidia viticida*
Rose chafer	*Macrodactylus subspinosus*
Rove beetle, big-jawed	*Oxyporus major*
Rove beetle, broad	*Coproporus ventriculus*
Rove beetle, cross-toothed	*Oxyporus femoralis*
Rove beetle, hairy	*Staphylinus maxillosus*
Rove beetle, obscure	*Gyrophaena vinula*
Rove beetle, paederine	*Sunius confluentus*
Rove beetle, painted	*Megalopsidia caelatus*
Rove beetle, parasitic	*Aleochara sp.*
Rove beetle, spiny-legged	*Bledius emarginatus*
Rove beetle, triangular	*Tachyporus jocosus*
Sap beetle, banded	*Glischrochilus fasciatus*
Sap beetle, dusky	*Carpophilus lugubris*
Sap beetle, narrow	*Conotelus obscurus*
Sap beetle, red	*Glischrochilus sp.*
Sap beetle, sculptured	*Phenolia grossa*
Sap beetle, six-spotted	*Prometopia sexmaculata*
Sap beetle, two-spotted	*Glischrochilas obtusus*
Sawfly, elm	*Climbex americanus*
Sawyer, Carolina	*Monochamus carolinensis*
Sawyer, white-spotted	*Monochamus scutellaris*

Common name	Scientific name
Scale, osytershell	*Lepidosaphes ulmi*
Scarab, scooped	*Onthophagus hecate*
Scarab, yellow-scaled	*Valgus seticollis*
Scavenger, dung	*Sphaeridium scarabeoides*
Scavenger beetle, giant	*Hydrophilus triangularis*
Scavenger beetle, minute water	*Hydrochus squamifer*
Scavenger beetle, narrow	*Tropisternus lateralis*
Scavenger beetle, water	*Berosus striatus*
Scolops, partridge	*Scolops perdix*
Scorpion, snow	*Boreus sp.*
Seed bug, long-necked	*Myodoca serripes*
Serica, chocolate	*Serica vespertina*
Shield-back, alternate	*Eurygaster alternata*
Shield-back, anchor	*Stiretrus anchorago*
Snail-eater, eastern	*Scaphinotus elevatus*
Snowflea	*Hypogastrura nivicola*
Soldier beetle, goldenrod	*Chauliognathus sp.*
Soldier beetle, hairy	*Podabrus tomentosus*
Soldier beetle, Pennsylvania	*Chauliognathus pennslvanicus*
Soldier beetle, two-lined	*Cantharis bilineatus*
Spider beetle, hairy	*Ptinus villiger*
Spittlebug, diamond-backed	*Lepyronia quadrangularis*
Spittlebug, dogwood	*Clastoptera proteus*
Spittlebug, meadow	*Philaenius spumarius*
Spittlebug, pine	*Aphrophora parallelus*
Spittlebug, Saratoga	*Aphrophora saratogensis*
Stag beetle, antelope	*Dorcus parallelus*
Stink bug, brown	*Eustichus servus*
Stink bug, green	*Acrosternum hilare*
Stink bug, Lugen's	*Mormidea lugens*
Stink bug, twice-stabbed	*Cosmopepla bimaculata*
Stink bug, two-spotted	*Perillus bioculatus*
Stonefly, common	*Family Perlodidae*
Stonefly, giant	*Family Pteronarcyidae*
Stonefly, green	*Family Chloroperlidae*
Stonefly, rolled-wing	*Family Leuctridae*
Stonefly, small winter	*Family Capniidae*
Stonefly, spring	*Family Nemouridae*
Stonefly, two-lined	*Isoperla patricia*
Stonefly, winter	*Family Taeniopterygidae*
Thrips, flower	*Frankliniella tritici*
Tiger beetle, blackish	*Cicindela unipuncata*
Tiger beetle, bronze	*Cicindela repanda*
Tiger beetle, dainty	*Cicindela lepida*
Tiger beetle, dark brown	*Cicindela rufiventris*
Tiger beetle, noble	*Cicindela formosa*
Tiger beetle, purple	*Cicindela purpurea*
Tiger beetle, river	*Cicindela tranquebarica*
Tiger beetle, sand	*Cicindela scutellaris*
Tiger beetle, six-spotted	*Cicindela sexgutttata*
Tiger beetle, Virginia big-headed	*Megacephala virginica*
Timberworm, oak	*Arrhenides minutus*
Toe-biter	*Belostoma flumineum*
Tortoise beetle, argus	*Chelymorpha cassidea*
Tortoise beetle, black-legged	*Metriona bivittata*
Tortoise beetle, golden	*Metriona bicolor*
Tortoise beetle, mottled	*Deloyala guttata*
Treehopper, buffalo	*Stictocephala bubalus*
Treehopper, oak	*Family Membracidae*
Tumble bug	*Canthon laevis*
Twig pruner	*Elaphidion villosum*
Wasp, burrowing	*Ptilothrix bombiformis*
Wasp, digger	*Sphex ichneumoneus*
Wasp, eastern sand	*Bembix spinolae*
Wasp, paper	*Polistes fuscatus*
Wasp, potter	*Eumenes fraternus*
Wasp, sand	*Prionyx atratus*
Water beetle, crawling	*Haliplus triopsis*
Water boatmen	*Family Corixidae*
Water bug, giant	*Lethocerus americanus*
Water bug, velvet	*Merragata hebroides*
Water scavenger beetle	*Family Hydrophilidae*
Water scorpion	*Family Nepidae*
Water strider	*Family Gerridae*
Water strider, small	*Rhagovelia obesa*
Water treader	*Family Mesoveliidae*
Weevil, black oak	*Curculio baculi*
Weevil, cranberry	*Anthonomus suturalis*
Weevil, hazelnut	*Curculio neoeorylus*
Weevil, northern pine	*Pissodes approximatus*
Weevil, Pale's	*Hylobius pales*
Weevil, pine	*Hylobius congener*
Weevil, pine root collar	*Hylobius radicis*
Weevil, red elm bark	*Magdalis armicollis*
Weevil, rice water	*Listrongtus oryzophilus*
Weevil, sawtooth sunflower	*Lixus fimbriolatus*
Weevil, square	*Attelabus nigripes*
Weevil, strawberry root	*Otiorhynchus ovatus*
Weevil, water	*Hyperodes sp.*
Weevil, white pine	*Pissodes strobi*
Whirligig, common	*Gyrinus borealis*
Whirligig, minute	*Gyrinus minutus*
Wireworm, community	*Melanotus communis*
Wireworm, tobacco	*Conoderus vespertinus*
Yellow jacket	*Vespula pennsylvanica*
Yellow jacket, northern	*Vespula vulgaris*

Glossary

Boreal. Referring to the cool regions of the Northern Hemisphere.

Circumboreal forest. The typical spruce–fir forest distributed around the high latitudes of the Northern Hemisphere.

Conglomerate. Rounded small pebbles or boulders cemented together by hardened clay.

Dalles/dells. A river gorge with nearly vertical walls.

Dominant. A plant or animal exerting considerable influence on a community.

Driftless area. The hilly area in western Wisconsin that escaped being covered with past glaciers (drift refers to rock debris ground and deposited by the glacier).

Drumlin. An oval-shaped hill of glacial till formed at the bottom of an ice sheet.

Duff. Incompletely decomposed organic material.

Edge(s). The portion of an ecosystem near its perimeter.

Endemic. Restricted to a particular geographic area.

Ephemeral pond. Pond lasting only a short period of time; also called vernal (springtime) or temporary pond.

Esker. A narrow ridge, often sinuous, composed of sand and gravel deposited by glacial melt water.

Exotic species. Not native; a species that has been introduced to an area.

Extirpated. Removed from an area.

Flowage. A water impoundment that creates a shallow lake or marsh behind a dam or dike.

Forb(s). A broad-leaved herbaceous plant; often referenced in prairie and savanna communities.

Glacial Lake Wisconsin. A large lake that covered much of Adams, Juneau, Jackson, and Wood counties when the Wisconsin glacier was melting.

Glacial till. Material carried and deposited directly by a glacier; till does not come from melt water.

Heath. Low-lying vegetation dominated by small-leaved shrubs.

Herb(s). A plant having nonwoody stems and which dies down annually; often referenced in forested communities.

Hummock. A rounded knoll or hillock, referring to the rounded tussocks of many sedges.

Invertebrates. Animals without backbones, such as insects, worms, and freshwater mussels.

Kettle (lake or hole). A depression in glacial drift caused by delayed melting of a block of ice.

Lekking grounds. An assembly area for courtship display.

Marl. A loose, wet soil deposit that contains calcium carbonate.

Mast. Nuts, especially beechnuts and acorns, that accumulate on the forest floor.

Metamorphic rock. Rock that has been changed by pressure or heat.

Microhabitat. A small, specialized habitat.

Monadnocks. Extremely hard, resistant rocks that survived the glaciers.

Moraine. Assorted sands, gravel, and rocky material deposited by a glacier; generally, moraines are irregular bands of hills that formed at the edge of a glacier.

Mycorrhyzial fungi. The fungi associated with the root system of a plant.

Oxbow lake. A lake that forms when a river straightens, leaving a lake in the former bend.

Patch(es). A relatively similar area of vegetation that differs from its surroundings.

Peat. An accumulation of partially decomposed plant material, mostly in water-logged habitats.

Perch(ed). A natural community or feature lying in a topographically higher position on the land than the majority of the natural community.

Periphyton. A community of plants and animals adhering to and forming a surface coating on rocks, plants, or other submerged objects.

Relict(s). Persistent remnants of a formerly widespread community existing in certain isolated areas.

Ridge and swale. High and low areas of underwater sand and gravel deposits that became land when the water levels dropped.

Sand blows. Open sandy areas with sparse vegetation. The loose sand can be blown by the wind to a different location.

Sedge. Any species of tufted plants that differ from grasses in having achenes (single-seeded fruits) and solid stems.

Seep. Water passing slowly through or oozing from the ground.

Specialists. A species adapted to live in only one or a few natural communities or a species found at only a few locations in the state.

State-endangered. In grave danger of being lost from the state (usually fewer than five populations).

State-threatened. At risk of becoming endangered in the state.

Succession. The gradual and predictable process of community change and replacement.

Succession, primary. The first inhabitants in the process of succession.

Succession, secondary. The species that replaces the primary inhabitants

Tension zone. A landscape where many species reach the limits of their range.

Terminal moraine. A moraine built at the edge of a glacier.

Tombolo. A sandbar connecting an island with the mainland (or a larger island).

Transitional. An area of passage from one state to another; in Wisconsin, the zone between the southern forests and the northern forests.

Tussock. A compact tuft of grass or sedge.

Witness tree. The original land surveyors in the mid-1800s marked the closest trees to the survey point. The four closest trees over four inches in diameter, one for each cardinal direction, were recorded as to distance from the survey point and species.

Bibliography

Note: References followed by an asterisk (*) contain the common and scientific names of entire groups of species that are listed throughout this book.

Alexander, R. D., A. E. Pace, and D. Otte. 1972. The singing insects of Michigan. *The Great Lakes Entomologist* 5, no. 2:33–63.

Allan, J. D. 1995. *Stream ecology: Structure and function of running waters.* London: Chapman & Hall.

Arnett, R. H., Jr., and R. L. Jacques Jr. 1985. *Insect life.* Englewood Cliffs, N.J.: Prentice Hall.

Arno, S. F. 1984. *Timberline: Mountain and arctic forest frontiers.* Seattle: The Mountaineers.

Bates, J. 1995. *Trailside botany.* Duluth, Minn.: Pfeifer-Hamilton.

Behler, J. L., and F. W. King. 1979. *The Audubon Society field guide to North American reptiles and amphibians.* New York: Knopf.

Bellrose, F. C. 1976. *Ducks, geese, and swans of North America.* Harrisburg, Pa.: Stackpole Books.

Benyus, J. M. 1989. *Northwoods wildlife.* Minoqua, Wisc.: Northword Press.

Bissonette, J. A. 1997. *Wildlife and landscape ecology: Effects of pattern and scale.* New York: Springer.

Black, R. F. 1974. *Geology of Ice Age National Scientific Reserve of Wisconsin.* Scientific Monograph Series, no. 2. Washington, D.C.: National Park Service.

Borror, D. J., and R. E. White. 1970. *A field guide to the insects of America north of Mexico.* Boston: Houghton Mifflin Co.

Brewer, J., and D. Winter. 1986. *Butterflies and moths.* Englewood Cliffs, N.J.: Prentice Hall.

Bull, J. L., and J. Farrand. 1977. *The Audubon Society field guide to North American birds, eastern region.* New York: Knopf.

Caduto, M. J. 1985. *Pond and brook.* Englewood Cliffs, N.J.: Prentice Hall.

Carroll, G. C., and D. T. Wicklow. 1992. *The fungal community: Its organization and role in the ecosystem,* 2d ed. New York: Marcel Dekker, Inc.

Chadde, S. 1998. *A Great Lakes wetland flora.* Calumet, Mich.: Pocketflora Press.

Clayton, M. 1998. *Photo atlas of the vascular plants,* 2d ed. [computer file]. Madison: University of Wisconsin Board of Regents.*

Cobb, B. 1963. *A field guide to the ferns.* Boston: Houghton Mifflin Co.

Collins S. L., and L. L. Wallace. 1990. *Fire in North American tallgrass prairie.* Norman: University of Oklahoma Press.

Courtenay, B., and H. H. Burdsall Jr. 1982. *A field guide to mushrooms and their relatives.* New York: Van Nostrand Reinhold Co.

Courtenay, B., and J. H. Zimmerman. 1972. *Wildflowers and weeds.* New York: Van Nostrand Reinhold Co.

Covell, C. 1984. *A field guide to the moths of eastern North America.* Boston: Houghton Mifflin Co.*

Crum, H. 1983. *Mosses of the Great Lakes forest.* Ann Arbor: University of Michigan Press.

Crum, H. 1988. *A focus on peatlands and peat mosses.* Ann Arbor: University of Michigan Press.

Curtis, J. T. 1959. *Vegetation of Wisconsin.* Madison: University of Wisconsin Press.

Daniel, G., and J. Sullivan. 1981. *A Sierra Club naturalist's guide to the North Woods of Michigan, Wisconsin, and Minnesota.* San Francisco: Sierra Club Books.

Davis, M. B. 1996. *Eastern old-growth forests.* Washington, D.C.: Island Press.

DeGraaf, R. M., and J. H. Rappole. 1995. *Neotropical migratory birds.* Ithaca, N.Y.: Comstock Cornell.

Dickinson, G., and K. Murphy. 1998. *Ecosystems.* London: Routledge.

Dillon, E. S., and L. S. Dillon. 1972. *A manual of common beetles of eastern North America,* vols. 1 and 2. New York: Dover.

Dunn, G. 1996. *Insects of the Great Lakes region.* Ann Arbor: University of Michigan Press.

Ebner, J. A. 1970. *The butterflies of Wisconsin.* Milwaukee Public Museum Popular Science Handbook Series, no. 12. Milwaukee: Milwaukee Public Museum.

Eggers, S. D., and D. M. Reed. 1997. *Wetland plants and plant communities of Minnesota and Wisconsin* [computer file]. St. Paul, Minn.: Northern Prairie Wildlife Research Center.

Fahey, J., Jr. 1999. *Field guide to the birds of North America.* Washington, D.C.: The National Geographic Society.*

Fassett, N. C. 1951. *Grasses of Wisconsin.* Madison: University of Wisconsin Press.

Fassett, N. C. 1957. *A manual of aquatic plants.* Madison: University of Wisconsin Press.

Fassett, N. C. 1978. *Spring flora of Wisconsin.* Madison: The University of Wisconsin Press.

Finan, A. S. 2000. *Wisconsin forests at the millennium: An assessment.* Madison: Wisconsin Department of Natural Resources.

Forman, R. T. T., and M. Godron. 1986. *Landscape ecology.* New York: John Wiley and Sons.

Gibson, D. J. 1989. Effects of animal disturbance on tallgrass prairie vegetation. *American Midland Naturalist* 121:144–154.

Graveland, J. R., R. van der Wal. 1994. Poor reproduction in forest passerines from decline of snail abundance on acidified soils. *Nature* 368:446–447.

Green, J. C. 1995. *Birds and forests.* St. Paul: Minnesota Department of Natural Resources.

Hart, S. 1998. Beetle mania: An attraction to fire. *Bioscience* 48, no. 1:3–5.

Helfer, J. 1987. *How to know the grasshoppers, crickets, cockroaches and their allies.* New York: Dover.*

Helmers, D. L. *Shorebird management manual.* Manomet, Mass.: Western Hemisphere Shorebird Reserve Network.

Hickin, N. E. 1963. *The insect factor in wood decay.* London: Hutchinson & Co.

Hoagman, W. 1994. *Great Lakes coastal plants: A field guide.* East Lansing: Michigan State University.

Hole, F. D. 1976. *Soils of Wisconsin.* Madison: University of Wisconsin Press.

Hole, F. D., and G. A. Nielsen. 1968. Soil genesis under prairie. In *Proceedings of a symposium on prairie and prairie restoration,* ed. P. Schramm, 28–34. Galesburg, Ill.: Knox College, Biological Field Station.

Jackson, H. H. T. 1961. *Mammals of Wisconsin.* Madison: University of Wisconsin Press.

James, S. W. 1988. The post-fire environment and earthworm populations in tallgrass prairie. *Ecology* 50:874–877.

Jones, J. K., Jr., and E. C. Burney. 1988. *Handbook of mammals of the north-central states.* Minneapolis: University of Minnesota Press. *

Jordan , W. R., III, M. E. Gilpin, and J. D. Aber. 1987. *Restoration ecology: A synthetic approach to ecological restoration.* Cambridge: Cambridge University Press.

Judziewicz, E. J., R. G. Koch. 1993. Flora of the Apostle Islands. *Michigan Botanist* 32, no. 2:43–189.

Kricher, J. C., and G. Morrison. 1988. *Eastern forests.* Boston: Houghton Mifflin Co.

Knobel, E. 1980. *Field guide to the grasses, sedges and rushes of the United States.* New York: Dover.

Ladd, D. 1995. *Tallgrass prairie wildflowers.* Helena, Mont.: Falcon Press Publishing.

Lange, K. I. 1989. *Ancient rocks and vanished glaciers: A natural history of Devil's Lake State Park, Wisconsin.* Baraboo, Wisc.: State of Wisconsin Department of Natural Resources.

Legler, K., D. Legler, and D. Westover. 1996. *Guide to common dragonflies of Wisconsin.* Sauk City, Wisc.: Karl Legler.*

Lellinger, D. 1985. *A field manual of the ferns and fern-allies of the United States and Canada.* Washington, D.C.: Smithsonian Institution Press.

Levi, H. W., L. R. Levi, and H. S. Zim. 1968. *A guide to spiders and their kin.* New York: Golden Press.*

Lincoff, G. 1981. *The Audubon Society field guide to North American mushrooms.* New York: Knopf.*

Lunn, E. 1982. *Plants of the Illinois dunelands.* Lake Forest, Ill.: The Illinois Dunesland Preservation Society.

Martin, A. C., H. S. Zim, and A. L. Nelson. 1951. *American wildlife and plants: A guide to wildlife food habits.* New York: Dover.

Martin, L. 1965. *The physical geography of Wisconsin.* Madison: University of Wisconsin Press.

Maser, C., and J. M. Trappe. 1984. *The seen and unseen world of the fallen tree.* General Technical Report PNW-164. Portland, Ore.: Pacific Northwest Forest and Range Experiment Station, USDA Forest Service.

Mathiak, H. 1979. *A river survey of unionid mussels of Wisconsin 1973–1977.* Horicon, Wisc.: Sand Shell Press.

McCarty, K. 1998. Landscape-scale restoration in Missouri savannas and woodlands. *Restoration and Management Notes* 16, no. 1:22–32.

McKnight, K. H., and V. B. McKnight. 1987. *A field guide to mushrooms, North America.* Boston: Houghton Mifflin Co.

McPherson, G. R. 1997. *Ecology and management of North American savannas.* Tucson: University of Arizona Press.

Medlin, J. J. 1996. *Michigan lichens.* Bulletin, no. 60. Bloomfield Hills, Mich.: Cranbrook Institute of Science.

Meyer, K. D. 1995. Swallow-tailed kite. In *The Birds of North America,* eds. A. Poole and F. Gill, No. 138, 1–24. Philadelphia: The Academy of National Sciences and Washington, D.C.: The American Ornithologists' Union.

Milne, L., and M. Milne. 1980. *The Audubon Society field guide to North American insects and spiders.* New York: Knopf.

Mitchell, R. T., and H. S. Zim. 1977. *Butterflies and moths.* New York: Golden Press.

Mohlenbrock, R. H. 1983. *Where have all the flowers gone.* New York: Macmillan Publishing Co.

Mossman, M. J., and K. I. Lange. 1982. *Breeding birds of the Baraboo Hills, Wisconsin.* Madison, Wisc.: Department of Natural Resources and Wisconsin Society for Ornithology.

Muir, J. 1965. *The story of my boyhood and youth.* Madison: University of Wisconsin Press.

Niering, W. A., and N. C. Olmstead. 1979. *The Audubon Society field guide to North American wildflowers, eastern.* New York: Knopf.

Opler, P. A., and V. Malikul. 1992. *A field guide to eastern butterflies.* Boston: Houghton Mifflin Co.*

Orr, C. C., and O. J. Dickinson. 1966. Nematodes in true prairie soils of Kansas *Transactions of the Kansas Academy of Science* 69:317–334.

Ostergren, R. C., and T. R. Vale. 1997. *Wisconsin land and life.* Madison: University of Wisconsin Press.

Peterson, R. T., and M. McKenney. 1968. *A field guide to the wildflowers of northeastern and north-central North America.* Boston: Houghton Mifflin Co.

Peterson, R. T., and V. M. Peterson. 1980. *A field guide to the birds.* Boston: Houghton Mifflin Co.

Petrides, G. 1972. *A field guide to the shrubs.* Boston: Houghton Mifflin Co.

Pettersson, R. B., et al. 1995. Invertebrate communities in boreal forest canopies as influenced by forestry and lichens with implications for passerine birds. *Biological Conservation* 74:57–63.

Pielou E. C. 1991. *After the ice age.* Chicago: University of Chicago Press.

Pratt, C. R. 1995. *Ecology: Complete course review.* Springhouse, Pa.: Springhouse.

Putnam, R. J. 1994. *Community ecology.* London: Chapman & Hall.

Putnam, R. J., and S. D. Wratten. 1984. *Principles of ecology.* Berkeley: University of California Press.

Pyle, R. 1981. *The Audubon Society field guide to North American butterflies.* New York: Knopf.

Pyne, S. J. 1982. *Fire in America: A cultural history of wildland and rural fire.* Seattle: University of Washington Press.

Reed, C. C. 1994. Species richness of insects on prairie flowers in southeastern Minnesota. In *Proceedings of the 14th Annual North American Prairie Conference,* ed. D.C. Hartnett, 103–116. Manhattan: Kansas State University.

Robbins, S. D., Jr. 1991 *Wisconsin birdlife.* Madison: University of Wisconsin Press.

Roth, C. E. 1984. *The plant observer's guidebook.* Englewood Cliffs, N.J.: Prentice Hall.

Sample, D. W., and M. J. Mossman. 1997. *Managing habitat for grassland birds: A guide for Wisconsin.* Madison: Wisconsin Department of Natural Resources.

Samson, F. B., and F. L. Knopf. 1996. *Prairie conservation.* Washington, D.C.: Island Press.

Schorger, A. W. 1982. *Wildlife in early Wisconsin.* Stevens Point, Wisc.: The Wildlife Society.

Schultz, G. 1986. *Wisconsin's foundations.* Dubuque, Iowa: Kendall/Hunt.

Sheviak, C. J., and M. J. Bowles. 1986. The prairie fringed orchids: A pollinator-isolated species pair. *Rhodora* 88:267–278.

Temple, S. A., and J. R. Cary 1987. *Wisconsin birds: A seasonal and geographical guide.* Madison: University of Wisconsin Press.

Terres, J. K. 1980. *The Audubon Society encyclopedia of North American birds.* New York: Knopf.

Vance, F. P., J. R. Jowsey, and J. S. McLean. 1984. *Wildflowers of the northern Great Plains.* Minneapolis: University of Minnesota Press.

Vogt, R. 1981. *Natural history of amphibians and reptiles of Wisconsin.* Milwaukee, Wisc.: Milwaukee Public Museum.*

Waldbauer, G. 1998. *The birder's bug book.* Cambridge, Mass.: Harvard University Press.

Watts, M. T. 1975. *Reading the landscape of America.* New York: Collier Books.

Weller, M. W. 1981. *Freshwater marshes: Ecology and wildlife management.* Minneapolis: University of Minnesota Press.

Whitacker J., Jr. 1980. *The Audubon Society field guide to North American mammals.* New York: Knopf.

White, R. E. 1983. *A Field guide to the beetles of North America.* Boston: Houghton Mifflin Co.

Williams, J. G., and A. E. Williams. 1983. *A field guide to the orchids of North America.* New York: Universe Books.

Index

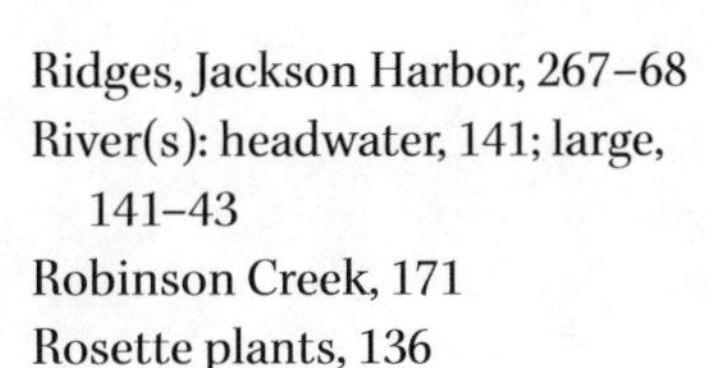